ECOSYSTEM CLASSIFICATION FOR ENVIRONMENTAL MANAGEMENT

Ecology & Environment

VOLUME 2

Ecosystem Classification for Environmental Management

edited by

Frans Klijn

Centre of Environmental Science,
Leiden University,
The Netherlands

KLUWER ACADEMIC PUBLISHERS
DORDRECHT / BOSTON / LONDON

Library of Congress Cataloging-in-Publication Data

```
Ecosystem classification for environmental management / edited by
  Frans Klijn.
      p.    cm. -- (Ecology & environment ; v. 2)
    Outgrowth of an international workshop held Dec. 1992 at Leiden
  University.
    Includes index.
    ISBN 0-7923-2917-1 (HB : acid free paper)
    1. Biotic communities--Classification--Congresses.
  2. Environmental protection--Congresses.  3. Ecosystem management-
  -Congresses.    I. Klijn, Frans.  II. Series.
  QH540.7.E36  1994
  574.5'012--dc20                                              94-19087
```

ISBN 0-7923-2917-1

Published by Kluwer Academic Publishers,
P.O. Box 17, 3300 AA Dordrecht, The Netherlands.

Kluwer Academic Publishers incorporates
the publishing programmes of
D. Reidel, Martinus Nijhoff, Dr W. Junk and MTP Press.

Sold and distributed in the U.S.A. and Canada
by Kluwer Academic Publishers,
101 Philip Drive, Norwell, MA 02061, U.S.A.

In all other countries, sold and distributed
by Kluwer Academic Publishers Group,
P.O. Box 322, 3300 AH Dordrecht, The Netherlands.

Printed on acid-free paper

All Rights Reserved
© 1994 Kluwer Academic Publishers
No part of the material protected by this copyright notice may be reproduced or
utilized in any form or by any means, electronic or mechanical,
including photocopying, recording or by any information storage and
retrieval system, without written permission from the copyright owner.

Printed in the Netherlands

Preface

When Lovelock published his 'Gaia', it was for many people quite a relief. We would not be able to destroy life on earth.
Lovelock illustrated this argument with a wealth of mechanistic feedback processes, as we know them to occur in ecosystems. These feedback processes would, somehow, lead the earth as a whole into a new equilibrium. An equilibrium with life within, be it in an entirely changed environment. This is, indeed, let us be earnest: a *functioning ecosystem*.
But what kind of ecosystem? The Gaia-hypothesis triggered a great deal of thought and discussion about what we actually require as an environment. *Biodiversity* as an abbreviation of *biotic diversity* has since become the focal point of societal concern.
But again, when we think about it, we are not only interested in the sheer number of species on earth. We also have other interests: nearby, in our backyards, in the surrounding countryside, and on the various locations where we would like to spend our holidays. We also want to preserve rare or characteristic species just for their own sake. In fact, we want species in viable populations to be part of communities that are self-maintaining in environments where they belong. We know we cannot ask for this without protecting their environment, which is also our environment. This is where the next fashionable term emerges: *sustainability*.
By combining these concepts, that nowadays form not only the basis of environmentalism but also of public concern, we come to the topic of this book. We want to sustain species diversity at several spatial scale levels, not only on the earth as a whole but, also, within our own reach. Since each environment can be regarded as an ecosystem, as long as a number of criteria are met, this is where the plural form *ecosystems* is needed. To sustain biodiversity, we must preserve *ecodiversity*: a varied pattern of ecosystems at the earth's surface composed by the interaction of abiotic, biotic and antropic forces.

Prerequisites to preserving ecodiversity are knowing the subject and putting some order in it for communication reasons. This requires classification, typification, and mapping. Especially the last, as we are concerned about the surface areas various ecosystem types occupy.

In comparison with the classification of species, ecosystems pose a number of special problems. First, the very concept of ecosystem already raises some discussion. Then, ecosystems are such intricate systems that we can approach the problem of their classification from many sides. Additionally, ecosystems can be recognized at many spatial scale levels.

So classifying ecosystems confronts us with a number of fundamental and practical dilemmas. At the same time, the recognition that not all ecosystems are equally susceptible to man-induced environmental change, nor equally valuable, requires that we put a great deal of effort into it. This is desperately needed for ensuring a more sound environmental management, which may be supported either by assessing the carrying capacity of ecosystems or by clearly presenting what the consequences of human actions will be by means of environmental impact assessments.

This book grew out of the initiative of the Centre of Environmental Science (CML) and the Rijksherbarium/ Hortus Botanicus (RHHB), both of Leiden University, to organise an international workshop on ecosystem classification for environmental management in December 1992.

The book treats the why and how of ecosystem classification and mapping as a prerequisite to environmentally sound management. Written by a number of renowned landscape ecologists from several West European countries, it discusses a wealth of theories, concepts, and methods from plant ecology, vegetation science, physical geography, and other environmental sciences, composed in such a way as to constitute practically applicable tools.

The book is divided into three parts: theory, approaches to classification, and applications. More specifically, the three parts focus on:
- an introduction to ecosystem classification and mapping from theoretical points of view;
- examples of ecosystem classifications from various methodological approaches; and
- the application of classifications of ecosystems and/or ecosystem components for predictive modelling, for nature valuation, and for the monitoring of nature values.

Most authors address more than one topic and freely wander from theory to practice and back. It will be obvious that classifying the various contributions into one of the three categories was almost as difficult as ecosystem classification itself.

In addition, the various authors represent a variety of scientific communities struggling with the subject, in other words: ecologists' diversity. This is reflected in the terminology and use of concepts that may differ in the various chapters.

Despite this diversity there was a general feeling during the workshop that we should attempt to reach as great an understanding as possible, since we share a concern for our one, but varied world. A first step toward such understanding requires that you, reader, reflect on the contents of this book. I sincerely hope it will enhance discussions among us.

Acknowledgements

The December 1992 workshop and the writing of this book were endorsed by the International Association for Landscape Ecology (IALE) and financially supported by WWF - Netherlands (WNF), the Netherlands' National Institute of Public Health and Environmental Protection (RIVM), the Netherlands' National Institute for Inland Water Management and Waste Water Treatment (RIZA), and the Leiden University Fund (LUF). We are grateful to these institutions for their support.

I also thank Professor Ies Zonneveld, Professor Helias Udo de Haes, and Dr. Ruud van der Meijden for the inspiring discussions during the organization of the workshop and the editing of this book.

Special thanks are due to Drs. Henk Bezemer, who, quietly in the background, played a crucial role in the practical organisation of the workshop and also did the technical editing of this book.

Frans Klijn
Leiden, 30 November 1993

Contributors

Geert de Blust
Institute of Nature Conservation
Kiewitdreef 5
B-3500 Hasselt
Flanders, Belgium

Jesper Brandt
Department of Geography and
Computer Science
Roskilde University Centre,
House 19.2
P.O. Box 260
DK-4000 Roskilde
Denmark

Robert G.H. Bunce
Inst. for Terrestrial Ecology (ITE)
Merlewood Research Station
Windermere Road
Grange-over-Sands
LA 11 6JU Cumbria
England

Frans A.M. Claessen
National Institute for Inland Water
Management and Waste Water
Treatment (RIZA)
P.O. Box 17
NL-8200 AA Lelystad
The Netherlands

Michel Godron
Laboratoire de Systematique et
d'Ecologie Mediterranéennes
(LSEM) - Institut de Botanique
Université Montpellier II
163, rue Auguste Broussonet
F-34000 - Montpellier
France

Kees (C.) L.G. Groen
Centre of Environmental Science
Leiden University
P.O. Box 9518
NL-2300 RA Leiden
The Netherlands

Wolfgang Haber
Lehrstuhl für Landschaftsökologie,
Weihenstephan
Technische Universität München
D-85350 Freising
Germany

Esbern Holmes
Department of Geography and
Computer Science
Roskilde University Centre,
House 19.2
P.O. Box 260
DK-4000 Roskilde
Denmark

x   CONTRIBUTORS

Frans Klijn
Centre of Environmental Science
Leiden University
P.O. Box 9518
NL-2300 RA Leiden
The Netherlands

Eckhart Kuijken
Institute of Nature Conservation
Kiewitdreef 5
B-3500 Hasselt
Flanders, Belgium

Dorthe Larsen
Department of Geography and
Computer Science
Roskilde University Centre,
House 19.2
P.O. Box 260
DK-4000 Roskilde
Denmark

Joris B. Latour
National Institute of Public Health
and Environmental Protection
(RIVM)
P.O. Box 1
NL-3720 BA Bilthoven
The Netherlands

Roman Lenz
GSF Forschungszentrum für
Umwelt und Gesundheid
P.O. Box 1129
D-85758 Oberschleißheim
Germany

Ruud van der Meijden
Rijksherbarium/ Hortus Botanicus
Leiden University
P.O. Box 9514
NL-2300 RA Leiden
The Netherlands

J. Gerard Nienhuis
National Institute of Public Health
and Environmental Protection
(RIVM)
P.O. Box 1
NL-3720 BA Bilthoven
The Netherlands

Desiré Paelinckx
Institute of Nature Conservation
Kiewitdreef 5
B-3500 Hasselt
Flanders, Belgium

Rudo Reiling
National Institute of Public Health
and Environmental Protection
(RIVM)
P.O. Box 1
NL-3720 BA Bilthoven
The Netherlands

Han (J.) Runhaar
Centre of Environmental Science
Leiden University
Postbus 9518
NL-2300 RA Leiden
The Netherlands

Helias A. Udo de Haes
Centre of Environmental Science
Leiden University
Postbus 9518
NL-2300 RA Leiden
The Netherlands

Jaap Wiertz
National Institute of Public Health
and Environmental Protection
(RIVM)
P.O. Box 1
NL-3720 BA Bilthoven
The Netherlands

J. Flip (P.) M. Witte
Department of Water Resources
Agricultural University
Nieuwe Kanaal 11
NL-6709 PA Wageningen
The Netherlands

Isaac S. Zonneveld
Vaarwerkhorst 63
NL-7531 HL Enschede
The Netherlands

Contents

Preface   v

Contributors   ix

PART 1 THEORY

1. Environmental Policy and Ecosystem Classification   1
   Helias A. Udo de Haes and Frans Klijn

2. Basic Principles of Classification   23
   Isaac S. Zonneveld

3. Systems Ecological Concepts for Environmental Planning   49
   Wolfgang Haber

4. The Natural Hierarchy of Ecological Systems   69
   Michel Godron

PART 2 APPROACHES TO CLASSIFICATION

5. Spatially Nested Ecosystems, Guidelines for Classification from a Hierarchical Perspective   85
   Frans Klijn

6. Ecosystem Classification by Budgets of Material: the Example of Forest Ecosystems Classified as Proton Budget Types   117
   Roman Lenz

7. The Use of Site Factors as Classification Characteristics for Ecotopes   139
   Han (J.) Runhaar and Helias A. Udo de Haes

8. The Application of Quantitative Methods of Classification to   173
   Strategic Ecological Survey in Britain
   Robert G.H. Bunce

PART 3 APPLICATIONS

9. A Flexible Multiple Stress Model: who needs a priori Classification?   183
   Joris B. Latour, Rudo Reiling and Jaap Wiertz

10. Ecosystem Classification and Hydro-ecological Modelling   199
    for National Water Management
    *Frans A.M. Claessen, Frans Klijn, J. Flip (P.) M. Witte and
    J. Gerard Nienhuis*

11. Up-to-date Information on Nature Quality for Environmental   223
    Management in Flanders
    Geert de Blust, Desiré Paelinckx and Eckhart Kuijken

12. Monitoring 'Small Biotopes'   251
    Jesper Brandt, Esbern Holmes and Dorthe Larsen

13. The Use of Floristic Data to establish the Occurrence and Quality   275
    of Ecosystems
    Kees (C.) L.G. Groen, Ruud van der Meijden and Han (J.) Runhaar

Index   291

Plates   295

Environmental policy and ecosystem classification 1

Helias A. Udo de Haes and Frans Klijn

ABSTRACT - Ecosystems can be defined in various ways, including or excluding man. For environmental policy, it is clarifying to restrict the ecosystem concept to the environment of human society, thus not including the society itself. The societal system then relates to the ecosystem in two ways. First, ecosystems are affected by man's activities, and, second, they fulfill societal needs. To illustrate this, we shall introduce various applications of the ecosystem approach in policy analyses.
Next, we discuss the nature of ecosystems in relation to a number of hierarchies. To this end, we distinguish hierarchies of system levels, of organizational levels, and of scale levels. The third hierarchy is especially relevant for the classification and mapping of ecosystems.
The ecosystem concept refers to both abstractions and concrete tangible wholes at the earth's surface. Concrete ecosystems may be called ecotopes or, more generally speaking, ecological land units. These are subject to classification and mapping, and hence are the major concern of this book. Two fundamentally different approaches to the classification and mapping of ecosystems will be introduced: a deductive approach starting from a theoretical framework and an inductive approach starting from empirical data.

Ecosystems and societal systems

Environmental policymaking requires information of various kinds. With regard to the state of the environment, it concerns the detection, prediction, and monitoring of changes. In this context, 'environment' is considered as the entire physical environment, consisting of abiotic and biotic components in a mutual relationship, or in other words, as an ecosystem.

Now we could regard man as a biotic component, too. This would imply that he is part of the ecosystem. However, man is also the cause of environmental problems by definition, he has the ability to foresee the consequences of his actions, and we may expect him to solve them. This places man in a special position beside being just another mammal. In addition, ecosystems are the object of study of the natural sciences. This means that by regarding man as part of the ecosystem, we would not cover all his aspects. We would not include his psychological, social, and cultural sides, which may be very important to understand his motives and actions. For these reasons, it is clarifying to distinguish between two related systems: a societal system on the one hand and an ecosystem as its environment on the other. This distinction is especially helpful for environmental policy analysis. We realize that it is an arbitrary distinction.

Environmental policy is mainly concerned with changes in the environment due to human activities. If these changes involve abiotic and biotic processes, they can be understood as ecological process-response relations. Sometimes, such processes are foreign to the ecosystem, such as toxification by xenobiotic substances, in other cases the processes occur in unnaturally high rates, such as acidification by anthropogenic acids.

The response of the ecosystem can be understood as a change of its characteristics. In turn, the changed ecosystem may form a less suitable or less attractive environment for society. Hence, a reverse influence of the environment on society must also be distinguished.

In this perception, the ecosystem can be visualized as the central module in a cause-effect chain as depicted in Figure 1.1, in which the ecosystem is related to society in two ways.

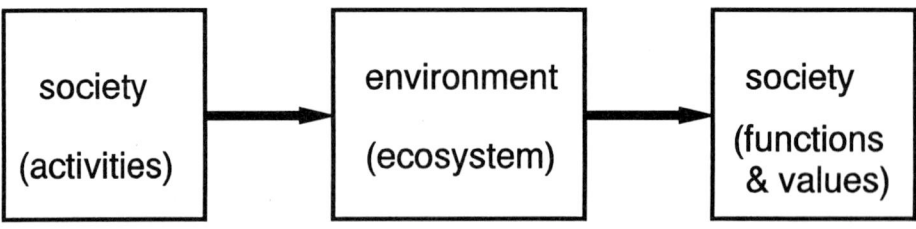

Figure 1.1 The ecosystem is related to society in two ways. First, man's activities influence the ecosystem's characteristics, second, the changed ecosystem forms a less suitable or attractive environment for society

In providing information on the state of the environment to policymakers, we may follow different approaches. Either we focus on individual characteristics of abiotic and biotic components of the environment, or we follow an ecosystem approach in an attempt to take into account all relations between these characteristics.

This book focuses on the ecosystem approach, which is gaining ground in environmental science and policy for a number of reasons:
- a growing recognition of the connection between abiotic and biotic components among environmental policymakers;[1]
- a growing awareness and interest in the value of ecosystems as 'wholes', both as a resource and for their own sake; and
- the higher efficiency of an integrated approach, which requires less information from direct surveys, thanks to correlations between ecosystem characteristics.[2]

This introductory chapter will predominantly be concerned with defining the object of classification more precisely. We shall also go further into the desired applications for environmental policy and we shall introduce two fundamentally different approaches to classification. Finally, the other chapters of this book will be introduced.

Abstract and concrete ecosystems

An ecosystem can be defined as an abstraction, *sensu* General Systems Theory (Von Bertalanffy, 1950), or as a concrete recognizable object: a tangible whole. In the abstract sense, it is a system of structurally related abiotic and biotic components that are also functionally related by physical, chemical, and biological processes (after Tansley, 1935, Chorley and Kennedy, 1971, Van der Maarel and Dauvellier, 1978, Odum, 1983 and others). It is obvious that abstract ecosystems are not the object of this book. They do not exist in reality, but only in the minds of researchers and in their papers.

---

[1] Disadvantages of a too narrow 'compartmental' approach emerged, for example, in the definition of quality standards for individual environmental compartments, such as air, soil, and groundwater. These did not match, due to the fact that the transmittal of problems to other compartments was not taken into account.

[2] This is especially important in relation to field survey and mapping. Attempts to combine information on individual components or characteristics *afterwards* often cause immense problems.

In this book, we limit ourselves to concrete recognizable ecosystems: real-space and real-time ecosystems. Such concrete ecosystems will have to match the definition given above, but extended by criteria on how to delimit their boundaries in space and time. Delimitation in time is seldom discussed, whereas delimitation of ecosystems in space is exactly the problem we are dealing with when questioning how to classify and map ecosystems.

Instead of going into this question here, we refer to other contributions in this book. In contrast, we shall discuss the question why ecosystem classification and mapping are relevant for environmental policy. It appears that the most important reason is found in the desire of policymakers to have access to quantified information related to surface areas. This requires mapping, which in turn requires classification. Indeed, the possibility of quantifying results in terms of surface area is the most prominent advantage of ecosystem classification and mapping over point data in monitoring. The quantified information desired by policymakers concerns the quality of the environment: the quality as affected by man's activities and the quality in terms of the significance for society. To understand this, we may glance at the simple effect-chain of Figure 1.1, focusing on the impact of man's activities on the ecosystem as well as on the reverse, namely, the significance of ecosystems for society.

Then we can recognize that man's activities affect different ecosystems in different intensities, due to differences in the susceptibility of ecosystems to various environmental hazards. Susceptibility is a function of the abiotic and biotic processes which are disturbed or triggered. This implies that the classification of ecosystems is a relevant tool for establishing how large a part of a country or region is susceptible to environmental hazards. It requires that the classification be suited for susceptibility assessments, and that the relevant characteristics for such an assessment can be estimated or quantified.

Secondly, not all ecosystems are valued as equally significant for society, whether it be for their life-supporting functions or for their nature values. As for the life-supporting functions, these have traditionally been covered by physical land evaluation as an applied branch of land classification (FAO, 1976; Beek, 1978). Land evaluation has gradually shifted away from mere soil classification toward a more integrated approach in the last decades, thus increasingly using ecological land classifications (Vink, 1975). The assessment of nature values is essentially a variation on land evaluation as far as the procedure is concerned. However, it often still struggles with the definition of nature conservation goals and the subsequent criteria to measure nature values (see, e.g. Nip et al., 1992). A nature value assessment is often based on surveys of biotic ecosystem components, vegetation, flora and/or fauna, because these are the most highly esteemed. It is obvious that ecosystem classifications for environmental conservation ought to be suitable for significance assessment. Especially if nature values have to be assessed, this requires biotic characteristics to be covered with a classification.

Static and dynamic approaches to environmental policy analysis

After having discussed the relation between the ecosystem and society in general, we may interpret the scheme of Figure 1.1 in geographical terms.
This requires that we first distinguish the spatial pattern of interventions or depositions, i.e. the loads (module 1 in Figure 1.1). Secondly, the ecosystem pattern must be known (module 2 in Figure 1.1). Finally, we must know the pattern of the land use or desired land use (module 3 in Figure 1.1) which determines the requirements on ecosystem quality from land utilization and/or nature conservation points of view.

We can then draw two schemes of application related to Figure 1.1. The first is of a 'static' nature, especially valuable as a means to detect potential or actual environmental hazards or the deterioration of environmental quality. The second is of a 'dynamic' nature meant for prediction.

The *static* scheme is given in Figure 1.2. We can discern three properties of ecosystems in this scheme, namely, susceptibility, significance and vulnerability. The susceptibility of ecosystems can be defined as the degree of change resulting from a hypothetical load. It is specific for a well-defined hazard, such as acidification or groundwater lowering. In many instances, it may also be necessary to specify the response variable, e.g., to distinguish between effects on the groundwater or topsoil quality. Groundwater quality is relevant for vegetations dependent on it or for public water supply, whereas a deteriorated topsoil quality may pose a risk to foodchains based on either plant life or soil fauna. Susceptibility is an ecosystem property, which may be determined

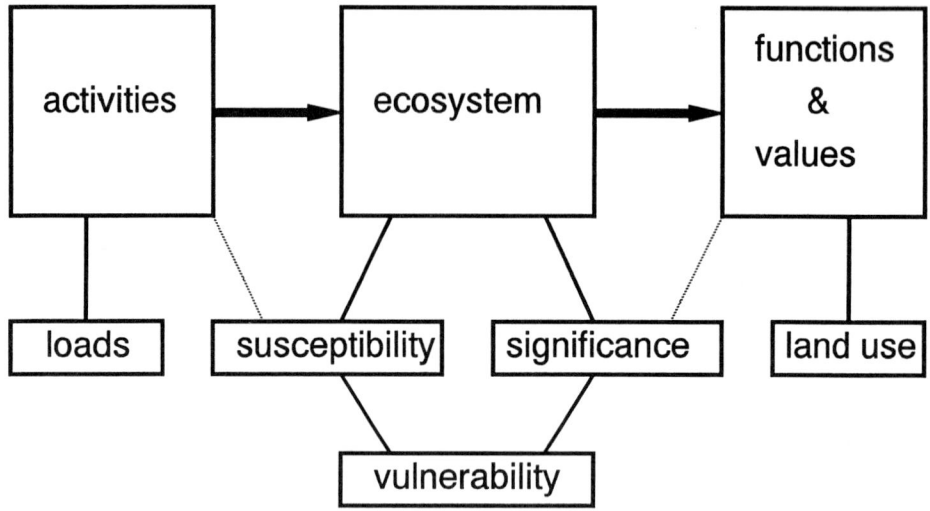

Figure 1.2 The general scheme of relations between society and ecosystem (Figure 1.1) may be interpreted in terms of patterns that can be mapped. Also, we may derive maps that are especially relevant for policy analysis, in this case of 'static' nature. So, in each rectangle we may read 'Map of ...'

in an objective way. It is based on facts only, i.e., sheer response as measured by natural science.

The significance of an ecosystem is the value attributed to it by society with respect to its present and future land use function, other life-supporting functions or nature value. It is, consequently, specific for a well-defined function, such as agricultural production, fisheries, public water supply or nature conservation. So, in assessing the significance of ecosystems we are also dealing with values instead of only with facts.

The vulnerability of ecosystems, finally, we define as the combination of susceptibility and significance (after Veelenturf, 1987). In fact, if we use the term vulnerability in connection with ecosystems, we are not only interested in the degree of change, but we also incorporate an answer to the question of how much we care.

The assessment of susceptibility, significance, and vulnerability resembles the well-known and widely accepted procedure of physical land evaluation (FAO, 1976; Beek, 1978) referred to earlier in connection with significance assessment. As for susceptibility, this can also be regarded as analogous to suitability, as defined in land evaluation. In fact, suitability assessment implies the comparison of land use requirements with relevant land qualities, whereas, analogically, susceptibility assessment implies the comparison of loads with carrying capacity determinants. Suitability assessment heavily leans on the definition of 'relevant and foreseeable land utilization types' (Vink, 1975), whereas susceptibility assessment is based on relevant and foreseeable environmental hazards. Susceptibility can be determined in a relatively simple way, as described in the FAO-Framework for land evaluation (FAO, 1976; see also Vink, 1975), resulting in relative classes. It may also be assessed in a more sophisticated way, resulting in the quantification of critical loads.

The *dynamic* scheme is shown in Figure 1.3. It is relevant in the context of predictive modelling. In this scheme, we can discern the forecasting of the effects in factual terms, the appraisal of the effects in normative terms, and the combination of both, which can be understood as environmental impact assessment. Often, the effect appraisal is inherently connected with the entire impact assessment.

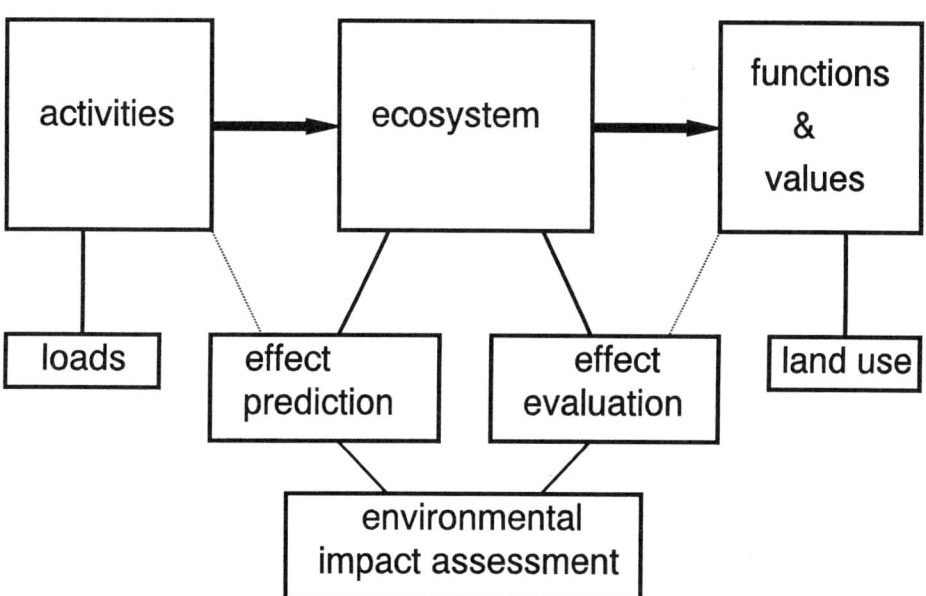

Figure 1.3 The general scheme of relations between society and ecosystem (Figure 1.1) can also be interpreted in 'dynamic' terms. In this case, the analysis starts by predicting changes in ecosystem characteristics

Obviously, the forecasting of effects requires *quantitative* knowledge of loads — the intensity of depositions or interventions — whereas for susceptibility assessment *qualitative* knowledge suffices. Also, the pattern of depositions and interventions must be known in quantitative terms, as well as the process-response relations. Therefore, effect-forecasting necessitates the specification of both the precise type and intensity of the load and the precise type and responsiveness of the effect variable.

In fact, the schemes concerning the static and dynamic approaches could be regarded as representing two families of applications. Families that could be named after the traditions they best fit in, namely, the family of *physical land evaluation* on the one hand, and the family of *environmental impact assessment* (EIA) on the other. Against this background, we could state that ecosystem classifications and maps for environmental policy ought to be applicable in land evaluations and/or environmental impact assessments.

System levels, organizational levels, and scale levels

After having defined the ecosystem in a general way, we should clarify its position in relation to a number of hierarchies frequently distinguished. In this context, three hierarchies are relevant with special reference to ecology: system levels, organizational levels and scale levels.

System levels were defined during the development of General System Theory (Von Bertalanffy, 1950). They were adapted by Chorley and Kennedy (1971) for use in physical geography. Since system theory developed in an attempt to achieve a unifying concept for as many sciences as possible, it is mainly concerned with the question of how to depict and understand natural wholes. This explains the levels recognized in a *structural hierarchy of system levels*, which run from morphological systems, via cascade systems, and process-response systems to control systems with some kind of natural or artificial intelligence. As examples of the latter category, Chorley and Kennedy mention cells, individual plants and animals, ecosystems, and man-environment systems.
System levels are relevant because they relate to different explanatory mechanisms. Hence, they closely correspond to different scientific disciplines.
The upper levels always include the foregoing levels. The choice for a certain level is partly determined by the object of study in the sense that each object

has a highest level at which it can be studied. In contrast, we do not need to regard an object on the highest possible level. So, while we may study a complex object in a simple way, it must be considered senseless to regard a simple object as a complex system. On the other hand, to *fully* understand life, it is not enough to regard it as a simple cascade system. A biotic system requires an explanation by biological mechanisms. Often, however, we are not interested in full understanding, but only in partial explanations.

Organizational levels have been distinguished by, among others, Miller (1975) and Haber (1982). They always concern concrete spatial and functional units: tangible wholes. The hierarchical levels are distinguished on the basis of increasing degree of organization, but the explanatory mechanisms remain largely within the same scientific field.

From Miller (1978) we can retrieve the following hierarchy of organizational levels in society: individual, group, organization, society, supranational system. These are relevant in the realm of social sciences. Haber (1982; 1992), varying on Miller (1975), defined a hierarchy of organizational levels for the whole universe, when positioning landscape ecology in the realm of ecological science. This hierarchy contained, among other things, the subsequent levels of organism, population, community, ecosystem, landscape/ecosystem-complex, human society-environment system, ecosphere, earth ('Gaia').[3]

There has been dispute about the sense of distinguishing the population and community levels (see Rowe, 1961; Schultz, 1967). As for population, it has been argued that this is not a 'tangible whole', and should belong to the organizational level of the organism. With regard to communities, when defining the ecosystem concept, Tansley (1935) already objected to these being considered as wholes loose from their abiotic environment.

These objections do not mean that populations or communities cannot be objects of study. They can, but perhaps they should not be considered as organizational levels in themselves, but as 'aspects' or subsystems of organizational levels. On the other hand, populations would fit nicely in the forementioned hierarchy of societal organization as distinguished by Miller (1978). It is obvious that the definition of organizational levels is arbitrary and a matter of taste. We shall not try to settle it here. But we do state that a hierarchy of various organizational levels, which is different from the hier-

---

[3] It is noticeable that there are some remarkable shifts in this list. The first shift occurs between community and ecosystem, when suddenly abiotic components are included, and the second one is caused by the appearance of man in the 'human society-environment system'.

archy of system levels, can be recognized. We would propose to distinguish at least cells, organs, organisms, ecosystems and (human) society-environment systems.

It is noticeable that in this hierarchy the ecosystem appears as a seperate and individual level. According to Feibleman's 'Laws of Integrative Levels' (Feibleman, 1954), any organizational level can only be understood by giving attention to the levels above and below (see also Rowe, 1961; O'Neill, 1988). A certain organizational level is made up of components of the level below, whereas its significance or function can be understood only in relation to the level above. This fits nicely with ecosystems, because they are made up of biotic and abiotic components, while their functional significance can only be understood fully in relation to the society-environment system.

The third hierarchy is the hierarchy of *scale levels*. It may concern both spatial and temporal scales, which are partly connected. The size and the rate of change of the object of study differ, but no fundamentally new aspects can be discerned, which means that the system level remains the same. Also, the concrete units remain at the same organizational level.

If we apply this to ecosystems and have a look at the definition we gave earlier: 'systems of structurally related abiotic and biotic components that are also functionally related by physical, chemical, and biological processes', it is clear that ecosystems may be defined on all scales of magnitude and complexity. This implies that ecosystems can be distinguished from very small up to the earth as a whole. The organizational level and system level would remain the same, viz. the ecosystem as the most complex example of a control system.

It is obvious that the distinction of scale levels is even more arbitrary than the distinction of the levels in the other two hierachies. We shall, however, not go into it further here, as Klijn will treat this subject later in the book.

In practice, these three hierarchies are often confused.[4] For example, why should the ecosphere be of a different organizational level than the ecosystem? Or why should a landscape? Recently, discussion on the Gaia-hypothesis — or

---

[4] It is noticeable that Chorley and Kennedy (1971, page 4) distinguish 'levels of complexity' *on top* of their main division of system levels, thus lengthening their list of system levels with self-maintaining systems, plants, animals, ecosystems, man, social systems, and human ecosystems. Elsewhere, they regard these systems as *examples* of control systems. Therefore, with the list on page 4 of their book they also add to confusion by mixing up system levels and something like organizational levels.

should we say hypotheses? — also revealed that Lovelock never intended to define *one controlling intelligence* behind his Gaia, but just pointed out that it would classify as a control system in terms of system levels.[5] In fact, we could maintain that Gaia is a very complex example of a control system, with many intelligences influencing its overall response, but not fundamentally different from an ecosystem. Its overall response merely appears being purposeful. This implies, that for our task of developing tools for environmental policy, we need not define a system level higher than that of a control system, nor an organizational level above the ecosystem level. It is only the scale level that differs.

In our opinion, the three hierachies should be regarded as three axes in a conceptual space. However, they are not entirely independent, which is obvious from the fact that the control system level is relevant for biotic organizational levels, such as cells or organisms, but not for the mineral composition of bedrock, for example, which can be understood perfectly by regarding it as a process-response system.

In conclusion, we repeat that ecosystems can be distinguished at many different spatial scale levels, which is important for questions of classification and mapping. Various authors will go into this in this book.

Ecosystems and ecotopes

The concrete ecosystems we are dealing with in this book have a certain spatial extension; this enables mapping. However, we need criteria on how to distinguish them and, secondly, how to map them.

Theoretically, the density of relations within an ecosystem, as compared to the density of relations with its surroundings, is the best criterion of delimiting the boundaries of an ecosystem. However, this principle is not sufficiently practicable. Common practice is much simpler, where ecosystems are defined on the basis of apparent homogeneity in comparison to their surroundings.

---

[5] One of the main reasons for confusion may be the different interpretations of 'intelligence'. Control systems are defined as being controlled by some intelligence. However, this includes both artificial 'intelligence', as in a thermostat, and natural 'intelligence', as in living organisms. Perhaps, it would help to distinguish more system levels by differentiating between *non-living* control systems, *living* control systems which are really self-maintaining and self-reproductive, and *thinking* control systems as living control systems that also act in a purposeful way. Man would classify as the latter, of course, and Gaia would do perfectly as a chaotic complex of the second.

Thus, we can define spatial units: ecological land units or ecotopes, as they have been defined by Tansley (1939) and also independently by Troll (see Troll, 1968; 1970).

Ecotopes are generally regarded as the smallest ecological land units that are relevant in landscape ecology. They are sometimes addressed as 'landscape cells'. Ecotopes are defined as homogeneous ecological units, the spatial expression of ecosystems predominantly determined by their structural characteristics. Van der Maarel and Dauvellier (1978) stress that ecotopes have 'a certain spatial extension'. From the examples of ecotopes these authors and Neef (1967; 1970) give, it appears that ecotopes should be relatively homogeneous with respect to vegetation structure. However, we stated above that concrete ecosystems can be distinguished on many spatial scale levels; this is less strict than the requirements of 'a certain spatial extension' and of homogeneity to vegetation structure permit.

Therefore, for general use irrespective of spatial scale, we argue to use the neutral *ecological land unit*, whereas the term *ecotope* should be used only for ecosystems with homogeneous vegetation structure, i.e., for one spatial scale level.

After having established the spatial scale of ecotopes, we may question whether they are ecosystems or merely the spatial expression of ecosystems. This question was put by Haber (oral communication) followed by the proposal to distinguish between an ecotope's contents and an ecotope's essence: the contents being an ecosystem in a structural and functional meaning, and the essence being the location on the earth's surface without any contents. This is a rather academic and dialectical distinction, resulting from a semantic analysis of the term ecotope. We shall regard ecotopes as ecosystems, whereas maps are the means to express their 'full topology' — their location in longitude, latitude, and altitude.

Runhaar and Udo de Haes discuss the nature of ecotopes more elaborately in this book (see also Veen, 1982; Zonneveld, 1989).

Many different approaches to defining ecological land units larger than ecotopes have been proposed, which can be divided into two groups. The first puts emphasis on the *specific heterogeneity* of certain land units. The second emphasizes *homogeneities* on *larger scale levels* by increasing the distance to the object of study, as if it were a pointillistic painting which requires stepping backwards. The first approach is followed by, e.g. Haase (1976; 1989) who defines *'ecochores'* at different spatial scale levels; the second by, e.g. Klijn (in this book).

Ecosystems and landscapes

So far, we have been speaking of ecosystems to be classified and not of landscapes, even though ecosystem classification and mapping is the major concern of people who call themselves landscape ecologists. Moreover, landscape ecology is sometimes defined as an ecosystem approach to landscapes. This would imply that landscapes are the 'tangible wholes' to be treated as objects. We shall try to elucidate the difference between ecosystems and landscapes.

Landscape ecology is based on the recognition of two groups of relations: vertical or 'topological' relations between abiotic and/or biotic components on the one hand, and horizontal or 'chorological' relations between land units on the other. Today the last group of relations seems to attract the most attention, as exemplified by the first volume of the Studies in Landscape Ecology of the International Association for Landscape Ecology (Vos and Opdam, 1993), which is entirely dedicated to these chorological relations.
The study of such relations, however, requires the distinction of ecological land units first. While no strongly deviating views on the ecotope concept appear, they do exist on the delimitation of larger spatial units. These result from a difference in emphasis on either topological or chorological relations.

As mentioned in the former section, the delimitation of the boundaries of ecological land units above the level of ecotopes can be based on either a *specific homogeneity* or a *specific heterogeneity*: recurrent patterns of ecotopes. When the homogeneity, which results from controlling factors in the topological dimension, is emphasized, there is no reason to deviate from the term ecosystem. Both the patterns Godron observes in his contribution to this book and the hierarchical classification scheme presented by Klijn are entirely based on such topological relations between ecosystem components. They recognize ecosystems on the earth's surface on the basis of relative homogeneity.

If, on the other hand, land is considered as a habitat for certain fauna species, it may be necessary to focus on the relatively autonomous movements of animals between sub-habitats or functional sites (Vos et al. 1982). Animals with large home-ranges especially use geographical areas which need to be heterogeneous. The ecological land units that are distinguished in such a 'habitat-approach' are defined on the basis of specific heterogeneity or specific pattern characteristics which differ from those in other 'landscapes'. This requires the recognition of 'landscapes' and a specific approach from a faunal, i.e., species point of view. In this respect, landscape can be defined as a mosaic of patches (Urban et al. 1987) and corridors within a matrix (Forman and Godron 1986) or a chorological conglomerate (Zonneveld 1984) of ecotopes.

Though this 'choric' approach *sensu* Haase (1973; 1976; 1989) belongs to landscape ecology, we would rather not regard it as an ecosystem approach, as it is species-oriented and, hence, essentially autoecological. This implies that we do not agree with the forementioned definition of landscape ecology being an ecosystem approach to landscapes: it is certainly an *ecological* approach but need not be a *systems* approach.

Deductive and inductive approaches to classification

In this introduction, we would also like to address one point of difference in classification approaches. It concerns the difference between deduction from theory and induction from data. In the next chapter, Zonneveld will treat many other points of difference.

A deductive approach starts with a theory, a hypothesis of how nature is built and functions. For ecosystem classification, such a theory could imply a notion of what causes the differences between ecosystems. It allows focusing on these supposedly differentiating factors from the beginning of the classification enterprise on. This results in relatively strict guidelines to classifying and mapping.

Of course, these theories do not emerge from nothing. They are based on previous research and experience. The main scientific task of this approach is testing the hypotheses. Starting from a general conception, the theory is adapted after trials. In this book, the classification framework presented by Klijn and the ecotopes classification of Runhaar represent such a deductive approach.

In an inductive approach, priority is put on the sampling of data, followed by applying relatively standard quantitative methods to arrive at a classification, e.g., TWINSPAN. It starts without any postulate about relations, and assumes that similarities and dissimilarities will emerge from the data.

However, the sampled data largely determine the outcome of whatever quantitative method is used. Hence, it is of the utmost importance to be very careful when sampling. The effect on the resulting classification of undersampling or oversampling of some ecosystems must not be underestimated. A 'wrong' sampling technique often results in the distinction of unique types. In addition, the choice of ecosystem characteristics surveyed may affect the outcome.

In this book, the contribution of Bunce is an example of this inductive approach starting from primary data. Others use secondary data, e.g., Latour et al. and Groen et al.

Conclusions

This book is about ecosystem classification for environmental policy. To this end, it is clarifying to distinguish the ecosystem from a societal system. The ecosystem can then be regarded as the environment of human society, consisting of related abiotic and biotic components. It may be negatively affected by human activities, but is also significant for society for its life-supporting functions and nature values.

In this book, we are only interested in ecosystems as concrete tangible wholes on the earth's surface, and not in ecosystems in abstraction.

The classification of concrete ecosystems is not a goal in itself. It is mainly a tool for environmental policy analyses. We distinguished two main families of applications: those of a static nature resembling physical land evaluation, and those of a more dynamic nature mainly associated with environmental impact assessment. Both families of applications have to meet the criteria of being related to causes of ecological change and the appraisal of ecosystem quality by society. Consequently, ecosystem classifications should preferably be related to environmental hazards and to environmental policy targets at the same time.

Ecosystems can be understood as systems in terms of General Systems Theory, forming one of the most complex examples of control systems. In addition, ecosystems form a specific level in a hierarchy of organizational levels. As such, the ecosystem can be regarded as the one and only relevant level between organisms and individual abiotic components on the level below, and the society-environment system on the level above.

Then, ecosystems may be distinguished at different spatial scale levels from very small up to the entire earth.

We distinguished between ecosystems as functional units and ecological land units as their spatial expression. For ecosystems of a certain spatial extension with homogeneous vegetation structure and relatively constant site factors, the term ecotope is traditionally used and seems appropriate. Larger spatial units that do not meet the criterion of homogeneous vegetation structure should be called ecological land units.

In our opinion, the term 'landscape' as an alternative for ecosystems at larger spatial scale levels is not really clarifying in the present context. It may, however, be a relevant concept in other contexts, such as for faunal habitat studies, for studies focusing on scenic aspects of landscapes, or for other studies for which a 'choric' approach is most appropriate.

Introduction to the other chapters

Finally, we should introduce the other chapters of this book to you. The following topics are addressed:
- a further introduction to ecosystem classification and mapping from theoretical points of view;
- examples of ecosystem classifications from various methodological approaches; and
- the application of classifications of ecosystems and/or ecosystem components for predictive modelling, for nature valuation, and for the monitoring of nature values.

These topics are covered both from a theoretical point of view and from experiences with the classification of ecosystems for various applications in different countries. Most authors address more than one topic and freely wander from theory to practice and back. Applications are never treated without also addressing theories or the practice of classifying ecosystems, nor are practical classifications discussed without regarding their applications. This can be understood as revealing the applied character of ecosystem classification, which is never a goal in itself.

We distinguished three parts in the book: theory, approaches to classification, and applications. However, from the above observations it should be clear that classifying the various contributions into one of these three categories was no less difficult than ecosystem classification itself. We ordered the chapters somewhat arbitrarily. The papers next to the beginning and end of each part manifest a gradual shift to the adjacent parts.

At this point, we must also emphasize that the following chapters are not tuned *in extenso* to achieve a unanimously supported nomenclature, use of concepts, or approach. The various contributions are a reflection of the variety of scientific communities struggling with the subject. Each author uses his own terminology, and expresses his own ideas. We think it is better to become acquainted with ecologists' diversity, meaning that this first general chapter is mainly intended as a reference for comparing the different points of view.

In chapter two, Zonneveld treats the goal and principles of classification as a means for systematic ordering. He goes into the use of diagnostic properties and guiding principles, and also introduces us to many different approaches to classification, typification, and mapping.

Chapter three, delivered by Haber, deals with the concepts of ecosystem and ecotope, and especially with the problems in applying the concepts in practice. Haber proposes to deal with application problems for practical environmental planning by a kind of 'scaling', which enables integrating knowledge on various organizational levels.

Godron, in the next chapter, discusses the patterns of ecosystems on the surface of the earth. He explains the controlling factors that determine these patterns and discusses which factors are easy or difficult to recognize. He stresses the role of spatial scale.

After these contributions that mainly set the stage, we gradually shift toward various approaches to classification.

In chapter five, Klijn argues for a deductive theoretical approach based on the recognition of pattern-controlling factors, as presented in the previous chapter. To this end, he postulates a framework for classification, related to a hierarchical ecosystem model. Thus he derives classification guidelines *sensu* Zonneveld for different spatial scale levels as also distinguished by Godron. He exemplifies some classifications with maps and applications.

In chapter six, Lenz introduces an apparently entirely different approach based on matter budgets, i.e. processes, rather than on structural characteristics of ecosystems. However, this also develops from a deductive theoretical approach which is comparable to the one in the previous chapter. Perhaps the approach is not so different after all, but rather complementary? Lenz illustrates the use of the classification with the assessment of critical loads of acidifying substances in a region of Germany.

Runhaar and Udo de Haes, in the seventh chapter, explicitly focus on the scale level of ecotopes. As the third representatives of a deductive approach, they present a classification scheme for ecotopes that was deliberately developed for environmental impact assessment. At first glance, their 'ecotope' appears to deviate from what it is commonly understood to be. However, they explain the rationale behind it, viz. nationwide applicability, and relate it to the more common conception.

Bunce wrote the eighth and final chapter in the part on approaches to classification. He explains how a land classification for the whole of Britain was made, following an inductive approach with environmental data from grid cells. He discusses the application of the resulting map, mainly to stratify further surveys, as well as a number of other recent applications.

Latour et al. open the third part of this book, i.e., on applications. In chapter nine they question whether ecosystem classifications are appropriate for modelling at all. Instead, they argue a species-centered approach, which would be more flexible and more closely tied to what society considers worthwhile to protect. The applicability of this species-centered approach is illustrated by a number of pilot studies, mainly carried out in the Netherlands.

In chapter ten, Claessen et al. apply the ecotope classification that was introduced by Runhaar and Udo de Haes in chapter seven, in a predictive model on the effects of changes in water management. In addition, a related classification of 'ecoseries' is used to account for abiotic processes. The model has succesfully been applied in a nationwide policy analysis for the Netherlands' policy on drinking water supply.

De Blust et al. begin a series of three chapters on environmental quality assessment with special emphasis on nature values. Based on experiences in Flanders, they discuss the advantages and disadvantages of maps, basically of biotically defined spatial units, on the one hand, and grid-cell data bases filled with information from floristic surveys on the other. They give recommendations on how to best combine such diverse information.

In chapter twelve, Brandt et al. draw from a wealth of experience on monitoring the dynamics of 'small biotopes' in Denmark. In agricultural landscapes, small (semi-)natural elements, such as hedgerows and pools, constitute major nature values. Monitoring such small elements requires its own rationale as to classification and data management. The experience of the authors with monitoring guarantees the importance of the 'lessons learned' they present in their chapter.

In the last chapter, Groen et al. discuss the possibilities of using floristic information to estimate the occurrence of ecosystems in grid cells. They argue that such floristic data can be updated easily and rapidly, which is favourable for monitoring purposes. However, the use of species composition as an indicator of ecosystem occurrence is not without problems, as they demonstrate by comparing the results based on floristic data with other data bases that focused on ecosystems from scratch.

References

Beek, K.J., 1978. *Land evaluation for agricultural development.* ILRI publication 23, Wageningen.
Chorley, R.J. and B.A. Kennedy, 1971. *Physical Geography. A Systems Approach.* Prentice-Hall International Inc., London, 370 pp.
FAO, 1976. *A framework for land evaluation.* FAO soils bulletin 32, Rome. Feibleman, J.K., 1954. Theory of integrative levels. *British Journal for the Philosophy of Science* 5: 59-66.
Forman, R.T.T. and M. Godron, 1986. *Landscape Ecology.* Wiley & Sons, New York.
Haase, G., 1973. Zur Ausgliederung von Raumeinheiten der chorischen und regionischen Dimension. *Petermann's Geogr. Mitt.* 117: 81-90.
Haase, G., 1976. Die Arealstruktur chorischer Naturräume. *Petermann's Geogr. Mitt.* 120: 130-135
Haase, G., 1989. Medium scale landscape classification in the German Democratic Republic. *Landscape Ecology* 3/1: 29-41.
Haber, W., 1982. Naturschutzprobleme als Herausforderung an die Forschung. *Natur und Landschaft* 57/1: 3-8.

Haber, W., 1992. Erfahrungen und Erkenntnisse aus 25 Jahren der Lehre und Forschung in Landschaftsökologie: Kann man ökologisch planen? In: F. Duhme, R. Lenz and L. Spandau (eds.): *25 Jahre Lehrstuhl für Landschaftsökologie in Weihenstephan mit Prof.Dr.Dr.h.c. W. Haber.* Festschrift, Weihenstephan.

Miller, J.G., 1975. The nature of living systems. *Behavioral Science* 20 (1975): 343-365.

Miller, J.G., 1978. *Living Systems.* McGraw-Hill Inc., 1102 pp.

Neef, E., 1967. *Die theoretische Grundlagen der Landschaftslehre.* Verlag H. Haack, Gotha-Leipzig.

Neef, E., 1970. Zu einigen Begriffen in der Ökologie. *Arch. Naturschutz Landschafsforsch.* 10/4: 233-240.

Nip, M.I., J.B. Latour, F. Klijn, P.K. Koster, C.L.G. Groen, H.A. Udo de Haes and H.A.M. De Kruijf, 1992. Environmental Quality Assessment of Ecodistricts: a Comprehensive Method for Environmental Policy. In: D.H. McKenzie, D.E. Hyatt and V.J. McDonald (eds.): *Proceedings of the International Symposium on Ecological Indicators, held Oct. 16-18, 1990 in Fort Lauderdale, Florida.* Elsevier Applied Science, London, etc., pp. 865-881.

Odum, H.T., 1983. *Systems Ecology, an introduction.* John Wiley & Sons, New York.

O'Neill, R.V., 1988. Hierarchy Theory and Global Change. In: T. Rosswall, R.G.Woodmansee, and P.G. Risser (eds.): *Scales and Global Change.* John Wiley & Sons, Chichester/London.

Rowe, J.S., 1961. The level-of-integration concept and ecology. *Ecology* 42/2: 420-427.

Schultz, A.M., 1967. The ecosystem as conceptual tool in the management of natural resources. In: S.V. Ciriacy-Wantrup and J.J. Parsons: *Natural resources: quality and quantity.* University of California Press, Berkely and Los Angeles.

Tansley, 1935. The use and abuse of vegetational concepts and terms. *Ecology* 16: 284-307.

Tansley, 1939. *The British isles and their vegetation.* Cambridge.

Troll, C. 1968. Landschaftsökologie. In: R. Tüxen (ed.): *Pflanzensoziologie und Landschaftsökologie.* Junk Publishers, The Hague.

Troll, C. 1970. Landschaftsökologie (geoecology) und Biocoenologie. Eine terminologische Studie. *Rev. Roum. Géol., Géophys. et Géogr.: Série de Géographie* 14/1: 9-18.

Urban, D.L., R.V. O'Neill and H.H. Shugart Jr., 1987. Landscape Ecology. A hierarchical perspective can help scientists understand spatial patterns. *Bioscience* 37/2: 119-127.

Van der Maarel, E. and P.L. Dauvellier, 1978. *Naar een Globaal Ecologisch Model voor de ruimtelijke ontwikkeling van Nederland.* Staatsuitgeverij, The Hague.

Veelenturf, P.W.M. (ed.), 1987. *Landschapsecologische kartering van Nederland, Fase 1.* Studierapport 39, Rijksplanologische Dienst. Staatsuitgeverij, The Hague.

Veen, A.W.L., 1982. Specifying the concept of landscape cell (ecotope) in terms of interacting physico-chemical processes and external vegetation characteristics. In: *Proc. Int. Congr. Neth. Soc. Landscape Ecol. Veldhoven, 1981.* Pudoc, Wageningen, 1992.

Vink, A.P.A., 1975. *Land Use in Advancing Agriculture.* Springer Verlag, Berlin etc., 394 pp.

Von Bertalanffy L., 1950. An Outline of General System Theory. *The British Journal for the Philosophy of Science* 1 (1950-1951): 134-165.

Vos, C.C. and P. Opdam (eds.), 1993. *Landscape ecology of a stressed environment.* Chapman & Hall, London.

Vos, W., W.B. Harms and A.H.F. Stortelder (eds.), 1982. *Vooronderzoek naar landschapsecologische relaties tussen ecosystemen.* De Dorschkamp, report 246, Wageningen.

Zonneveld, I.S., 1984. Landschapsbeeld en landschapsecologie. *Landschap* 1: 5-9.

Zonneveld, I.S., 1989. The land unit. A fundamental concept in landscape ecology, and its applications. *Landscape Ecology* 3: 67-86.

Basic principles of classification 2

Isaac S. Zonneveld

ABSTRACT - Classification is a systematic ordering of the object of research, in this case, ecosystems at the earth's surface or, in other words: landscape units as 'holons'. As for general principles of classification, we can learn a lot from the best-known classification, the taxonomical classification of species. This has functioned as an example for similar classifications of land attributes, such as soil or vegetation.
For a classification of ecosystems, we must select diagnostic characteristics from the large number of ecosystem properties, for which selection guiding principles are an aid. Examples are given of the most important guiding principles and possible diagnostic characteristics are discussed. Also, it is argued that typifications of land attributes are the best diagnostic characteristics for ecosystem classification.
Two different approaches to classification are compared, viz. by agglomeration, which leads to the most pure typification, and by sub-division, which is always connected with mapping. These two approaches are related to two different hierachies. A hierarchy of agglomeration is related to classification by agglomeration with abstract boundaries in a typification, whereas a hierarchy in space is related to classification by sub-division with concrete boundaries in the field. A map's legend, for example, is essentially a classification by sub-division. However, units can be described by means of the units of a typification.
Finally, it is questioned whether a world embracing typification is worthwhile to strive for. It is concluded that for most applied surveys an ad-hoc classification by sub-division is the most appropriate, but using land attribute typifications for defining the legend units.

Introduction

This book is about methods of classifying ecosystems in such a way that the results will best serve conservation and environmental policy matters.

By 'ecosystems' we mean bodies at the earth's surface. They represent landscape units or land units: 'holons' (see Naveh and Lieberman, 1983) that, depending on the spatial scale, range from ecotopes as the smallest to units of larger size. In other words, we are dealing with a landscape ecological subject.

In scientific jargon 'classification'[1] means a systematic ordering of the object of research, that is, putting the objects in 'pigeon holes' from which they can easily be retrieved. This necessarily includes the labelling of the units with a code or name. A hierarchic structure in the system highly facilitates the 'storing' and 'retrieval' of the material. Classification is a scientific discipline bound to strict logic and is a basic requirement to grasp the object of study.

The classical object of scientific classification is organisms, (the first action of man in paradise, Genesis 2:19). Organisms are clearly defined individuals. They are classical examples of individual wholes. These wholes may show morphological resemblance based on 'consanguinity', hence, real relationship. The taxonomist therefore claims to develop a natural system which shows, to a certain extent, correlation with the phylogeny, the development lines of the organisms, and the degrees of genetic kinship. Although in practice the evolutionary relationship cannot always be verified exactly, the evolutionary considerations are used as guiding principles for the selection of properties that can be used to characterize and recognize units: taxa, and to design the hierarchical structure of the system, with families, orders, classes, etc. These properties, or diagnostic characteristics, are always purely morpho-metric, which means measurable on the concrete object. Thanks to its great success and overwhelming influence on the biological sciences this biological taxonomy became the great example for systematic ordering in science.

Consequently, for land attributes such as soils or vegetation, (syn-)taxonomic classification systems have also been developed with a similar character. This, however, requires the presumption that in soils and vegetation, basic wholes can be distinguished of a certain individuality or, in other words, that the

---

[1] Elsewhere the term classification is used with slightly different meanings. Evaluation may be part of it, but ordering in one way or another is always included.

theory of Smuts (1926) about hierarchic wholes would hold for soils and vegetation. This means that soils and vegetation are systems in the sense of systems theory (see Von Bertalanffy, 1968). Indeed, in soil science and vegetation science this is assumed, as is illustrated by the emergence of the concepts *(phyto-)coenon* for a vegetation whole, and *pedon* for the basic soil unit.

The main difference from the taxonomy of organisms, however, is that soils and vegetation form a continuum at the earth's surface. Hence, the distinction of basic individuals as such already forms a problem. It is obvious that the same holds for ecosystems at landscape scale.

In contrast to biology, which deals with individual organisms, geography, with its strong orientation on space and spatial forms, follows other ways of thinking in the systematic ordering of its subject. Its ordering tends to separate what is different by cutting off, rather than combine what is similar by clustering. It is obvious that in its classification approach, landscape ecology, as a daughter science of both biology and geography, is influenced by both parents.

In the following we will discuss the need and possibilities for developing classification systems for landscape units, such as ecosystems, and the role of classification systems for other land components or attributes, such as vegetation, soil, and landform.

Guiding principles, properties, and the selection of diagnostic characteristics

Classification is *abstraction* (Van Melsen, 1955). This means that for reasons of study, we select only some of the many properties of a concrete tangible object and use these to describe abstract units that are supposed to represent reality. Classification, as explained in the former section, supposes a system character of the subject, except in the simplest cases of single value survey. The selected properties to be used as diagnostic characteristics do not need to be necessarily important, in the functional sense. They are chosen predominantly because of their recognizability, measurability, etc. They do not necessarily determine the 'character' or 'nature' of the subject, they only indicate it.[2]

So, in each classification one always distinguishes: 1) guiding principles; 2) properties; and 3) diagnostic characteristics.

1) *Guiding principles* are the principles or rules according to which the diagnostic properties are chosen and calibrated, and the hierarchy of a classification is assessed. These rules can be derived from internal functions of the system, external relations with the surroundings, or based on the genesis. They depend on the purpose of the classification.[3]

2) *Properties* are all aspects, attributes, or characteristics of the object to be classified, irrespective of their value for function, recognition, or classification.

3) *Diagnostic characteristics* are selected properties, abstracted to describe units. They must principally be measurable, i.e., 'morphometric' properties of the object of classification, and they must be relatively constant in time. They must never be derived from the environment of the subject.[4] Preferably, it should be possible to qualify or quantify diagnostic characteristics with readily available means. This may be by laboratory tests, but only in a minority of the cases; the majority should be diagnostic characteristics that can be observed in the field.

Systems must be characterized by system properties. As will be explained later, the classification of ecosystems as 'holistic entities' can be done by

---

[2] The English word 'characteristic' commonly used in this context, may lead to confusion. In Dutch, a diagnostic characteristic is called a 'kenmerk': 'distinction mark', but in the sense of a natural property, not as a label added afterward. The terms 'karakter' in Dutch or 'Charakter' in German exclusively point to internal functional aspects and not to externally visible parts of the classified object. The term characteristic can also easily be confused with 'property', which is a more neutral term. If we want to indicate a property that has to be used for distinguishing one unit from another, we should always specify it as a 'diagnostic characteristic'.

[3] The term 'criterion' is avoided as much as possible. Its literal meaning is 'measuring rod', which may point to a guiding principle as well as to a 'type of diagnostic characteristics'.

[4] This may be clear from an example of vegetation survey. If the vegetation should be used to indicate environmental conditions, diagnostic characteristics derived from the environment would cause circle reasoning.

using guiding principles and diagnostic characteristics accordingly. In practice, however, it may often be useful to use classifications of ecosystem components as well: the 'land attributes',[5] such as bedrock type, landform, climate, water, soil, vegetation, fauna and man. Classification units of the land attributes are used to characterize the 'holistic' land units.

Therefore, in the following examples of guiding principles and diagnostic characteristics, examples from classifications of the most relevant land attributes will also be included.

Examples of guiding principles and diagnostic characteristics

A general guiding principle for classification is the *convergence of evidence*. This means that in one object, or group of objects, various properties converge or coincide, in contrast to other objects. This points to the individuality of that object and reflects its system character. Diagnostic characteristics should be selected from those converging properties.

In addition, diagnostic characteristics should preferably be *system properties* and not properties of just the composing parts, i.e., of units of a hierarchically lower rank. The latter may only be used to define specific property *combinations*. For example, species composition can be regarded as a system property of vegetation, but a single species cannot.

The concepts of guiding principles and diagnostic properties may be clarified further by referring to existing classifications of soil, vegetation, geomorphology, other land attributes and, finally, ecosystems.

In soil classification one may, for example, select diagnostic characteristics from an 'ecological guiding principle', i.e., quality of the soil as rooting environment. Properties such as mineral content, permeability for water and air may then be selected. Such a guiding principle is especially useful when the soil classification is going to be used further in land use planning or landscape ecology in general.

However, for a better scientific understanding a 'genetic guiding principle' may be preferred, i.e., focussing on the soil genesis. Hence, processes of weathering, lessivage, homogenisation, etc. may lead the selection of diagnos-

---

[5] In landscape ecology the components in the vertical direction, such as atmosphere, vegetation, fauna, man, water and rock, are often indicated with the term 'land(scape) attribute'. In the horizontal direction we use the term 'land(scape) unit'.

tic characteristics. Properties related to this principle may be the stratification in soil horizons with different colour, consistency, texture, etc.

Fortunately, there is a certain correspondence between sets of characteristics selected according to 'ecology' and 'genesis', respectively, because of the system character of the soil. Thus, podzolic features point to poorness in nutrients, whereas intermediate weathering stages may indicate an abundance of available minerals. On the other hand, features like colour, which have no ecological significance but are a side effect of the genetic history, may serve as very clear diagnostic characteristics.

The above-mentioned requirement that diagnostic characteristics should be morphometrical and easy to measure poses a special complication, especially when using the ecological guiding principle. Difficulties arise because the really important operational factors, such as nutrient availability and vapour pressure in relation to intake into the roots, are usually not measurable. Van Wirdum (1981) distinguishes between so-called operational, conditional, positional and, later, also hereditary factors. Contrary to operational factors, conditional and positional factors are often clearly visible and can be easily used as diagnostic characteristics, again due to the system character of the object. Soil texture is a good example of a conditional factor strongly influencing the structure, consistency, nutrient exchange capacity, and water retention capacity of the soil. As an example of a positional factor, a colluvial soil at a footslope will likely receive more run-off water than soils up-slope.

For vegetation classification, a 'functional guiding principle' primarily refers to the useful qualities of the vegetation for use by man, by providing, for example, wood, useful species, vegetation cover, or slope stability.

In addition, the indication of environmental conditions is a very important commitment of vegetation science both for scientific and applied purposes. It means assistance in understanding the ecological function of other land attributes. This represents the 'ecological guiding principle'.[6]

As in soil classification, we can distinguish a 'genetic guiding principle' in vegetation classification, i.e., the development stage from instable pioneer to more stable succession stages, or according to various degrees of human impact or management. It can guide the selection of diagnostic characteristics such as structure and floristic species composition.

---

[6] The terms 'functional' and 'ecological' are somewhat arbitrary. The so-called ecological properties of soils are functional for land use. Also, one might state that the indicative value of vegetation, even though it follows from an 'ecological guiding principle', has an important function as indicator of environmental quality.

The vertical structure of vegetation cover is a very good example of a morphometric property that can be used as a diagnostic characteristic. The same holds for the floristic species composition.
Although stable properties are most appropriate as diagnostic characteristics, cyclic changes, if seasonal, may be used as such. This is especially true when relatively frequent sequential remote sensing is possible as is the case with satellite recording. The so-called multi-temporal satellite images are a means to record differences in the phenological cycles of the vegetation cover and, thus, to discriminate between different communities.

For geomorphological classifications, properties of landforms that are directly functional for land use are, for example, the degree and form of slopes, sedimentation, erosion pattern and intensity, drainage patterns, etc.
In contrast to this functional approach, the purely scientific aim of geomorphology is the understanding of the landforms surrounding us. It is clear that from this point of view the genesis of these landforms is the most logical guiding principle to select diagnostic characteristics. Therefore slope degree, shape of the slope, and horizontal configuration of slope types (pattern) are often used as diagnostic characteristics.

As for other land attributes, it suffices to be brief. For geological classifications, lithological differences are generally more important than orogenesis as a guiding principle. The contribution of atmosphere or climate, hydrology, animal kingdom, and man are much more dynamic. These attributes are especially relevant for their influence on other land attributes. Of these attributes some static properties, averages of seasonally fluctuating properties, but also processes as such can be used as diagnostic characteristics.
For man's influence, reclamation history or related cultural developments could be guiding principles. Land use is the main diagnostic characteristic.

For ecosystems or landscape units as wholes, the guiding principles for selecting diagnostic characteristics may be strongly influenced by the purpose of the classification.
Also, the degree of human influence on the landscape is important. In purely natural landscapes geomorphologic and climatic factors may be most decisive in combination with soil forming processes, whereas in cultural landscapes the reclamation history may be more important.

Combinations of many properties of the land attributes can be used as diagnostic characteristics, but shape, size and configuration of pattern elements can be used as well.

The special problems of classifying landscapes as wholes will be discussed in a later section, after a further treatment of the procedure of classification.

Two procedures for creating classes: subdivision or agglomeration

In the former section we highlighted the nature of diagnostic characteristics by discussing what kind of properties can be used for classification purposes and according to which principles they may be selected. Now we shall look at two different procedures that lead to classes of different character. These are specified below.

- Classification by *sub-division* or *descending* classification. Begun by regarding the subject as one unit at the highest level of division, and then cutting it into subsequently more and more segments based on differences.

- Classification by *agglomeration* or *ascending* classification. Begun by regarding the smallest spatial units as homogeneous, and then clustering these according to similarities. This may also be called 'typification'.

The latter is the purest abstraction in which the shape and place of units at the earth's surface do not count. In practice, one often finds a combination of both principles. A legend of a not-too-detailed map, in which units are described by means of a classification by agglomeration, such as the Braun-Blanquet syntaxonomy or the US Soil Taxonomy, is necessarily such an integration. In both procedures the previously mentioned principle of convergence of evidence plays an important role in selecting the boundaries between the classes and the different levels of division.

The universal use of the first approach is in mapping: the simplified depicting of reality on a much more detailed scale, accompanied by a description of the units in a legend which may have hierarchic levels. Classifications of landform, soil, and vegetation of such a character have been designed, especially as a base for mapping.

The best examples of typification are, as mentioned, the taxonomic classifications of plants, animals, soils and vegetations. Also, the system of classifying minerals is built on this principle. It appears, interestingly, that in landscape ecology courses, students with a geographical background have more problems with a system based on agglomeration of basic (land) units. Biologically trained students, in contrast, have problems thinking in systems by subdivision, because they are used to their clustering system. The dominant approach in each discipline is clearly reflected by this.

Classification by sub-division

Classification by sub-division first depicts the whole subject on a scale that can be taken in at a glance. For landscapes or land attributes such as soil, vegetation or landform this means a reduction in size with the help of aerial photography or other geodetic and cartographic means. Then, based on this comprehensive view, the image is cut into segments while looking for convergence of evidence. This, of course, is done stepwise, thus resulting in various division levels that can be regarded as forming a hierarchy. The resulting pieces at the various levels are then described. The final result has *essentially the character of a map legend.* Even if the map is not presented, the units always refer to sections at the earth's surface. The higher in hierarchic rank, the more unique they are. At the lower division levels, similar patches may occur in different mosaics belonging to different units at a higher division level in the hierarchy. This is an essential difference with classification by agglomeration, where a lower unit can principally occur only once in the hierarchic system.
The sub-division is executed according to certain guiding principles. Regarding the question of which level of division is appropriate to stop further splitting, we maintain that the use of the principle of convergence of evidence will guarantee the most 'natural' sub-division.

The guiding principles for selecting diagnostic characteristics in pure classifications of this type differ for each hierarchic level, which, in this case, is determined by the size of the survey area and the mapping scale. Consequently, in the near-to-total geospheric dimension the only land attribute that gives a clear picture is climate. Therefore, classifications of soils and vegetation at a global scale, are based on the guiding principle that climate influences soil genesis, which is a guarantee that the continuum of the earth's land surface can be cut into large, homogenous sections. At this scale, the concept of 'zonality' is connected with this divisive classification procedure. For more details see the contributions by Godron or Klijn in this book.

The informative value of classes achieved by sub-division is limited strongly by the difference between shape, form, and size of the patches at the earth's surface on the one hand, and those on the map which depends on the mapping scale on the other. This considerably reduces the usefulness of classification by sub-division in a pure form.

Often, relatively simple maps with single-value classes are the result. Therefore, classification by sub-division is best suited for global scales, where only few properties are relevant, namely those related to climate or major landforms that are homogeneous over large areas. It is also applicable for detailed maps with a special purpose, namely, for those cases where the area can be 'copied' metre by metre and each individual patch of the mosaic can be described as an individual unit. For intermediate dimensions classification by agglomeration is more suited.

However, if mapping is involved, sub-division always logically appears. As mentioned before it is the basis of provisional aerial photo-interpretation and it is inherent in establishing a hierarchy in a map legend. This also holds if a classification by agglomeration is used to define the object, such as vegetation. In such a case, the map patches are described in the legend representing the units resulting from classification by agglomeration. On detailed scales these patches are usually rather homogeneous and can be well characterized by one type. On less detailed scales, however, the patches on the map rather represent mosaics of types in reality. These can be very well described as complexes of the taxonomic units. The majority of soil maps and most vegetation surveys using floristic typification, e.g, according to the Braun-Blanquet classification, follow this principle (see Zonneveld 1988b).[7]

---

[7] It should be emphasized that using a higher hierarchical level from a classification by agglomeration does not solve the problem because this hierarchy is independent of scale. No larger spatial units will result. See also sections 5 and 6 or Klijn, chapter 5 in this book.

Classification by agglomeration, or typification

Classification by agglomeration starts by clustering the smallest units according to similarity into 'types': typification.[8] In plant and animal taxonomy, such a clustering concerns the individual plants and animals. In abiotic spheres it may, for example, concern individual minerals or sand grains. If we deal with continua at the earth's surface, it may concern the smallest structural units such as vegetation patches, soil bodies, or land units that can be recognized in the field or in aerial photographs.

From all the properties of the subject that were described, the diagnostic characteristics are retrieved by statistical methods, varying from the classical 'table method' to computerised multivariate analyses such as 'cluster analysis' or 'ordination' as examples of techniques especially applied in vegetation classification[9] (see Jongman et al., 1987; Gauch 1985; Whittaker, 1973).

In comparison with the classification of organisms, the classification of spatial units such as land units, soil bodies or vegetation poses an extra problem, namely, its being continuous in the field. This 'concrete continuum' makes it less easy to take samples.

But even more important than this 'concrete continuum' is the continuous character in abstraction. Even in the classification of organisms, which can always be separated into concrete individuals, it is not always possible to find clear boundaries. In abstraction, there may be gradual transitions in properties between populations, thus making the drawing of boundaries arbitrary. Such related populations then form so called 'clines'. These may show considerable differences at the extreme edges of the area of distribution but with all possible transitions in between.[10]

---

[8] Also, the term 'typology' exists, but this is also used for units resulting from classification by sub-division. Therefore we shall use 'typification', which primarily points to unambiguously characterizing the units.

[9] In the last decades, rapid developments took place in this field and these methods are now frequently used for vegetation typification. Among these methods one may distinguish between so-called clustering methods and ordination methods. Sometimes, the clustering methods are erroneously addressed as 'classification methods'. However, both are means to achieve classification.

[10] There are, for example, certain bird species with circumpolar distribution of which the individuals living at the extreme ends of the relatively narrow longitudinal area differ at species level, i.e., cannot mate. In contrast, normal genetic exchange is possible between individuals from not too remote areas.

This phenomenon is even more distinct in the classification of land attributes. These already have much less individuality in reality, because they are a continuous stratum at the earth's surface. If this is not clearly realized, it may give rise to fruitless disputes as it has, e.g., done on vegetation classification in the past. Therefore, in the next section we shall first discuss this phenomenon with vegetation classification as example. In the last section of this chapter, the application of both methods of classification will be discussed for ecosystems as wholes.

On concrete and abstract boundaries in classification and the selection of sample plots

Empirically, we know that plants composing a vegetation are not randomly distributed. There is a relation with the abiotic and biotic environment.
Let us suppose there is a place with N clearly distinguished environments, such as wet places, places with deep well-drained soil, ponds and outcrops of bare rock, each large enough to house a sufficient number of individual plants. Let us also suppose that each of these environments is absolutely homogenous, that samples from within the boundaries give equal results, that there are no intermediate environments, and, hence, that only sharp boundaries exist.
Then we could describe the vegetation on randomly distributed sample plots of sufficient size. Provided the number of samples is large enough, cluster analysis will give N clusters, each characterized by a specific species combination.

If we would have made a preferential sampling as a result of bias, such as a preference for a very beautiful or rare species, or a prejudice against some species, a chance would exist that not all N types were found. It is clear that only random sampling will give an objective picture. However, since samples are not points but instead occupy a certain area, a purely random sampling could imply the inclusion of samples from boundaries. So, by randomly scattering the samples, boundaries may be included, resulting in descriptions suggesting transitions that do not exist in reality. It depends on the size and form of the samples whether this phenomenon has a serious impact on the classification.

The vegetation pattern may have an influence on the required number of samples. Let us suppose that the total area is rather uniformly occupied by one vegetation type, forming a kind of 'matrix', whereas the other N-1 types together occupy only about 10%, resembling a 'dot pattern'. In that case, 90% of the samples will describe the matrix type. For a statistically sound description, the remaining 10% must be covered by a sufficient number of samples. In the case of a very irregular mosaic, even more samples are necessary than in a more regular one. Examples of such 'disbalanced' mosaics are the average desert with less than 10% wadi area in a homogeneous matrix, or the flat savanna with about 10% gallery forests. Purely random sampling would result in an oversampling of the 'matrix' area in comparison to the much more important wadi vegetations or gallery forests. It is obvious that for practical reasons one cannot waste so much effort and time. An overall reduction in sample number, however, would result in an undersampling of the most interesting smaller areas, and consequently a very vague distinction between clusters.

The solution to this sampling problem is *stratified random sampling* based on remote sensing, preferably aerial photography. This is a sufficiently objective means to distinguish small landscape patches from the matrix, such as the wadis and gallery forests of the forementioned examples, with their different soils and vegetations. This is objective because the properties that will be selected as diagnostic characteristics, such as individual species in the case of vegetation mapping or soil profiles in the case of soil survey, cannot be recognized on the photos. Only the general landscape pattern as such can be recognized. The sampling can then be carried out randomly but stratified, i.e., random within the strata. One should note that this use of aerial photography, the so-called landscape approach, is not primarily a mapping aid for drawing boundaries but rather an aid for non-biased sampling! (See further Zonneveld, 1988b.) Of course, at the same time, this photo-interpretation is the natural means to delineate the basic land units for classifying ecosystems as wholes. In the stratification it is also incorporated that the samples should be far from obvious boundaries. Thus, only real, gradual transitions will be sampled, as samples will never be located on sharp boundaries, yielding unrealistic descriptions. We have learned that, even with these precautions, well-defined clusters are rarely found. The hypothetical example of Box 2.1 may clarify this (see also Zonneveld, 1974).

36 I.S. ZONNEVELD

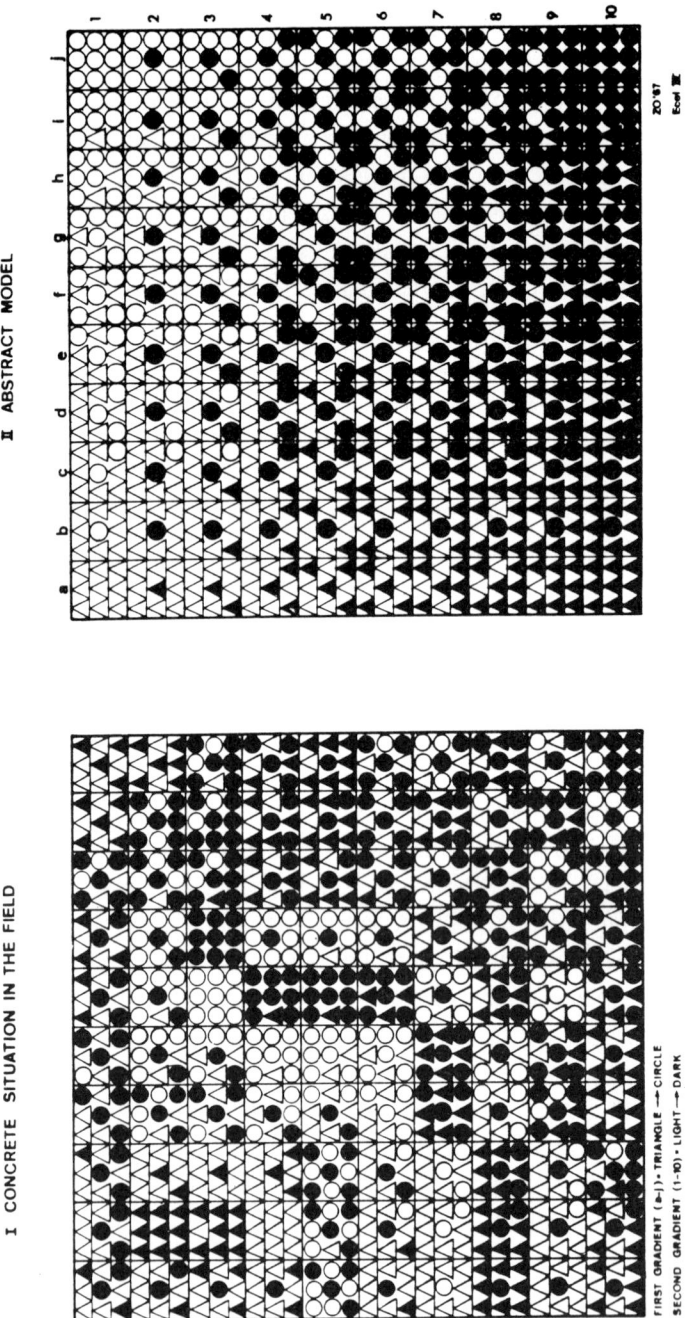

Figure 2.1 Concrete and abstract model of a continuum with two gradients. First gradient (a-j) = triangle → circle; second gradient (1-10) = light → dark

Box 2.1

Figure 2.1 shows a simplified image of the vegetation of the 'whole world'. By way of schematization the world is square, the quasi-endless number of patches is reduced to just one hundred patches of the same quadratic shape and size. The different plant compositions are expressed by only four symbols: black circle, white circle, black triangle and white triangle.

On the left (a) the real vegetation pattern is depicted. None of the patches is exactly the same, which is in perfect accordance with reality. Nevertheless, the picture shows a kind of mosaic in which boundary lines can apparently be drawn. It would be easy to make a provisional photo-interpretation as a base for a vegetation map of this area with a pattern revealing clear environmental differences indicated by the vegetation. We then only need to make a vegetation classification to define the legend units. For this purpose all 100 patches are described by their floristic composition. Multivariate analysis is then carried out with the results of this sampling. The outcome is depicted on the right (b). There the same patches occur, but now ordered, or 'ordinated', according to morphometric parameters. Only two axes appear: one from black to white and the other from triangle to circle. It appears that in the abstraction no boundaries occur. There are no clusters. There is only a beautiful ordination along two axes.

We are now free to cut our own boundaries in this ordination figure. In science, this is a common process of schematisation, but it also expresses a very common human need in daily life, namely, cutting steps in an otherwise unaccessible slope. We could also compare this to the stepwise subdivisions in price categories of theatre seats, which do not fully express the gradient in distance from the stage.

Back to our example. It appears that the four quarters would be the minimum number of classes, but we could just as well cut each gradient in three, i.e., the whole square into nine sub-squares to yield a fine classification as a legend to the vegetation map of the left figure.

Furthermore, it is not difficult to understand that if we had not used all the patches but, for example, only the northern half, holes certainly would have appeared in the right hand picture. Obvious clusters would have resulted and, also, sharp community descriptions. On the contrary, if we had taken the southern half instead of the northern, we would have also got clusters but different ones.

In reality, things may not be as bad as this simple example suggests, because there are many more patches and plants involved. The tendency, however, remains true. In many cases, the abstract boundaries between plant associations can be regarded as arbitrary, because they fully depend on the way of stratification and sampling.

Although we would not argue at abandoning this type of floristic classification, which proved to be a magnificent method especially for regional use all over the world, we maintain that our square world example shows that boundaries between types often must be assessed arbitrarily after large areas have been studied.

Hierarchy and classification

On several occasions we used the word hierarchy in connection with classification. We can distinguish a hierarchy in agglomeration and a hierarchy in space. And, again, a third hierarchy of a different character may be distinguished, namely, a hierarchy in dominance of relations.

The *hierarchy in agglomeration* points to the hierarchical levels in classifications by agglomeration. It refers, for example, to the so-called 'syn-taxonomical' levels in vegetation science, with the community as the basic level: the 'association' as the smallest phyto-coenon of Braun-Blanquet or the 'Zurich-Montpellier school'. Associations are combined to form 'alliances' on the basis of similarity in species composition. These 'alliances', in turn, are combined in 'orders' and these, finally, in 'classes'.
This hierarchy has no relation to space. For example, in the mountains several classes may occur within only a few hectares. In very homogeneous areas, such as flat desert, levels below even the association level may cover many square kilometres. In the taxonomic soil classification systems, the situation is similar, with the soil type, based on the minimum 'pedon', as the core of the abstract classification. So there is essentially no relation to the scale of maps or the hierarchy of their legends.

The second hierarchy, *hierarchy in space*, is characteristic for map legends and all classifications by subdivision with *chorological* character. It is the horizontal pattern that determines what belongs to which unit. Such a unit may be a complex mosaic consisting of syn-taxonomically different units. In maps based on pure classification by subdivision such a heterogeneous patch may be classified as one 'type'. In vegetation and soil surveys using existing taxonomies based on clustering, such map patches are described as complexes of (syn-)taxonomic units.[11]

---

[11] In some soil surveys complex mapping units are confusingly called 'associations' whenever they have a regular and recurrent pattern. This should not be confused with the association distinguished in vegetation science, which is the basic 'coenon' comparable with the basic soil type in soil science.
In vegetation science, such complex mapping units are sometimes adressed as 'sigma-syntaxonomic' units, such as 'sigma-associations', 'sigma-alliances' or 'sigma-orders'. However, this contributes to confusion by introducing unnecessarily complex 'scientific nomenclature'. It is much simpler and more clear to speak of a 'complex of communities': A + B, etc. The 'sigma-syntaxonomy' is not recommended for landscape ecological research, and certainly not for practical applications.

The third hierarchy is of a different kind. Van der Maarel and Dauvellier (1978) and, also, Bakker, Klijn, and Van Zadelhoff (1981) postulate that there is a hierarchy in the influence of spheres, such as atmosphere, hydrosphere, pedosphere, etc. Klijn will treat this hierarchy elsewhere in this book, in an attempt to transform it into a guideline for classification and mapping. Zonneveld (1989) and Van Gils (1989), however, maintain that such a guideline cannot be strictly applied, because the classification and mapping of ecosystems fully depend on the aim of the survey and the character of the landscape at issue.

Classification of landscape units as holons

The examples in the foregoing sections were chosen from the taxonomical classifications of plants and animals, from the classification of minerals and, also, from the classifications of the land attributes, vegetation and soil, but not from landscape units as wholes yet.

To this end, we should first integrate the 'pedon' and 'coenon' of the former sections with atmosphere, rock, relief, animals, and human artifacts into a 'holon' as mentioned in the introduction (see also Naveh and Lieberman, 1983; Zonneveld, 1979, 1988b, 1989).

The selection of diagnostic characteristics

Landscape units as holons can theoretically be the object of classification by agglomeration as was described for soil and vegetation. In fact, landscape units at any spatial scale are the tangible expression of the holons, which can be recognized both in the field and by the interpretation of aerial photos or other remotely sensed images.

Landscape units should be classified by means of diagnostic characteristics, which could be either land attribute classification units as properties of the holons or properties of a number of selected single land attributes. We shall first treat the use of properties of single land attributes for the classification of holons and give some examples. Then we shall go into the use of land attribute classification units as diagnostic characteristics.

The sampling of landscape units may concern properties of all land attributes, which can then be used for a multivariate analysis resulting in a landscape classification by agglomeration. For a good multivariate analysis as a basis for classification, it is required that the parameters should be comparable items.

In this context, the success of the floristic vegetation classification can be explained as due to the fact that only one kind of property is used, viz. floristic species composition. Moreover, biotic parameters contain a wealth of information because they are indicative for their environment.[12]

To base such a classification by multivariate analysis on soil properties is already more difficult, because most single soil parameters are not comparable and they are less easy to determine.[13]

This causes the practical problem of what parameters to include when dealing with landscapes, and how to balance or counteract the negative effects of the incomparability of the various parameters. One might, for example, introduce a certain weighing to prevent one parameter from entirely determining the result of the multivariate analysis.

It is clear that multivariate analysis only facilitates the handling of data; it is no guarantee for higher objectivity. Both the selection of parameters to be included and the evaluation of the result still largely depend on subjective reasoning.

Vos and Stortelder (1992), for example, classified ecotopes as smallest landscape units in two steps. Their classification was based on stratified randomly sampled vegetation data, analysed by means of cluster analysis with TWINSPAN. They then used the discriminant analysis programme DISCRIM (Ter Braak, 1982, 1986) to find the relation between vegetation and site factors, especially humus form.[14]

An example of an attempt to accommodate the incomparability of parameters is the sophisticated method of ecosystem classification that was introduced by Kwakernaak (1982). He used parameters from at least three land attributes, viz. landform, soil, and vegetation, for which he tackled the problem of incomparability of the various parameters, by using 'information' in the sense of general systems theory as a common denominator. First, the author distin-

---

[12] In such classifications by species composition only, the vegetation structure is either fully neglected or used for stratifying the samples before any statistical treatment, for example, by distinguishing between forests, grassland, arable land, etc. (see Westhoff and Den Held, 1969).

[13] Soil parameters can be colour, texture, thickness of horizons, structure or humus form, but, also, parameters to be determined by laboratory tests, such as pH, cation exchange capacity, nutrient availability, vapour pressure, etc.

[14] These authors also describe a procedure to arrive at higher level mapping units by the agglomeration of ecotopes which are arranged in recurrent patterns, thus forming larger spatial units: the chorae.

guished between so-called 'differentiating' or 'key' variables, such as altitude, slope aspect, slope form and inclination, on the one hand, and dependent variables such as soil, vegetation, erosion processes and land management, on the other. The key variables were used to determine the land units and were excluded from the calculations. The dependent variables, in contrast, were subjected to cluster analysis, after conversion into 'variance of information values'. The results of the various cluster analyses were then compared both mutually and with the key variables. Finally, the results of this correspondence analysis were used to balance the influence of the variables on the classification.
For practical applications, especially when data are lacking and relatively quick surveys are needed, such sophisticated methods are too complex.

The above-mentioned methods, as far as they try to integrate properties of soil, water, relief and vegetation, do not make maximal use of the system character of the land attributes. For instead of using the properties of these land attributes, we could also use the classification units of land attributes as properties of the holons.
In terms of the General Systems Theory (Von Bertalanffy, 1968) and Smuts's (1926) holism, land attributes can be regarded as subsystems, representing wholes on the hierarchical level below the holistic landscape units. Not using the system characteristics of this level means a denial of an enormous reservoir of knowledge about each of these subsystems from climatology, geology, geomorphology, soil science, hydrology and vegetation science. Apart from knowledge about genesis, chorology and chronology, it also includes already existing knowledge about ecological relations, which is reflected in the respective classification systems. This justifies the conclusion that it would be sound to use the classification units for land attributes as diagnostic characteristics for a classification of land units as holons. Fortunately, this is also the most feasible, and it enhances the co-operation between specialists from various sciences as well.
Especially classifications for geomorphology, soils, and vegetation are used as diagnostic characteristics in land classification practice. In contrast, classifications of climate, geology, hydrology, and human interference are used only on few occasions. More often, these attributes are merely implicitly covered. They should be derived from the first three by means of their indicative value. It is self-evident that the earlier explained requirements should be met, in particular, those about 'convergence of evidence' and the one about diagnostic characteristics that should be 'system properties'. This means that a land unit

as a whole should never be characterized by one land attribute only. Nevertheless, it has often been stated that land ecosystems can be defined by means of only one land attribute, such as soil or vegetation or even landform, because these would be sufficiently indicative for all the others. This reasoning might hold for purely natural and undisturbed landscapes. In reality, however, the correlation between the various land attributes is feeble due to human disturbance. So, we maintain that, although the indicative value of land attributes for other land attributes can be used advantageously, a classification of land units as wholes should be based on a combination of land attributes.

A first example in this context is the 'holistic' land unit survey developed by Zonneveld (1979, 1988 a, b, c, 1989). It starts with the making of a provisional map on the basis of aerial photograph interpretation, for which the surveyor focuses on landscape system properties rather than on properties of single land attributes such as soil or vegetation. These are, by convergence of evidence, relief, land use and vegetation structure. Then, the properties of soil, vegetation and geomorphology are determined in the field and used to describe the legend. The classification of each attribute may not be based on detailed assessments but on some important properties only, depending on the knowledge, time, and experience of the surveyor and the purpose of the survey. The mutual indication of the land attributes enables the avoidance of some time-consuming effort in this respect. Consequently, this type of land survey may require no more time than the survey of only one land attribute, and so saves time (Zonneveld, 1979, 1989).

Other examples are the related classification by Doing (1974) and the more purely geomorphologically oriented terrain classifications of Van Zuidam and Van Zuidam-Cancelado (1979). The so-called 'Land system' surveys as practised by CSIRO in Australia and DOS in Britain (see Christian and Steward, 1968), as well as the Ecological (Biophysical) Land Classification in Canada (see Thie and Ironside, 1976) are examples of classifications by subdivision that focus on the holon character of landscape units and for which land attribute classifications are used as diagnostic characteristics.

An alternative to starting with integrated land units from the provisional mapping onwards, is to integrate the land attributes into land units after having surveyed, mapped and classified soil, vegetation, landform, hydrology, land use, etc. separately. This may be the best method for areas, for which many data are available in map form. GIS techniques may help in the integration (see Zonneveld, 1979; Figure 2.2). For scientific purposes, however, it is preferable to distinguish landscape units as wholes beforehand.

BASIC PRINCIPLES OF CLASSIFICATION 43

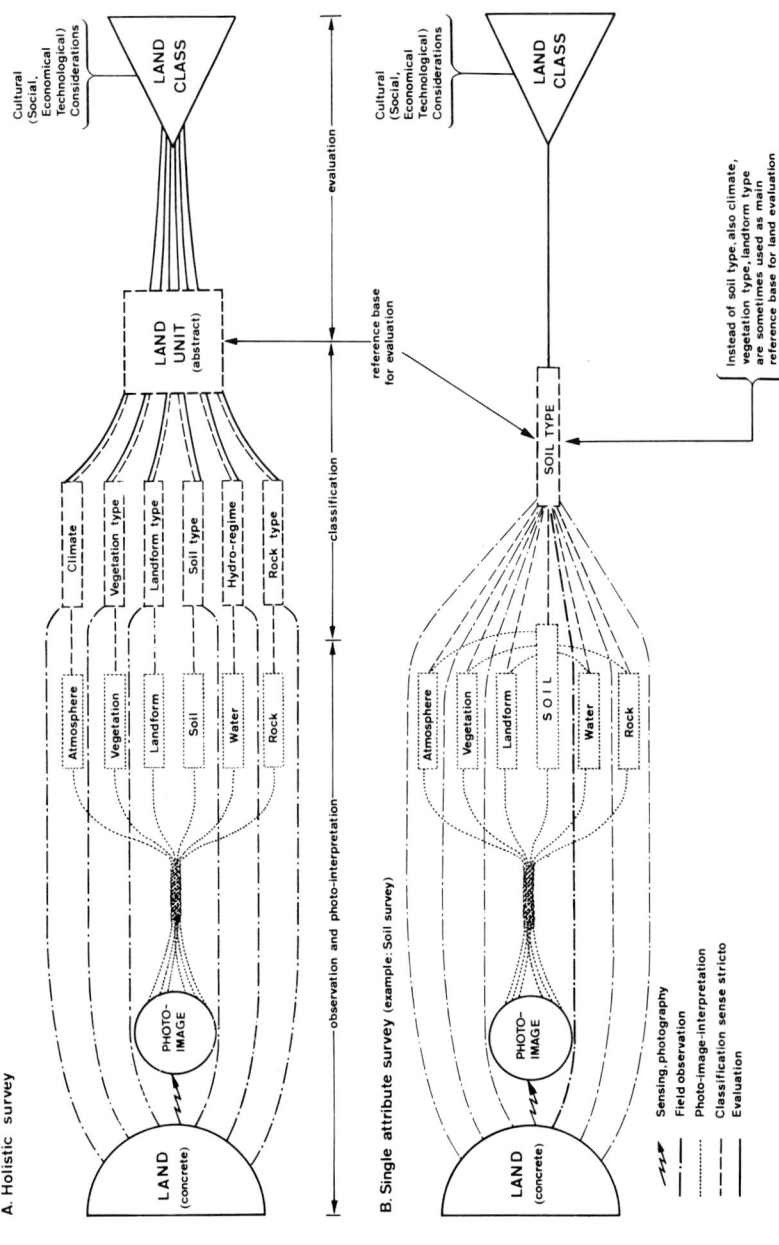

Figure 2.2 Comparison of holistic land survey and single attribute survey for land evaluation

Classification by agglomeration or by subdivision?

Still the question remains unanswered whether it is worthwhile to strive for a pure classification by agglomeration, i.e., typification, which may be hierarchically structured but independent of space. Or should we stick to classification by sub-division only, thus emphasizing the spatial hierarchy?

In Figure 2.3 the relation is illustrated between the units at different levels of a classification by sub-division in accordance with the 'hierarchy of space' on the one hand, and those from a non-spatial classification in accordance with the 'hierarchy of agglomeration' on the other. The basic unit in both hierar-

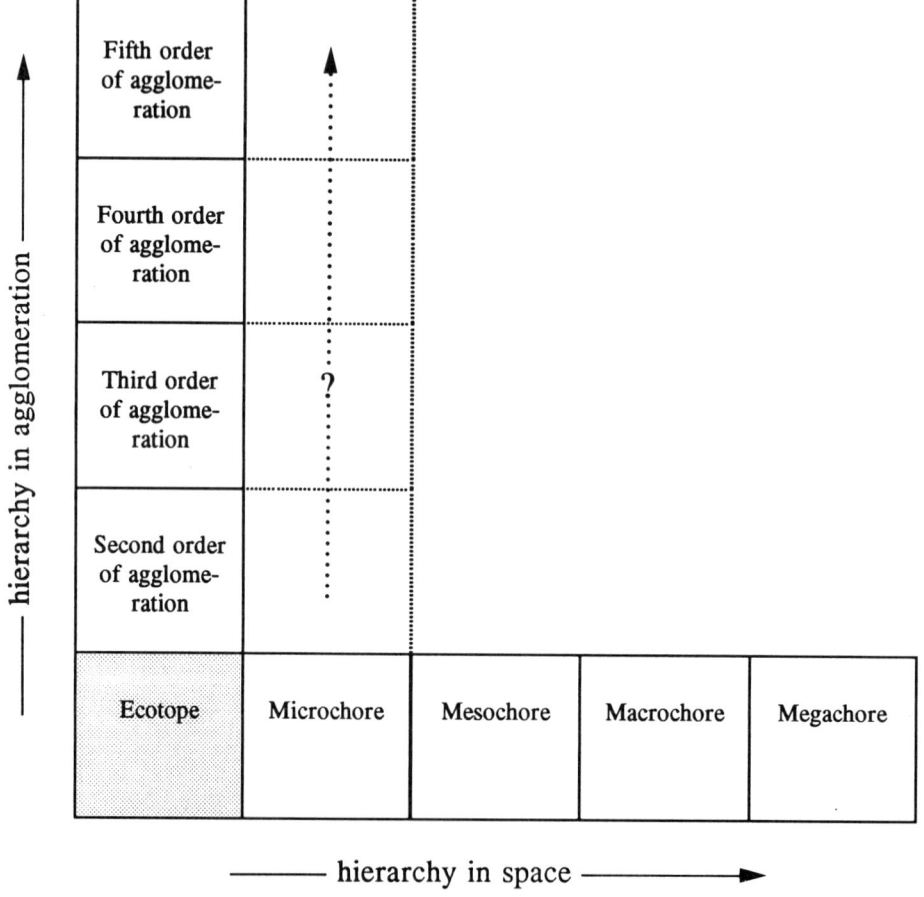

Figure 2.3 The relation between the levels of a hierarchy of space, a chorological hierarchy, and the levels of a hierarchy of agglomeration, a typification hierarchy

chies is the smallest area at the earth's surface that can still be called a landscape unit or ecosystem: an ecotope. In the vertical direction a typification hierarchy is indicated. In the horizontal direction we see a hierarchy of 'chorae' or 'chores': spatial units of increasing size ranging from 'microchore' to 'megachore'.

In both cases the higher units are complexes of ecotopes. In the vertical direction these are abstract units, in the horizontal direction concrete, larger spatial units. In the typification hierarchy, an ecotope type can be part of only one unit at a hierarchically higher level. In the spatial hierarchy, however, the same ecotope type may occur as an element of various mosaics at the same hierarchical level, i.e., as a component of different mapping units.[15]

Attempts to develop typifications for landscape units as wholes are known for confined regions, such as the Netherlands, Britain, or Russia. However, world-embracing classifications, as they exist for soil, landform, or vegetation, are not yet known for landscape units or ecosystems. Forman and Godron (1986) discussed the desirability to design a true phylogenetic typification for landscapes, but conclude that this will not yet be possible because of the complexity of the object. Attempts in this direction may certainly serve the development of the science, but any attempt is complicated considerably by the enormous influence of man, who becomes more and more independent from the natural factors. It would certainly be easier if only purely natural landscapes existed. Especially the increased temporal variability induced by man makes it practically impossible to design a logical and applicable worldwide hierarchical typification system.

Moreover, for most applied surveys it suffices to describe the higher levels in the legend, the chores, as complexes of ecotopes. If these are also described by means of the existing classifications of the various land attributes, the latter will provide the link for comparisons with more remote areas, which is the main reason to strive for an abstract classification by agglomeration. Therefore, we can conclude that it is at least doubtful whether it is worthwhile to attempt to develop one world-embracing pure typification, independent of space.

---

[15] For example, patches with similar vegetation and soil due to a dominant factor, such as dung concentration or termite activity, may occur in two different land units of higher hierarchical level dominated by range land but with different soil and vegetation.

References

Bakker, T.W.M , J.A. Klijn and F.J Van Zadelhoff, 1981. *Nederlandse kustduinen: Landschapsecologie.* Pudoc, Wageningen, 144 pp.

Christian, C.S. and G.A. Steward, 1968. Methodology of Integrated Surveys. In: *Aerial Surveys and integrated studies.* Proc. Toulouse Conf. 1964. UNESCO, Paris, pp. 233-280.

Doing, H., 1974. *Landschapsecologie van de duinstreek tussen Wassenaar en IJmuiden.* Mededelingen Landbouwhogeschool no. 72/4, Wageningen.

Forman, R.T.T. and M. Godron, 1986. *Landscape Ecology.* Wiley and Sons, New York, 619 pp.

Gauch, H.G. Jr., 1985. *Multivariate analysis in community ecology.* Cambridge University Press, Cambridge, 298 pp.

Jongman, R.H.G , C.J.F. ter Braak and O.F.R Van Tongeren, 1987. *Data analysis in community and landscape ecology.* Pudoc, Wageningen.

Kwakernaak, C., 1982. *Landscape Ecology of a Prealpine Area.* PhD thesis. Publ. Fys. Geogr. Bodemk. Lab., University of Amsterdam, no. 33.

Naveh, Z. and A.S. Lieberman, 1983. *Landscape Ecology, Theory and Application.* Springer-Verlag New York etc., 356 pp.

Smuts, J.C., 1926. *Holism and Evolution.* 2nd Edition 1971. Viking Press, New York.

Ter Braak, C.J.F., 1982. *DISCRIM. A modification of TWINSPAN (Hill, 1979) to construct simple discriminant functions and to classify attributes, given a hierarchical classification of samples.* IWIS-TNO, Wageningen.

Ter Braak, C.J.F., 1986. Interpreting a hierarchical classification with simple discriminant functions: an ecological example. In: E. Diday et al. (eds.), *Data analysis and informatics, IV.* Elsevier, Amsterdam etc., pp. 11-21.

Thie, J. and G. Ironside (eds.), 1976. *Ecological (biophysical) Land Classification in Canada.* Ecol. Landclass. Series no 1. Lands Directorate Environment, Canada.

Van der Maarel E. and P.L. Dauvellier, 1978. *Naar een Globaal Ecologisch Model voor de ruimtelijke ontwikkeling van Nederland.* Staatsuitgeverij, The Hague.

Van Gils, H.M., 1989. Legends of landscape ecological maps. *ITC journal*, 1: 41-48.

Van Melsen, A.G.M, 1955. *Natuurfilosofie.* Uitg.mij. N.V. Standaard-Boekhandel, Amsterdam, 360 pp.

Van Wirdum, G., 1981. Design for a land ecological survey of nature protection. In: *Proc. Int. Congr. Neth. Soc. Landscape Ecology, Veldhoven.* Pudoc, Wageningen, pp. 245-251.

Van Zuidam, R.A. and F.I. Van Zuidam-Cancelado, 1979. *Terrain analysis and classification using aerial photographs. A geomorphological approach.* ITC textbook VII.6. ITC, Enschede, 310 pp.

Von Bertalanffy, L., 1968. *General Systems Theory: Foundations, Development and Applications.* George Brasiler, New York.

Vos, W. and A. Stortelder, 1992. *Vanishing Tuscan landscapes, landscape ecology of a Sub-mediterranean - montane area (Solano Basin, Tuscany, Italy).* Pudoc, Wageningen, 402 pp.

Westhoff, V. and A.J. den Held, 1969. *Plantengemeenschappen in Nederland.* N.V. Thieme and Cie, Zutphen, 324 pp.

Whittaker, R.H., 1973 (ed.). *Ordination and Classification of Communities. Handbook of Vegetation Science, Vol. 5,* Junk Publishers, The Hague, 738 pp.

Zonneveld. I.S., 1974. On abstract and concrete boundaries, arranging and classification. In: Sommer and Tüxen (eds.), *Tatsachen und Probleme der Grenzen in der Vegetation.* Proceedings symposion IAVS, 1968, Rinteln. Cramer Verlag 3301, Lehre, pp. 17-43.

Zonneveld, I.S., 1979. *Land Evaluation and Land(scape)Science.* Vol. VII.4 ITC-textbook of Photo- Interpretation. ITC, Enschede. 134 pp.

Zonneveld, I.S., 1988a. Examples of vegetation Maps, Their Legends and Ecological Diagrams. In: A.W. Küchler and I.S. Zonneveld, *Vegetation Mapping. Handbook of Vegetation Science, Volume 10.* Kluwer Academic Publ. Dordrecht etc., chapter 11, 135-147.

Zonneveld, I.S., 1988b. The ITC-Method of Mapping Natural and Semi-natural Vegetation. in: A.W. Küchler and I.S. Zonneveld, *Vegetation Mapping. Handbook of Vegetation Science, Volume 10.* Kluwer Academic Publ. Dordrecht etc., chapter 29, 401-427.

Zonneveld I.S., 1988c. Landscape ecology and its application. In: M.R. Moss, (ed.), *Landscape Ecology and Management.* Proceedings of the first Symposium of the Canadian Society for Landscape Ecology and Management, Guelph, May 1987.

Zonneveld, I.S., 1989. The land unit - A fundamental concept in landscape ecology and its applications. *Landscape Ecology* 3/2: 67-89.

System ecological concepts for environmental planning 3

Wolfgang Haber

ABSTRACT - Environmental management and planning requires specific approaches and methods derived from landscape ecology and ecosystem theory. The concepts of ecosystem and ecotope have proved very practical if carefully defined. Whereas the ecosystem has its place in the hierarchy of organizational levels — where it represents the first level which fully integrates non-living and living factors — the ecotope is the smallest homogeneous spatial component of a landscape. Problems in applying both concepts are discussed, and a special system approach symbolized by a pyramid is developed. It combines both reductionistic and integrative methods and appears appropriate in dealing with the complexity of environmental systems.

Introduction

The environment of the many different organisms on earth, including humans, is notoriously complex and difficult to explore. Its subdivision and classification, however, is a necessity for all measures required to utilize, manage, protect, develop or change the environment, all of which involves planning. A generally agreed classification suited to as many purposes as possible would be most welcome, but is difficult to achieve — because complex phenomena allow many different approaches to classification. In addition, classifiers tend to stick to their own preference, some even considering classification a goal in itself.

The need for environmental planning is a special challenge for environmental classification. It must be based on ecological concepts, which should be scientifically sound and operational, and should contribute to ecological theory, or have at least a heuristic value. During 25 years of work at the Landscape Ecology Department in Weihenstephan, a number of ecological concepts were developed, adopted from others, examined, modified, or abolished (Haber, 1992; Duhme et al., 1992). From this experience, the concepts of ecosystem and ecotope have proved most useful and practical. However, they need to be carefully defined.

The ecosystem concept

The ecosystem as level of organization or integration

When Tansley (1935) introduced the ecosystem concept, he stated that organisms cannot be separated from their specific environment with which they form one physical system. These systems, he continued, are the basic units of nature on the earth's surface for an ecologist and can be called ecosystems; they occur in many different forms and sizes and represent one distinct category among the physical systems of the universe, which extend from the universe as a whole down to the atom.

With this last remark, Tansley pointed to levels of organization, later developed as a distinct hierarchy by Egler (1970), who called them 'levels of integration', and adopted by Miller (1975) and many others. We modified and expanded the — still simplistic — hierarchy as shown in Figure 3.1. Each level can be regarded and investigated as a system of its own. To each level corresponds a certain spatial and temporal dimension or 'scale'. Each level is also defined by its 'emergent properties' which distinguish it from the lower levels — the properties of which, of course, it comprises. The notion of emergent properties has been rejected by Harper (1982) and Fenchel (1987), but has been proven correct for even the subatomic level by physicists. The realm of ecology and of much of environmental science is derived from this hierarchy encompassing several — in our case seven — organizational levels, with the ecosystem as the central one. Many other scientific disciplines are devoted to only one single level, e.g. cytology, or molecular biology, and even within ecology there is a clear tendency to restrict research to only one level, for instance, population ecology. But the integrative goal of ecology is

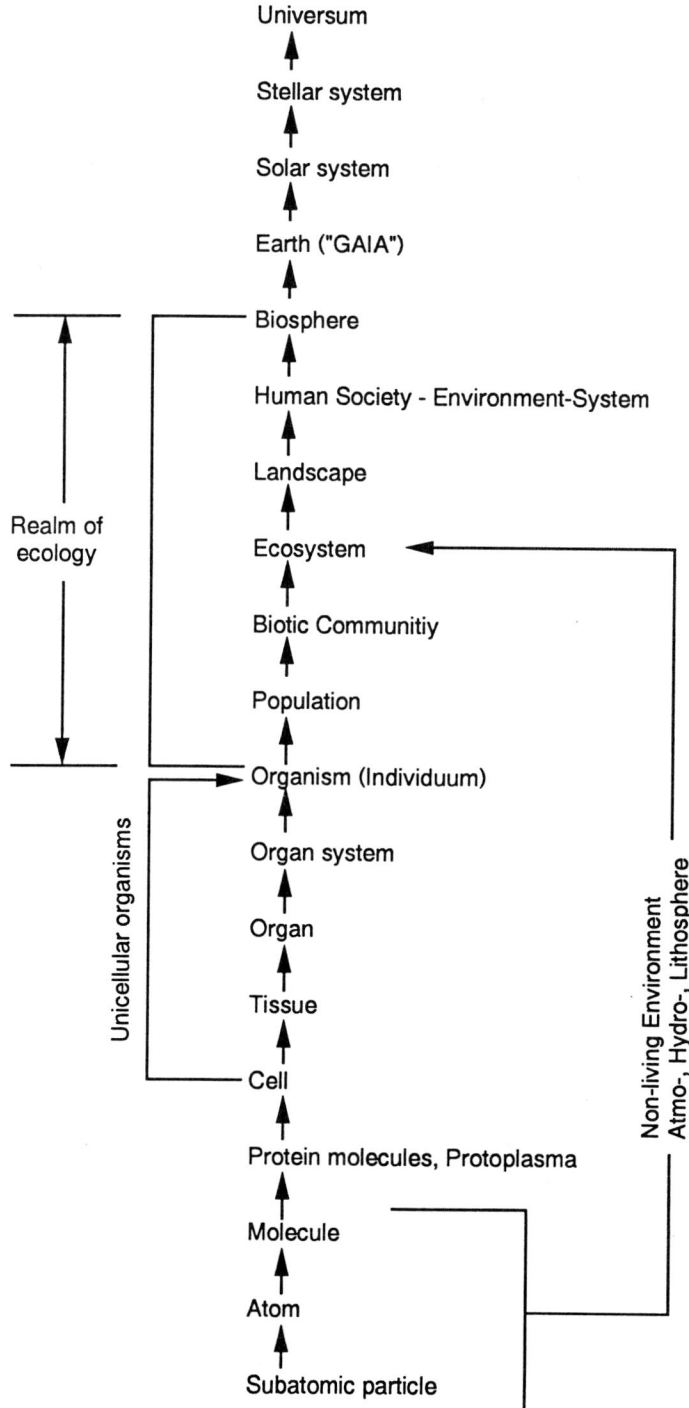

Figure 3.1 Hierarchy of levels of organization of non-living and living matter (from Haber 1993)

then disregarded. It is the very essence of ecology to relate different levels by 'downscaling' and 'upscaling', and to find out what, e.g. a certain environmental impact will mean for different levels — for which the results may be quite different.

Dimensions of the ecosystem

The place of the ecosystem in the hierarchy of organizational levels denotes that an ecosystem is a part or subdivision of a landscape as the next higher level, so its spatial dimension is restricted to a certain physical size or surface. By contrast, an understanding exists of ecosystems being of any size from the whole biosphere down to a small hedgerow between two fields. This is certainly not a workable concept. However, Ellenberg (1973, see also Ellenberg and Mueller-Dombois, 1974, p. 17), referring to this apparent lack of a given spatial dimension of ecosystems, introduced an elaborate classification of ecosystems into five categories, namely mega-, macro-, meso-, micro- and nano-ecosystems, that was as logical as it was comprehensive, but which did not find any wide application.

The ecosystem within the environmental spheres

There is another, much simpler hierarchical subdivision of the general environment, i.e. into 'environmental spheres', which was recommended and used by Van Leeuwen (1980). He distinguished the cosmosphere, atmosphere, hydrosphere, and lithosphere as the non-living environmental spheres, followed by the biosphere and pedosphere as living spheres (Figure 3.2). This hierarchical classification is important for the ranking of the effects of abiotic environmental or ecological factors, and should always be observed in environmental research and planning. We placed H. Walter's 'Standortsfaktoren' (Walter, 1986) into this hierarchy, which proved very valuable. Therefore, the environmental sphere approach was incorporated into the level-of-organization hierarchy (Figure 3.1). As the biotic organization levels from the molecular up to the community level represent a predominantly biological sequence, the non-living ecological factors are formally introduced into the hierarchy of organizational levels at the ecosystem level, thus making this level a particular and ecologically critical one. Ecosystems are the smallest components of the biosphere that can be regarded as systems themselves.

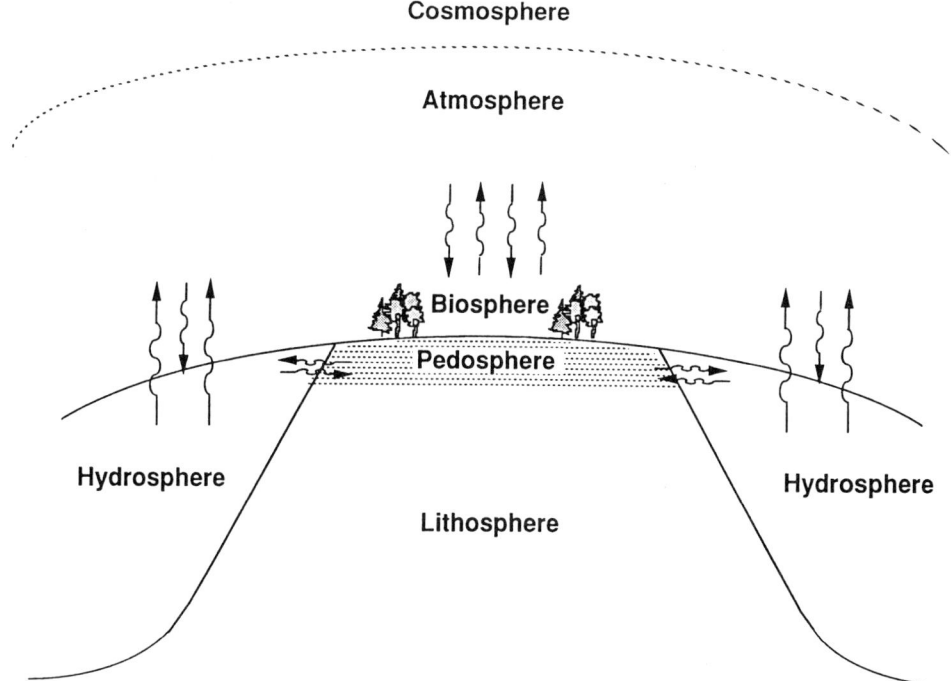

Figure 3.2 The environmental spheres and their hierarchy (Haber, 1993, after Van Leeuwen, 1980)

For understanding and explaining ecosystems from a functional point of view, we use the well-known functional scheme of a natural ecosystem modified from Ellenberg (1986); there is a terrestrial and an aquatic version (Figure 3.3). It does not need any further explanation here.

Problems in applying the ecosystem concept

The ecosystem concept is derived, as mentioned before, from an interaction of non-living and living components. The living components may be detached from the non-living ones and investigated as a biotic community with their own interactions. But it does not make sense — at least for an ecologist — to detach the non-living components from the ecosystem concept. They only typify the site of the ecosystem and point to the concept of ecotope, to be discussed below.

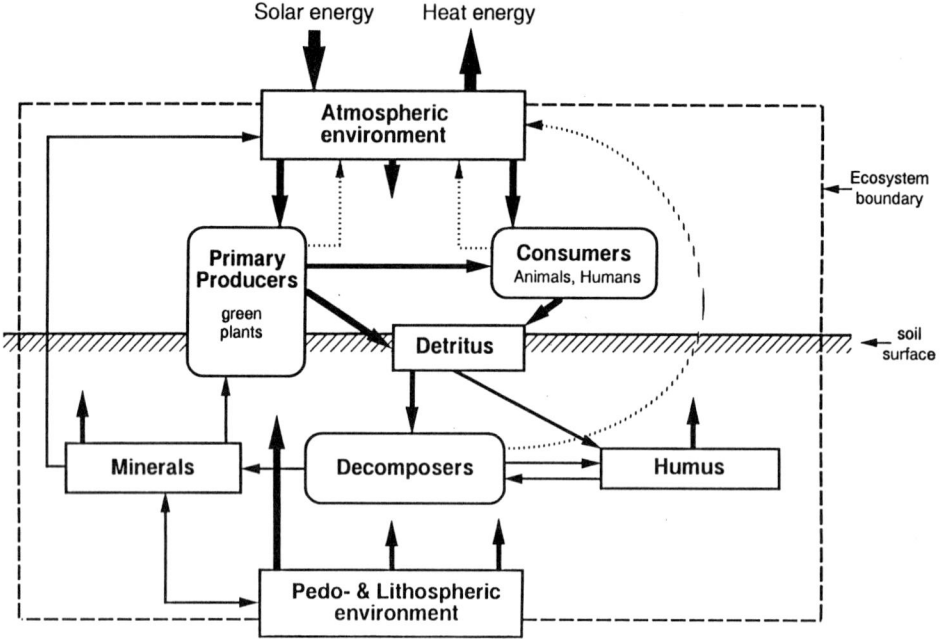

Figure 3.3 Simple functional model of a natural terrestrial ecosystem (adapted from Ellenberg 1986)

There is a problem with the ecosystem concept because it represents both an abstract unit (ecosystem type) and a concrete entity. For example, Kaule (1974) mapped and described a number of raised bogs in the pre-alpine region of southern Bavaria, each representing a distinct individual ecosystem, but all of them belonging to the same ecosystem-type of 'raised bogs'. In environmental planning, one is mostly dealing with concrete ecosystems, but sometimes, e.g. in conservation planning, with ecosystem types, too. To make a distinction and to avoid confusion, it has been proposed to call the concrete ecosystem 'ecotope' (cf. Naveh, 1984) (see below).

Another problem with the ecosystem concept is related to the fact that terrestrial ecosystems are defined (and also named) by vegetation characteristics. Animal ecologists have some difficulties in fitting animal communities and their biotopes, or animal habitats, into the ecosystem pattern, especially when working with larger vertebrates. The delineation of ecosystems clearly cannot be based on mobile organisms like freely moving animals, but only on immobile, firmly rooted and easily visible plants forming vegetation complexes. Yet, least part of the life cycle of every animal species can be assigned to a certain definable location — often with a specific vegetation structure — and, consequently, to a place in an ecosystem.

The ecotope concept

Origin and definition

As mentioned above, in environmental planning it is concrete sites that are to be dealt with or decided upon. Thus, many planners prefer a site approach. This desire can be met by the ecotope concept. Its origin is the discipline of landscape ecology. The first landscape ecologist was Alexander von Humboldt (1769-1859) who gave the first (and still valid) definition of 'landscape', but who did not mention landscape ecology, because 'ecology' was coined only six years after his death by Haeckel. The term 'landscape ecology' was introduced by Carl Troll in 1939. Both Humboldt and Troll were biology-minded and biologically trained geographers or landscape ecologists, respectively, which cannot be said of some younger landscape ecologists. It was Troll who coined the term 'ecotope' in 1950. His aim was to recognize the smallest component parts of the complex entity of a landscape. For these 'landscape cells' or 'tiles' ('Fliesen'[1]), as they were also called, he required spatial homogeneity which was basically defined by abiotic criteria, in particular physical and chemical properties of the substrate (bedrock) such as porosity, texture, pH, calcium content, silica content, etc. These properties constitute a small geographical land unit called physiotope or geotope. It may be colonized by organisms which are adapted to, and gradually transform, the physiotope by interacting with the physico-chemical properties. This interaction of living and non-living components constitutes an ecosystem and changes the physiotope into an ecotope. This change is manifested by phenomena such as humus and soil formation, the establishment of a special microclimate, of long-living plant structures and the creation of new ecological niches, to mention only a few.

Therefore, an ecotope is a concrete ecosystem at a given and defined site (cf. Haber, 1990a; 1993). There is some confusion caused by confounding 'eco-

---

[1] Schmithüsen (1948) introduced the term 'Fliesen' into his German explanation of the landscape mosaic which he liked to compare with a tiled floor or wall of a house or room ('Fliesengefüge'). Troll (1968) rejected 'Fliese', which he considered unsuitable for international discussions because it is difficult to translate and even to pronounce.

tope' with 'biotope'. Biotope means, by definition, the location (topos = place) of a biotic community, that is the living part of an ecosystem. So far, ecotopes and biotopes would coincide, but the approaches are different: one comes from landscape ecology, the other from community ecology. The ecotope approach yields the ecosystem more operational for planning purposes and results in a better fit in the hierarchy of organizational levels (see Figure 3.1). We tried to transform the functional ecosystem scheme (Figure 3.3) into a corresponding ecotope scheme (Figure 3.4), in which the key ecological processes are indicated by various arrows.

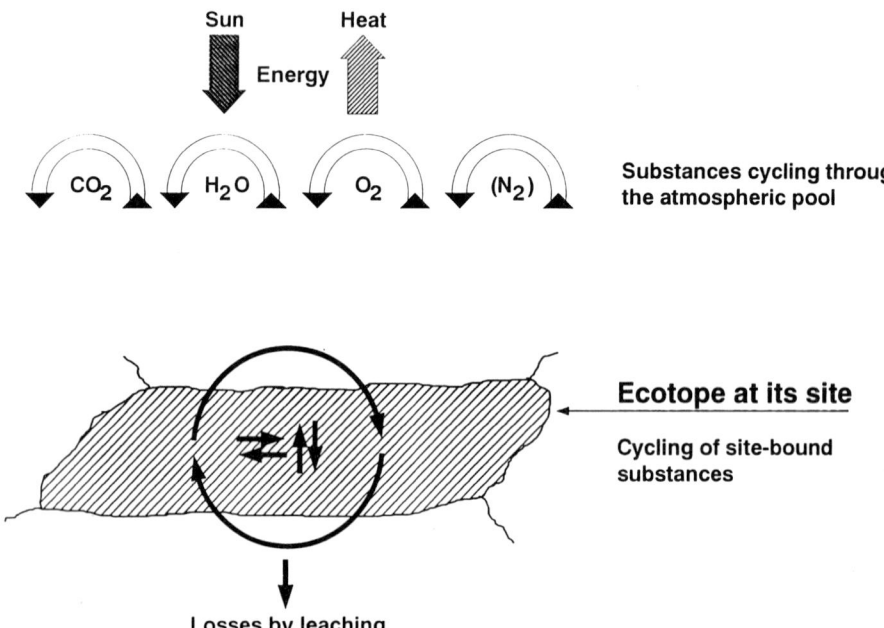

Figure 3.4 Model of a single ecotope at a given site as the basic component of a landscape. Note two types of matter cycling or flux: one through the atmosphere, the other bound to the site (circular pair of arrows). Vertical pair of arrows: relationships between organisms and site. Horizontal pair of arrows: relationships between organisms (from Haber, 1990b)

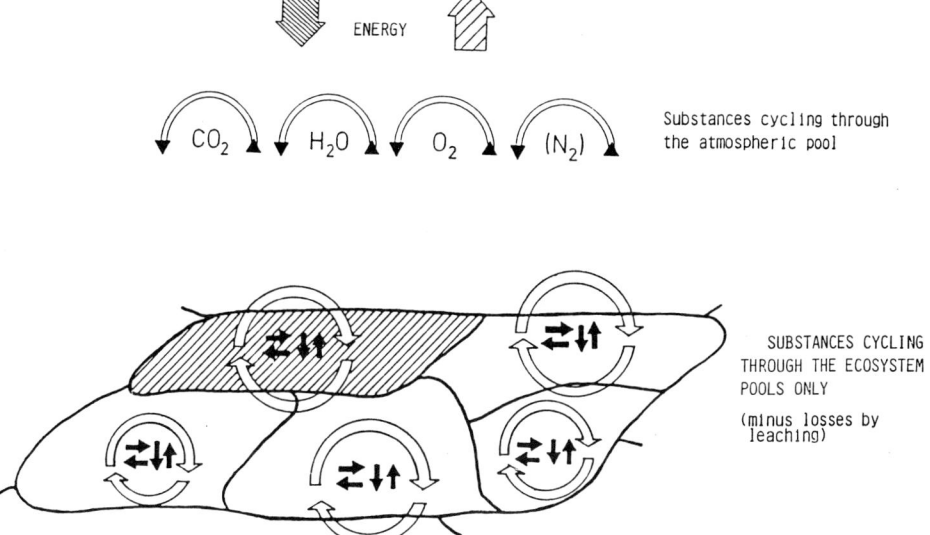

Figure 3.5 Ecotope pattern on level terrain. (From Haber, 1990b.) For explanations see Figure 3.4

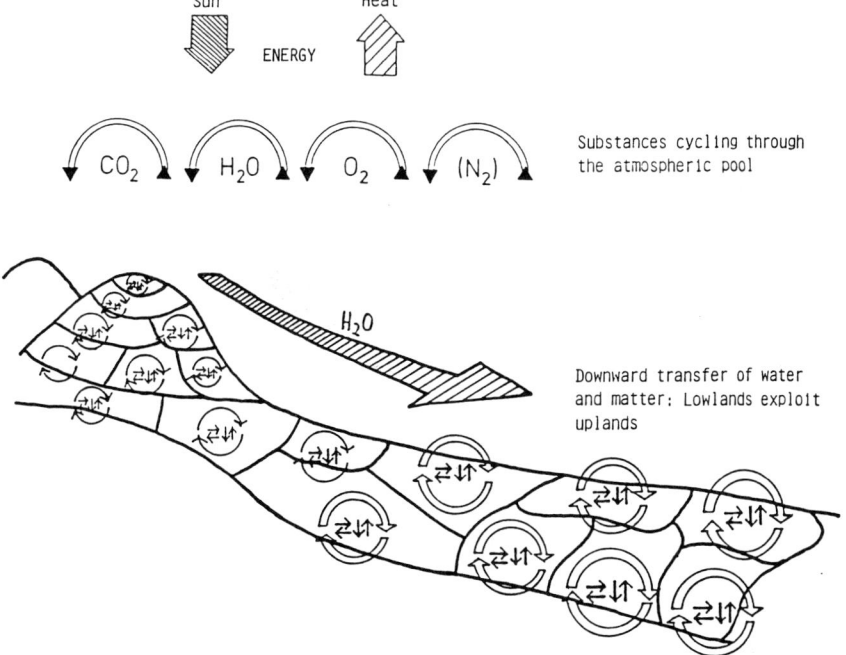

Figure 3.6 Ecotope pattern along a slope. For explanations see Figure 3.4. (From Haber, 1990b)

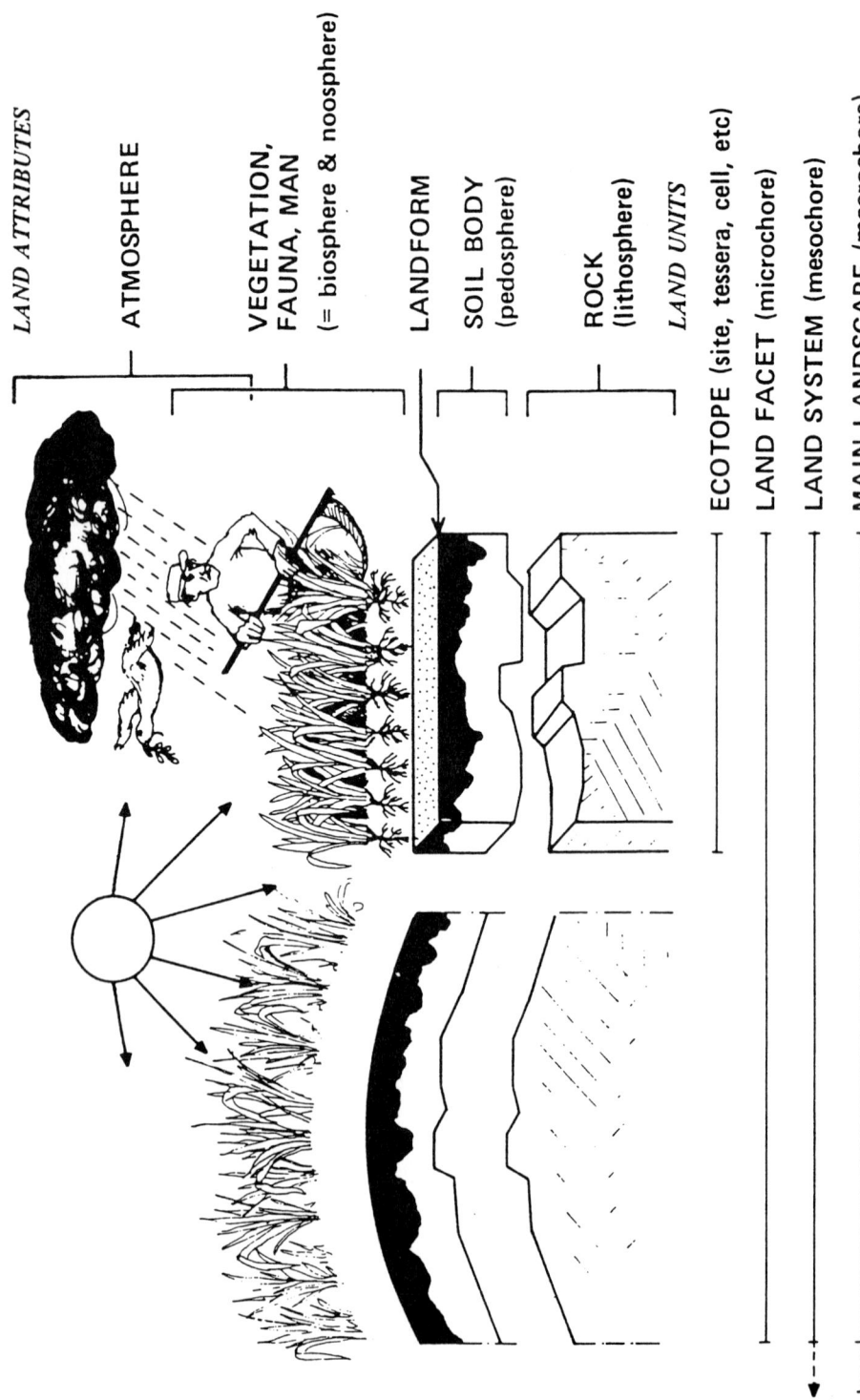

Figure 3.7 Spatial dimensions of landscape ecological units, based on the ecotope (from Zonneveld, 1990)

The ecotope pattern in the landscape

A landscape — also called ecosystem-complex or ecotope-complex, or, in geographical terms, a 'chore' — is considered as a pattern or 'mosaic' of ecotopes. Controlled by the general climatic, geological and relief conditions this is always a typical 'set' of ecotopes. The ecotope pattern of a natural landscape is largely determined by the physico-chemical properties of the bedrock which, however, may be altered or blurred by soil formation. Basically, there are two general ecotope pattern types:
1. on level terrain, where the lateral near-surface connections between ecotopes are few (Figure 3.5),
2. on inclined terrain, where the downward movement of water and substrate results in strong lateral connections and colluvial accumulations at foothills and in floodplains (Figure 3.6).

With pattern type no. 1, vertical interactions between ecosystem components dominate over horizontal interactions between ecotopes, whereas with type no. 2, horizontal relations between ecotopes are much more important. These are generally directed 'top-down', i.e. from the most elevated ecotopes to those downhill.

Of course, human land-use will profoundly influence or alter this spatial pattern. A homogeneous ecotope may be cleared of its vegetation cover (thus virtually 'reduced' to a physiotope), then subdivided into two or more parcels differently utilized: fields, meadows, planted forests or housing areas. This would result in the disruption of the original homogeneity and in a set of 'new', anthropogenic ecotopes and/or ecosystems, respectively.

The best graphical representation of an ecotope in a landscape with all attributes was given by Zonneveld (1990) and is reproduced in Figure 3.7.

Problems in applying the ecotope concept

This combined ecosystem/ecotope concept (Figure 3.8) has proved practical for environmental planning and management, at best in fine-grained landscapes. A disadvantage is that it does not easily lend itself to a comprehensive classification, nor does it fit into existing classifications. We devised, following a suggestion of Westhoff (1968), a simple classification of 'Main Ecosystem Types' according to decreasing naturalness (Table 3.1). This does not

Table 3.1 Main ecosystem or land-use types arranged according to decreasing naturalness or increasing artificiality

A.	Bio-Ecosystems	Dominance of natural components and biological processes.
A.1	Natural Ecosystems	Without direct human influence. Capable of self-regulation.
A.2	Near-natural Ecosystems	Influenced by humans but similar to A.1. Little changed after human abandonment. Capable of self-regulation.
A.3	Semi-natural Ecosystems	Resulting from human use of A.1 and A.2, but not (intentionally) created. Change significantly after human abandonment. Limited capability of self-regulation. Management required.
A.4	Anthropogenic (biotic) Ecosystems	Intentionally created by humans. Fully dependent on human control and management.
B.	Techno-Ecosystems Examples: Settlements (villages, cities) Traffic systems Industrial complexes	Anthropogenic (technical) systems: Dominance of technical structures (artefacts) and processes. Intentionally created by humans for industrial, economic or cultural activities. Dependent on human control and on the surrounding and interspersed bio-ecosystems.

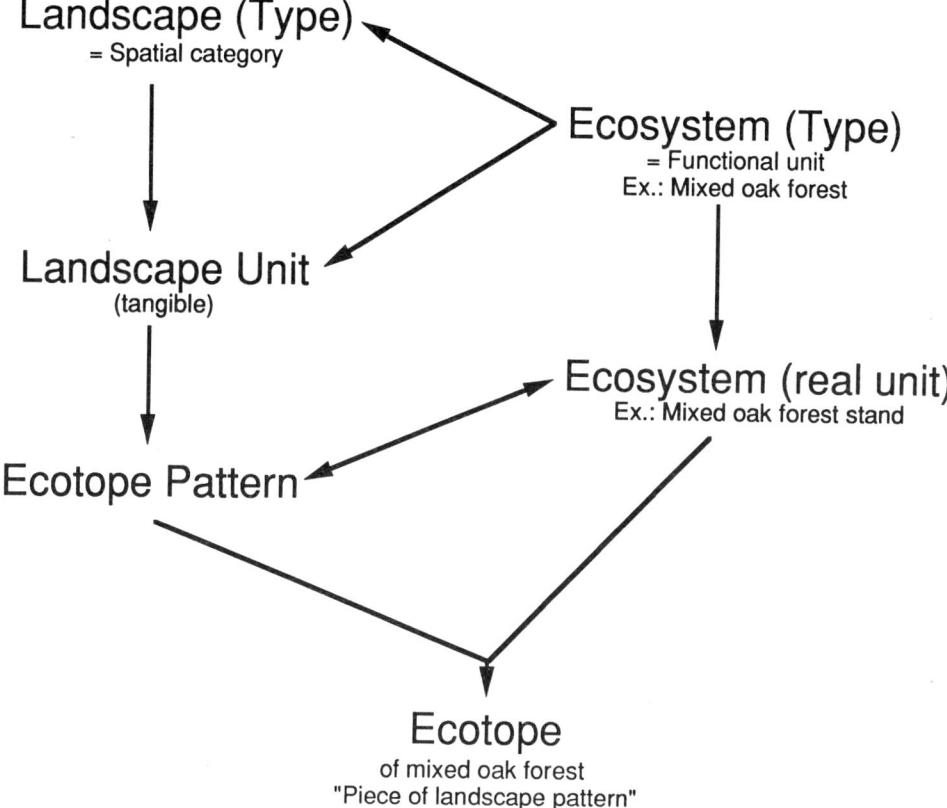

Figure 3.8 Landscape and ecosystem approach in landscape ecology

preclude the use of other classification systems, such as Ellenberg's (1973) mentioned before. And, also, it should be recalled that the available syntaxonomic vegetation classifications, in particular the continental Braun-Blanquetian system, offer excellent possibilities for ecosystem/ecotope classification. They are sometimes discredited for a too rigorous syntaxonomic emphasis, neglecting ecological connections or viewpoints; however, if used with less narrow-mindedness, the phytosociological approach is one of the most valuable tools for environmental management and planning.

The continental school of phytosociology has provided landscape ecology and environmental planning with an unrivalled, reliable basis of comprehensive ecological information (cf. Ellenberg, 1980; Westhoff, 1979), because of its thorough and detailed investigation, ecological interpretation, and floristic classification of Central European vegetation. This is especially the case after

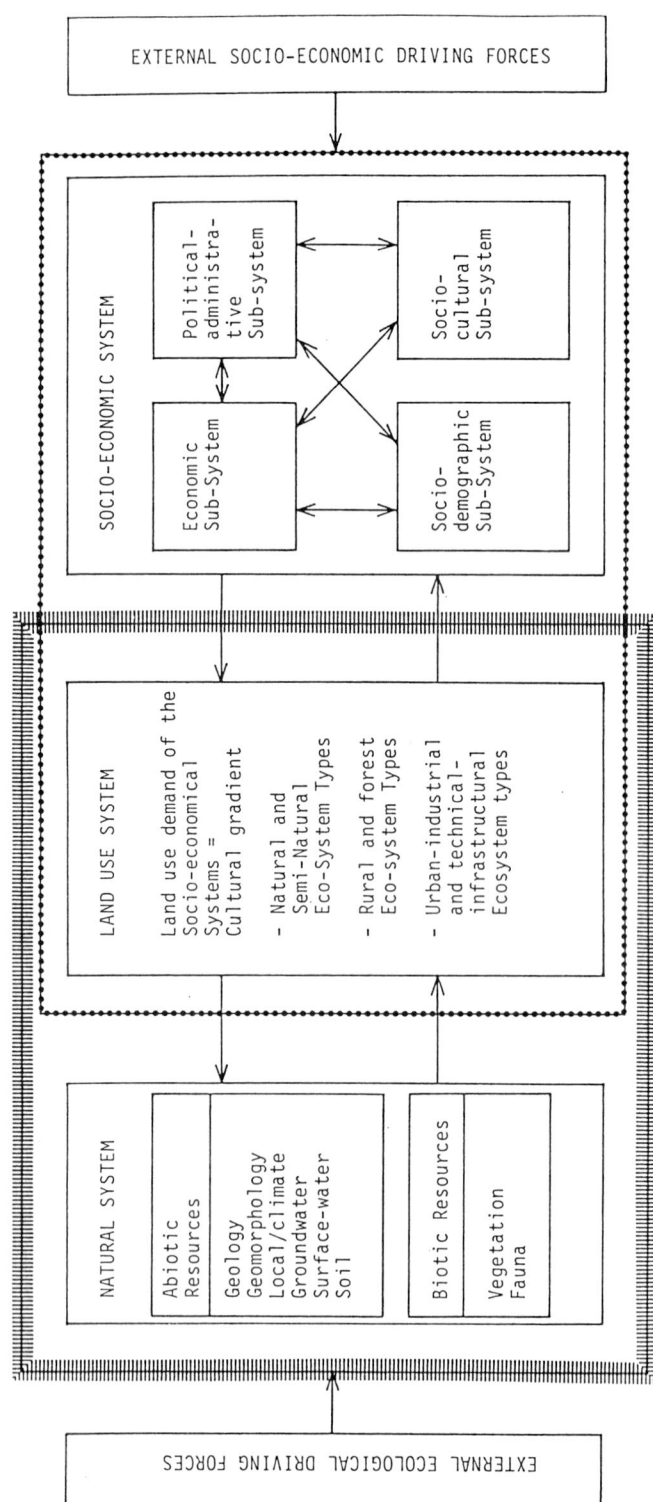

Figure 3.9 Simple model of a regional ecological-economic system. Explanation in the text. (Adapted from Messerli and Messerli, 1979)

a gradual shift away from 'pure' descriptive phytosociology and toward vegetation ecology (Pfadenhauer, 1992).

A system approach for environmental planning

Environmental management and planning needs additional system concepts to comply with new requirements and prescriptions. Looking at the hierarchy of organizational levels shown in Figure 3.1, much activity shifted from the ecosystem levels upward to the level of landscape and the society-environment-system. We can only briefly mention the concepts developed for these levels, but not treat them in any detail (see Haber, 1990a; 1990b; Haber et al., 1991; Tobias, 1991; Lenz and Schall, 1991; Kerner et al., 1991).

A very useful concept, originally developed by Messerli and Messerli (1979) for the Swiss Man and the Biosphere Programme (MAB 6), is the regional ecologic-economic system (Figure 3.9). It is a threefold system with the natural (eco-)system on its left side and the socio-economic system on its right side, representing the organization of nature and human society, respectively. The influence and imprint of the latter upon the former has produced the land use system, shown in the middle of the figure, which is nothing but our cultural landscape. Its gradient from natural to urban-industrial ecosystem-types corresponds to the classification shown in Table 3.1. Of course there are also external inputs and outputs, for example, air pollutants or government subsidies entering the regional system, and wastewaters or export goods leaving it.

To transform such a regional ecologic-economic system into an environmental planning or management model requires additional concepts allowing predictions and simulations. For such applications, a key problem is data availability and processing, characterised by the 'point-area dilemma'. One can get exact data from measurements only at a limited number of points that are expected to be representative for a given area. But for information about the whole area, one has to extrapolate from these points and, therefore, loses reliability. To avoid this and to be more exact, the number of measurement points may be increased, but this often requires a disproportionate expenditure of work and time, and one runs into another dilemma: the space-time dilemma. One can get reliable quantified data either in a spatial context or in a temporal sequence, but not in both dimensions.

To overcome these dilemmas, we have worked out a 'pyramid model' shown in Figure 3.10 (Haber, 1990a; Kerner et al., 1991). The bottom of the pyramid represents the cultural landscape or the regional system as shown in Figure 3.9. Here the 'ecological reality' with all its structures and processes is assessed and recorded, quantified wherever possible by measuring, counting, or weighing. However, this can only be done with high precision at a few carefully selected points (black dots in Figure 3.10).

To get an overall assessment of the whole region, the results of the point measurements have to be extrapolated and aggregated. This happens on the next higher level of the pyramid, called the 'spatial level'. The data are stored and processed in a geographical information system producing all kinds of maps and pictures of the region, input-output-comparisons and even plans — but only for specific points in time.

The dynamics of the region or of its components caused by human and other biotic activities, changes in inputs and outputs, etc., have to be assessed by introducing feedback processes. These require still more data aggregation and a higher degree of abstraction, and lead to a third working level called the 'temporal level'. Here, also, 'time charts' can be produced, showing where and when what changes will occur in the region under given impacts.

The last and uppermost level is the 'strategic level' where principal trends or changes in the regional system are estimated, using scenario or simulation techniques. Even erratic or catastrophic events can be simulated, of course in a more speculative way, but always supported by data from the lower levels.

The tapering of the pyramid symbolizes the unavoidable decrease in precision and reliability of both data and evidence. On each level, different methods of data processing are required, and different results are produced. But all data are derived from the same data set gained on the bottom level. Continuous validation of all results, especially of those having a predictive character, is necessary. It is achieved by iterative comparisons between levels and, in particular, with the lower levels. We call this procedure 'up- and down-scaling' or 'coupling of levels'. It prevents overemphasizing results of single-level approaches and methods. Thus, the pyramid concept combines reductionistic and integrative methods, which is very important for dealing with complex systems.

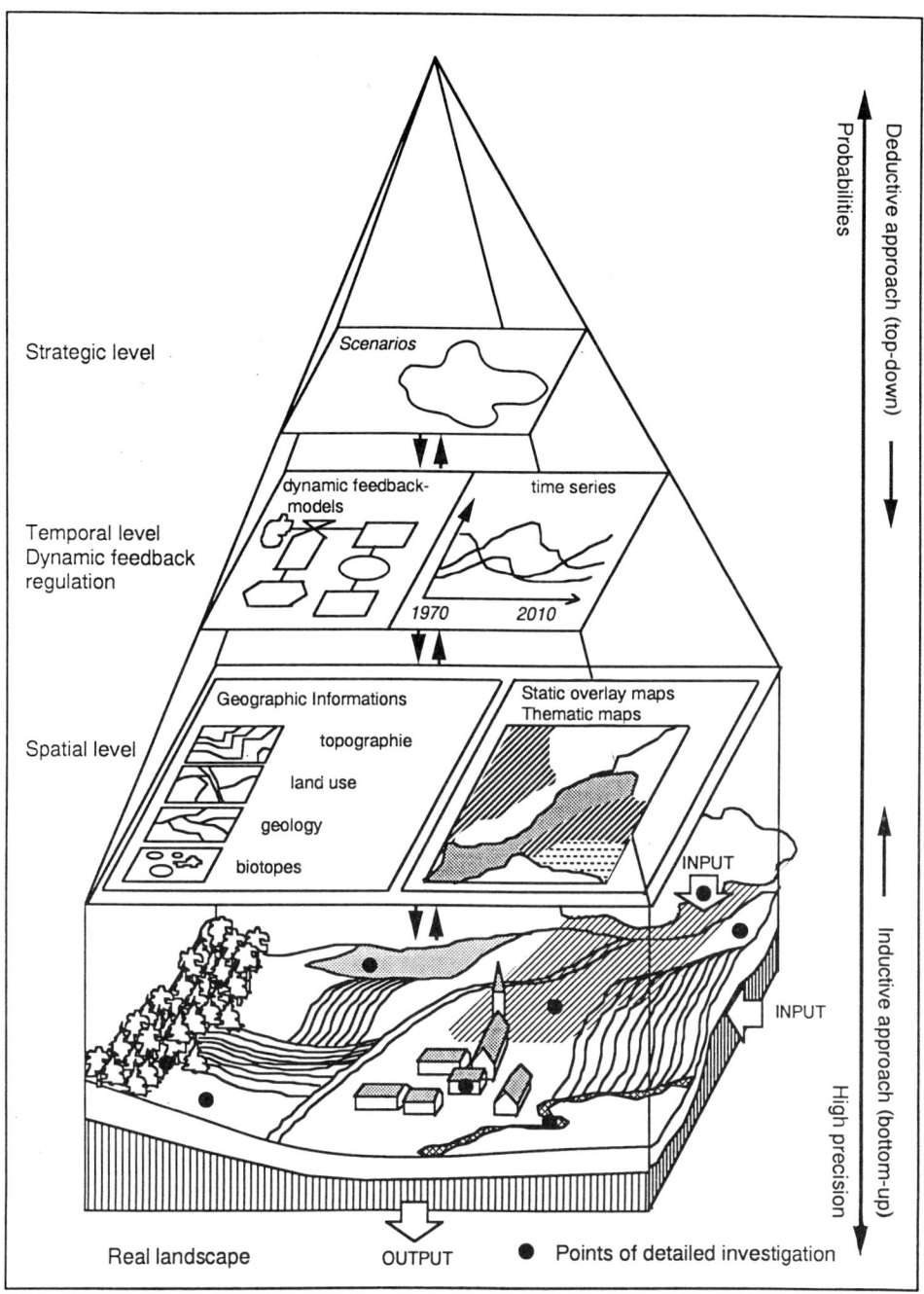

Figure 3.10 Different scale-adapted approaches to environmental management and planning of a given landscape or regional system. Explanations in the text. (From Kerner et al., 1991)

References

Duhme, F., R. Lenz, and L. Spandau (Eds.), 1992. *25 Jahre Lehrstuhl für Landschaftsökologie in Weihenstephan mit Prof. Dr. Dr.h.c. W. Haber* (Festschrift). Freunde der Landschaftsökologie Weihenstephan e.V., Freising.

Egler, F.E., 1970. *The way of science. A philosophy of ecology for the layman.* Hafner, New York.

Ellenberg, H., 1973. Versuch einer Klassifikation der Ökosysteme nach funktionalen Gesichtspunkten. In: H. Ellenberg (Ed.), *Ökosystemforschung.* Springer, Berlin /Heidelberg, pp. 235-265.

Ellenberg, H., 1986. *Vegetation Mitteleuropas mit den Alpen.* 4. Auflage. Ulmer, Stuttgart.

Ellenberg, H., and D. Mueller-Dombois, 1974. *Aims and methods of vegetation ecology.* Wiley, New York/London.

Fenchel, T., 1987. *Ecology - potentials and limitations.* Ecology Institute, Oldendorf/ Luhe. (Excellence in Ecology, Vol. 1.)

Haber, W., 1990a. Using landscape ecology in planning and management. In: I.S. Zonneveld and R.T.T. Forman (Eds.), *Changing landscapes: an ecological perspective.* Springer, New York/Berlin, pp. 217-232.

Haber, W., 1990b. Basic concepts of landscape ecology and their application in land management. *Physiology and Ecology Japan* 27 (Special Issue 'Ecology for Tomorrow', edited by H. Kawanabe, T. Ohgushi, M. Higashi), pp. 131-146.

Haber, W., 1992. Erfahrungen und Erkenntnisse aus 25 Jahren der Lehre und Forschung in Landschaftsökologie: Kann man ökologisch planen? In: F. Duhme, R. Lenz and L. Spandau (Eds.), *25 Jahre Lehrstuhl für Landschaftsökologie mit Prof. Dr. Dr.h.c. W. Haber* (Festschrift). Freunde der Landschaftsökologie Weihenstephan e.V., Freising, pp. 1-28.

Haber, W., 1993. *Ökologische Grundlagen des Umweltschutzes.* Economica, Bonn. (Umweltschutz - Grundlagen und Praxis Band 1.)

Haber, W., R. Lenz, P. Schall, R. Bachhuber, W.D. Grossmann, K. Tobias and H.F. Kerner, 1991. Prüfung von Hypothesen zum Waldsterben mit Einsatz dynamischer Feedbackmodelle und flächenbezogener Bilanzierungsrechnung für vier Schwerpunktforschungsräume der Bundesrepublik Deutschland. *Berichte Forschungszentrum Waldökosysteme* (Göttingen), Reihe B, Band 20.

Harper, J.L., 1982. After description. In: E.I. Newman, *The plant community as a working mechanism.* Blackwells, Oxford, pp. 11-26.

Kaule, G., 1974. *Die Übergangs- und Hochmoore Süddeutschlands und der Vogesen.* J. Cramer, Lehre (Dissertationes Botanicae Band 28).

Kerner, H.F., L. Spandau, J.G. Köppel and T. Wachter, 1991. Methoden zur angewandten Ökosystemforschung, entwickelt im MAB-Projekt 6 'Ökosystemforschung Berchtesgaden' (Werkstattbericht). In: *MAB-Mitteilungen 35*, hrsg.v. Deutsches Nationalkomitee 'Der Mensch und die Biosphäre' (MAB). Bundesministerium für Umwelt, Naturschutz und Reaktorsicherheit, Bonn. 2 volumes.

Lenz, R. and P. Schall, 1991. Theorie und Modellierung von Waldschadensprozessen im Fichtelgebirge -ihre hierarchische Strukturierung und technologische Anwendung. *Verh.Ges.f.Ökologie* 19/3 (Osnabrück 1989), pp. 647-661.

Messerli, B., and P. Messerli, 1979. *Wirtschaftliche Entwicklung und ökologische Belastbarkeit im Berggebiet*. Fachbeiträge zur Schweizerischen MAB-Information (Bern) Nr. 1. 20 pp. (also in: *Geographica Helvetica* 33: 203-210, 1978).

Miller, G.T., 1975. *Living in the environment: an introduction to environmental science*. Wadsworth, Belmont/Calif. (7th edition 1992.)

Naveh, Z., 1984. Conceptual and theoretical basis of landscape ecology as a human ecosystem science. In: Z. Naveh and A.S. Lieberman (Eds.), *Landscape ecology: Theory and application*. Springer, New York, pp. 26-105.

Pfadenhauer, J., 1993. *Vegetationsökologie - ein Skriptum*. IHW-Verlag, Eching (Bayern).

Schmithüsen, J., 1948. 'Fliesengefüge der Landschaft' und 'Ökotop'. Vorschläge zur begrifflichen Ordnung und zur Nomenklatur in der Landschaftsforschung. *Berichte zur deutschen Landeskunde* 5: 74-83.

Tansley, A.G., 1935. The use and abuse of vegetational concepts and terms. *Ecology* 16: 284-307.

Tobias, K., 1991. Konzeptionelle Grundlagen zur angewandten Ökosystemforschung. *Beiträge z.Umweltgestaltung* Band A 128. Erich Schmidt Verlag, Berlin.

Troll, C., 1939. Luftbildplan und ökologische Bodenforschung. *Zeitschrift der Gesellschaft für Erdkunde* 1939: 241-311. Berlin.

Troll, C., 1950. Die geographische Landschaft und ihre Erforschung. *Studium generale* 3, No. 4/5: 163-181 (also in: *Erdkundliches Wissen* 11: 14-51, 1966).

Troll, C., 1968. Discussion remark. In: R. Tüxen (Ed.), *Pflanzensoziologie und Landschaftsökologie*, Junk Publishers, The Hague, p. 42.

Van Leeuwen, C.G., 1980. *Ekologie I*. Delft, Faculty of Architecture TUD (Reader, unpubl.).

Walter, H., 1986. *Allgemeine Geobotanik*. 3. Auflage. Ulmer, Stuttgart.

Westhoff, V., 1968. Die 'ausgeräumte' Landschaft. Biologische Verarmung und Bereicherung der Kulturlandschaften. In: K. Buchwald and W. Engelhardt (Eds.), *Handbuch für Landschaftspflege und Naturschutz*, Band 2, pp. 1-10. - Die Reste der Naturlandschaft und ihre Pflege. In: *ibidem*, Band 3, pp. 251-165. BLV, München.

Westhoff, V., 1979. Phytosociology in the Netherlands: History, present state, future. In: M.J.A. Werger (Ed.), *The study of vegetation*, The Hague, pp. 81-121.

Zonneveld, I.S., 1990. Scope and concepts of landscape ecology as an emerging science. In: I.S. Zonneveld and R.T.T. Forman (Eds.), *Changing landscapes: an ecological perspective*. Springer, New York/Berlin, pp. 3-20.

The natural hierarchy of ecological systems

4

Michel Godron

ABSTRACT - For ecological research the scale of phenomena becomes increasingly important, both for reasons of sampling and for interpreting the results. At present, it seems possible to determine which biophysical factors are dominant in ecological zones, regions, sectors and/or districts, and stations or sites. The hierarchy of these ecological units then becomes a natural one, and their being spatially nested in each other becomes more logically founded.

Introduction

As soon as ecology became a science, spatially nested units of vegetation were defined and named. Two international congresses of botany (1901 and 1910) then gave excellent classifications that can be integrated profitably in a contemporary synthesis. However, the proposed classifications were only descriptive and did not explain, for example, why 'regions' should be distinguished. Consequently, they would today be regarded as intuitive classifications, because their basis was not made explicit.
At present, it is possible to trace the functional causes that justify the then-proposed spatial hierarchy. In this contribution we shall present these causes, which can be regarded as natural laws. Our objective is not a discussion on semantics, but a reasoning on the causes of the phenomena, as expressed at the beginning of modern science: *scire bene, scire causas.*

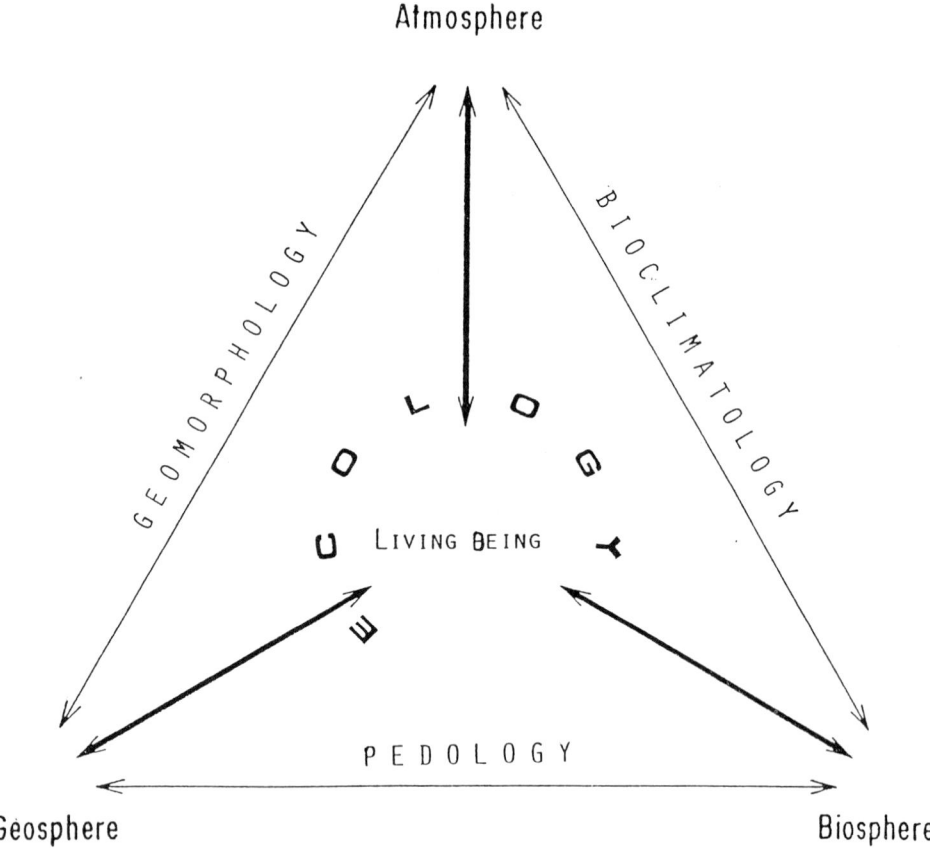

Figure 4.1 Relations between ecology and three basic environmental sciences

If we understand the individual causes and their interactions, an agreement on the names of units, such as ecotopes or stations, will be easier. The main causes of the differentiation of spatially nested ecological units shall be examined for three different scale levels successively, namely, zones and regions, sectors and/or districts, and stations or sites.

The law of zonality and the upper levels of the spatial hierarchy

The first ecological division of Earth is into oceans and continents. Here, we confine ourselves to examining patterns on continents only, which result from the combined action of various discriminant environmental factors, namely, climate, relief and geology. These three main factors (Figure 4.1) are used for making ecological subdivisions. However, it is necessary to put some order in the multitude of resulting spatial units.

On any continent we can distinguish four main types of tectonic structures: cordilleras or alpine chains, old shields, metamorphic borders and volcanic bodies (Figure 4.2). Are they good ecological units? Clearly not, if we look at examples. The Alps are ecologically very different from Patagonia or from

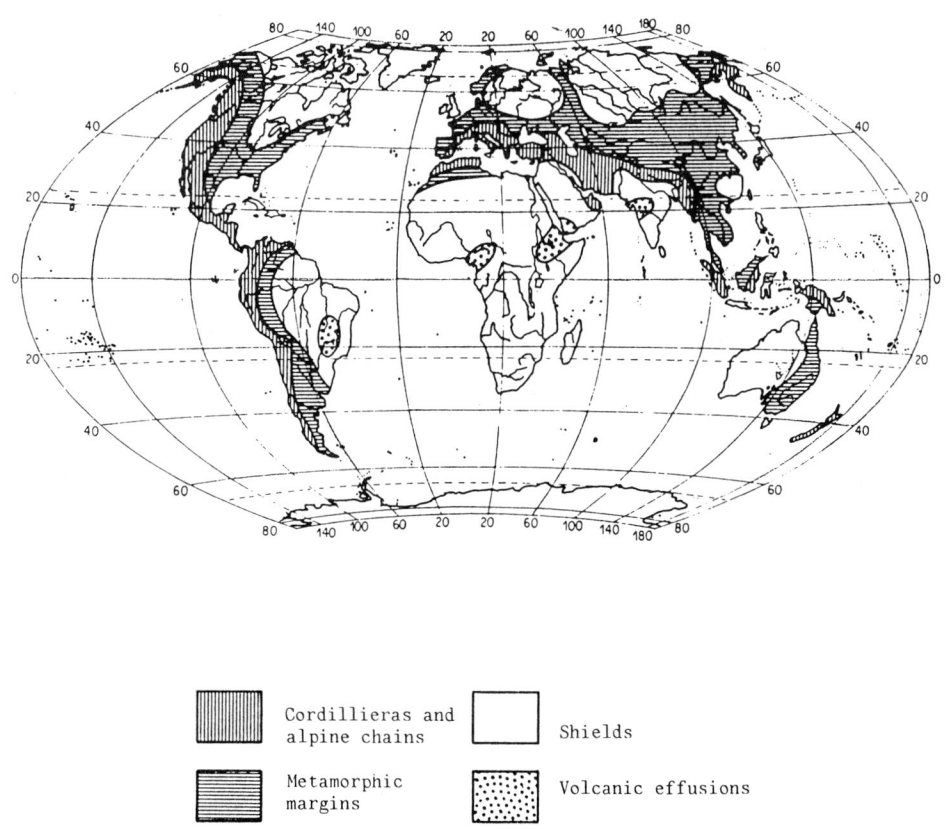

Figure 4.2 Main tectonical structures

Peruvian cordilleras. In the same way, the Canadian shield has very few ecological characteristics in common with the Sahara, and so on.

Would other aspects of geology have a more discriminant nature? We maintain they do not, because neither stratigraphy nor lithology would yield the same ecological unit in Europe and Patagonia.

In general, the major ecological pattern on Earth that can be observed on the scale of continents is related essentially to large climatic differences. The corresponding units are the so-called ecological zones.

Ecological zones

Thalès de Milet (about 600 B.C.) was the first scientist who noticed the primordial importance of climate, and he bequeathed us the word *zone* to name the resulting spatial units. During the Middle Ages and Renaissance, no one improved that first classification.

At the beginning of the 17th century, Boussingault observed that peat bogs are linked to cold climates. A. von Humboldt, L. Lesquereux and others came up with similar observations, thus further elaborating this kind of reasoning. However, the general law explaining the large zonal pattern was discovered by B. Dokuchaev (1879, 1883, 1902) and his enthusiastic disciples N. Sibirtsev and C. Glinka. They observed that, from the Black Sea to the Arctic Ocean, ecologically relatively homogeneous units occurred that were controlled almost entirely by latitude, because no mountain chain blocked the vast plains.

For each of these zones the climate determines the vegetation, and the joined action of climate and vegetation in turn determines the soil development (Table 4.1).

Table 4.1 Ecological zones are determined mainly by climate and reflected by dominant natural vegetation types and soil types

Climate type	Natural vegetation	Soil
semi-arid	steppe	tchernozem
continental, mesothermic soil	deciduous forest	brown
continental, cold	taïga (coniferous forest)	podzol
continental, very cold soil	tundra	arctic

THE NATURAL HIERARCHY OF ECOLOGICAL SYSTEMS 73

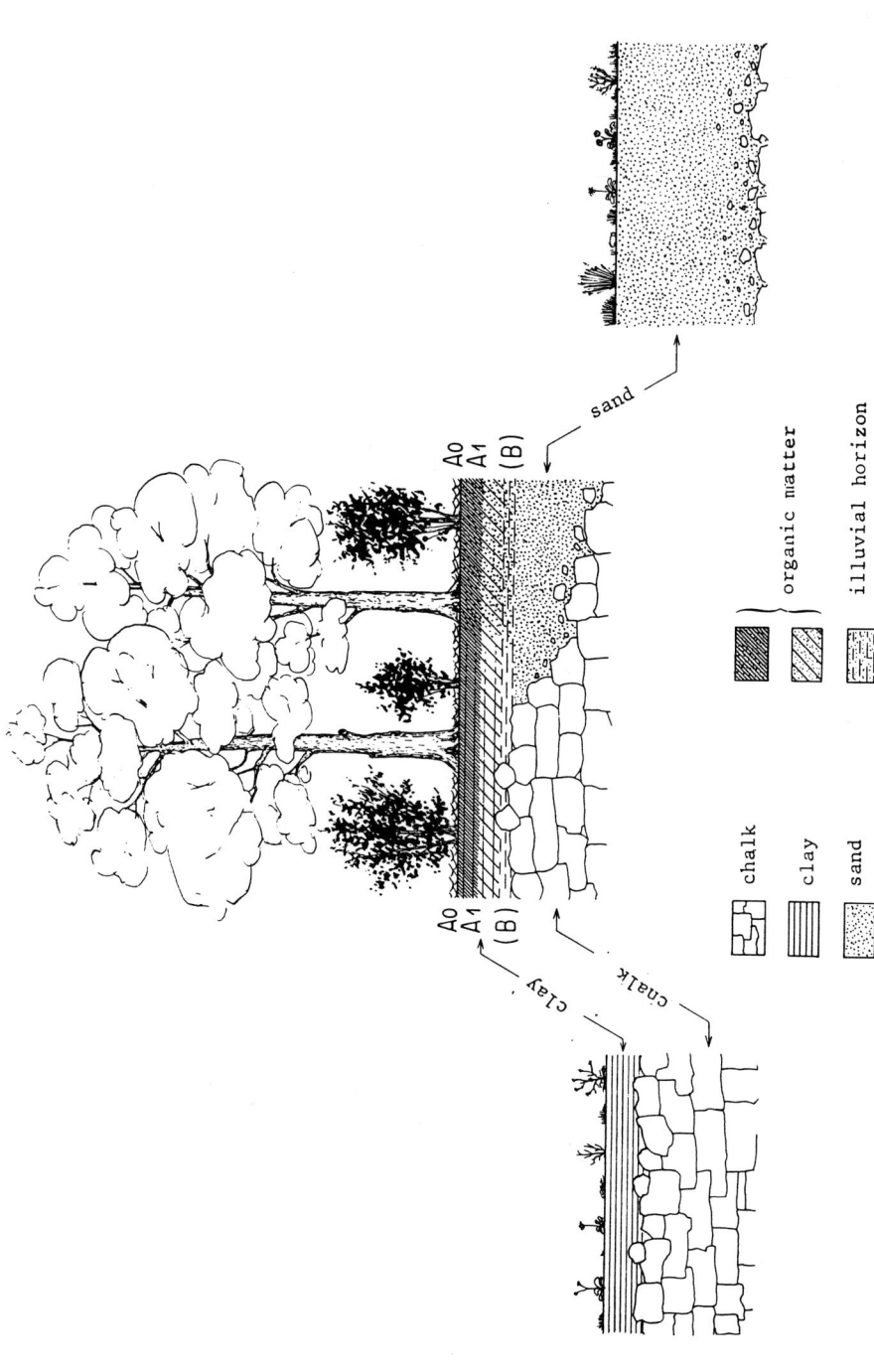

Figure 4.3 Convergent evolution of two young ecosystems in a temperate climate, towards one forest type independent of the parent material

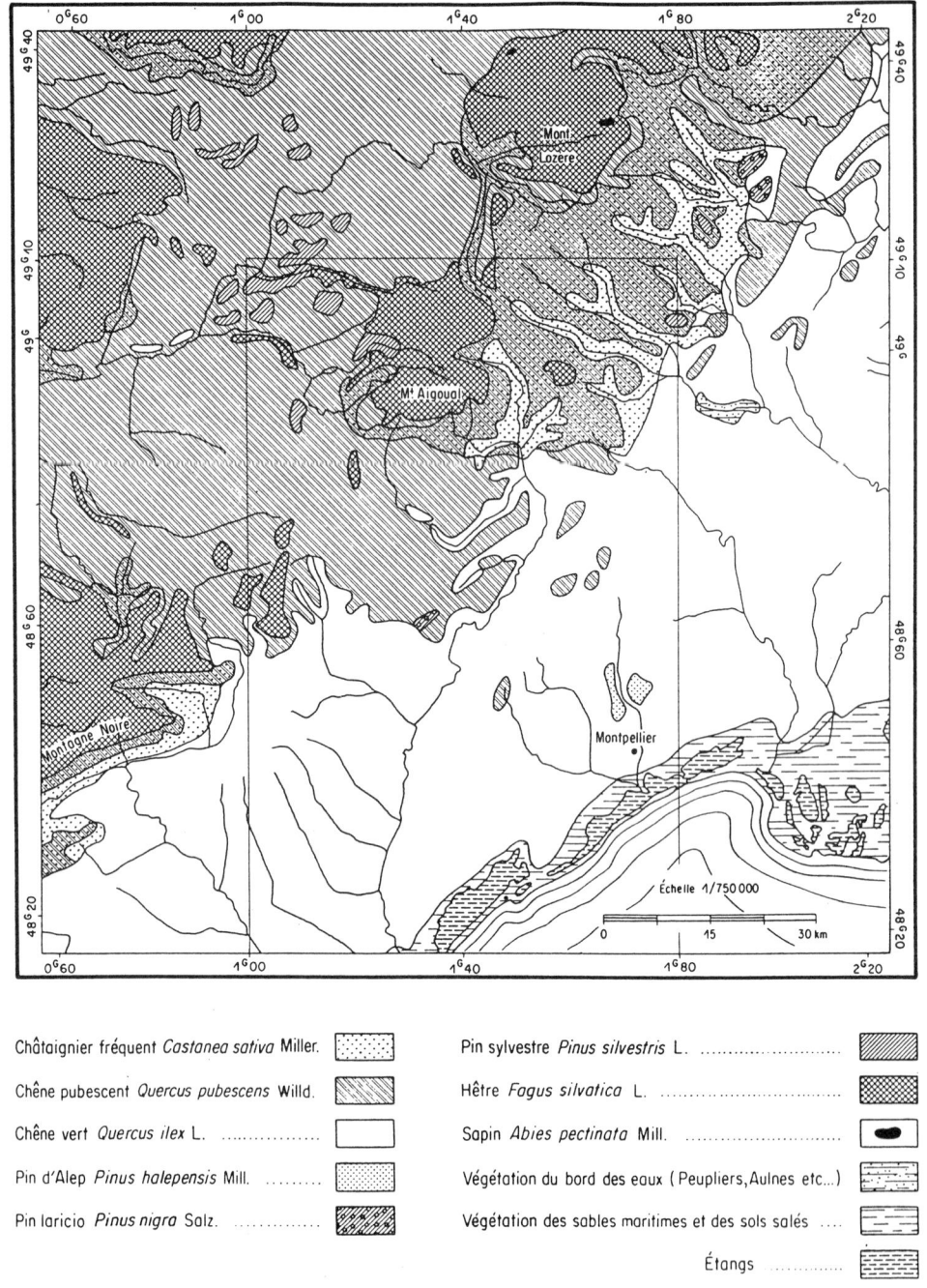

Figure 4.4 Vegetation belts of Languedoc, according to Gaussen (1938)

For a better understanding of these interactions, we shall have a brief look at the series of causes that explain these correspondences. We shall do so by comparing the development of two very different mineral soils, one on quartzeous sand and the other on clay (Figure 4.3), due to a progressive colonization by vegetation and subsequent succession.

In the beginning, the sand is excessively drained and poor in cations. The clay, in contrast, is poorly drained. Progressively, as vegetation produces litter that is converted into humus, which is incorporated in the soil, the sand can retain more moisture, whereas the clay acquires a better developed structure and becomes better aerated. As a result, the two soils become more alike and the different vegetations become more similar during the natural succession. In other words, there is a convergence in the development of these two ecological systems. This gives the law of zonality:

In each climate, the developments of both the soil and the biocenoses converge towards a characteristic 'family' of types.

This law is very general and valid all over the world, reflected by the fact that ecologists regard climate as the first environmental factor to take into account at the highest level of the spatial hierarchy.

Ecological regions

In fact, climate is not entirely homogeneous inside a zone. Both relief and distance to the ocean are additional causes of spatial differentiation, because they affect the climate. Thus, 'étages de végétation' are formed, which mainly depend on winter temperature (Figure 4.4). In each of these altitudinal belts the law of zonality applies, but there are reasons that oblige us to define a hierarchical level just below the level of ecological zones: *ecological regions*.

One of the best definitions of ecological regions is given by Ch. Flahault (1901), who wrote: 'large territories where the same climate reigns and that carry the same vegetation'. He specified that he was referring to large vegetation units, such as taiga, tundra, savannah, etc. as already classified by De Candolle (1812) and Schouw (1832).

Again, it is necessary to specify what type of climate we have in mind. To this end, it suffices to examine how climatologists proceed. They use all the available measurements and calculate the gradients between data collected in meteorological stations hundreds of kilometres apart. Consequently, climatic maps are usually produced at rather small mapping scales of at least 1:200,000.

In practice, for specifying the boundary limits of a climatic region, it is necessary to collect data from a score of meteorological stations, then to compare them and compute statistical regressions between altitude, temperatures, rainfall, evaporatranspiration, etc., in order to interpolate between the stations. The boundaries between the 'étages de végétation' can then be based on the correlations that have been found between altitude and bio-climatical factors such as rainfall, humidity, etc.

Also, on large plains or large plateaus, which cover hundreds or thousands of square kilometers, as is the case in eastern Canada or African Sahel, it is possible to distinguish ecological regions in relation to altitude, distance to the ocean, etc.

In this context, the next law reads:

An ecological region is a territory for which statistical regressions between physiographic characteristics such as altitude and distance to the ocean and bio-climatical factors have the same co-efficients.

Ecological sectors, districts, and landscapes

After climate, it is the geosphere that has a prominent influence. Therefore, we must look at the ecological units connected with geomorphology and soil. Manil (1963) named these units 'secteurs écologiques', which may be translated as *ecological sectors*.

Let us consider a valley in an area with homogeneous bedrock. If geomorphological processes are constant from the top of the divides to the river, the whole valley can be regarded as an ecological sector. If, instead, terraces have been formed by the river, during glaciations in temperate regions or during wet periods in the tropics, these terraces constitute autonomous ecological sectors, where the pedogenesis is not the same as on the upper slopes.

If we want to stress that the terraces have characteristics in common with the slopes, we may indicate the whole valley as a 'family of ecological sectors'. Alternatively, we may also distinguish a level intermediate between the region and the sector, for example a *district* (see Klijn, elsewhere in this book).

Now we may question if there is a natural law corresponding to these hierarchical level(s). We should say yes, because with the development of geomorphological dynamics by Tricart (1965) it appears that the four processes of morphogenesis, viz. weathering, erosion (comprising ablation), transport and deposition are sufficiently linked to justify a law:

Each ecological sector is characterized by a set of geomorphological processes that depend on the regional climate, either directly or indirectly by its influence on the vegetation cover.

The scale of these processes is intermediate between the scale of regions and the scale of ecological stations discussed in the next section.

Finally, we want to remark that it is at this same spatial scale level that *landscapes* can be mapped. We recall that in some recent publications on landscape ecology (e.g. Forman and Godron, 1986), a landscape is considered as a heterogeneous kilometers-wide unit, characterized by a recurrent pattern of a number of homogeneous units which could qualify as biocenoses, ecotopes or ecosystems, depending on the point of view. In other words: patches and corridors within a further homogeneous matrix. This definition enables mapping landscapes directly and specifying the contents of each landscape in the accompanying legend.
It appears that in practice a landscape often corresponds with an ecological sector, a family of ecological sectors, or manmade elements, for example, as in urban landscapes. Sometimes a landscape corresponds with an 'étage de végétation'.

Ecological stations or sites

In the spirit of our predecessors at the botanical congresses in the beginning of the century, we define the *ecological station* or *site* as the elementary unit characterized by a homogeneous vegetation which corresponds to a biocenosis (Moebius, 1877). Some authors use the word *ecotope* for this level of the spatial hierarchy (compare Udo de Haes and Klijn or Haber in previous chapters of this book), in which most studies on ecosystems were carried out.

In relation to the classification hierarchy of phytosociological units, Braun-Blanquet (1925) expressed that an association corresponds to 'une écologie particulière et autonome', i.e., an ecological station.

However, the upper levels of the phytosociological classification hierarchy, i.e. class, order and alliance, are characterized by species that can be regarded as actually being ecological species groups in the sense of Duvignaud or Ellenberg. For example, the *Quercion ilicis* is characterized by a large number of species confined to the mediterranean climate. These species constitute a distinct ecological group which is found in the association named *Quercetum ilicis*, but which is also found in other mediterranean associations.

Consequently, these species groups could be related to the larger units of the spatial hierarchy of ecological units as treated above.

As Figure 4.3 illustrates, vegetation development is a factor causing zonal homogenization. It also seems evident that vegetation grows higher and higher after a forest cutting or after the abandonment of a field. Therefore, we think it wise to make a brief excursion to explain the cause of this permanent tendency to grow, to see whether there is a controlling law and to examine the relevance of such a law for the ecological station concept.

A terrestrial plant is a thermodynamic machine that needs solar energy, water, nutrients, and CO_2. The simplest plant is a film covering the soil and exposing its face to sun and air. Actually, lichens function in such a way. They are able to live on bare rocks or extremely poor soils. The plants that take over from lichens are mosses, invading as soon as a 'microsoil' exists, followed by herbaceous plants and then bushes and finally trees. This succession is well known, but its cause is rarely explained: it is only competition for solar energy that pushes plants to grow higher than their neighbours, a tendency that is so strongly written in the genetic memory of a species that it is now a law:

On each point of earth, vegetation is bound to ascend as high as site factors, such as water availability, nutrient availability, salinity, etc., allow the local flora to grow.

The result of this law is a vertical distribution of biomass that can be described by a logarithmic curve. Figure 4.5, on the left, gives the curve for a forest in New Quebec, but Falinski (1973) found a similar vertical phytomass distribution in the natural forest of Bialowiecza. This curve is now the expression of an equilibrium between the forces that tie vegetation to the

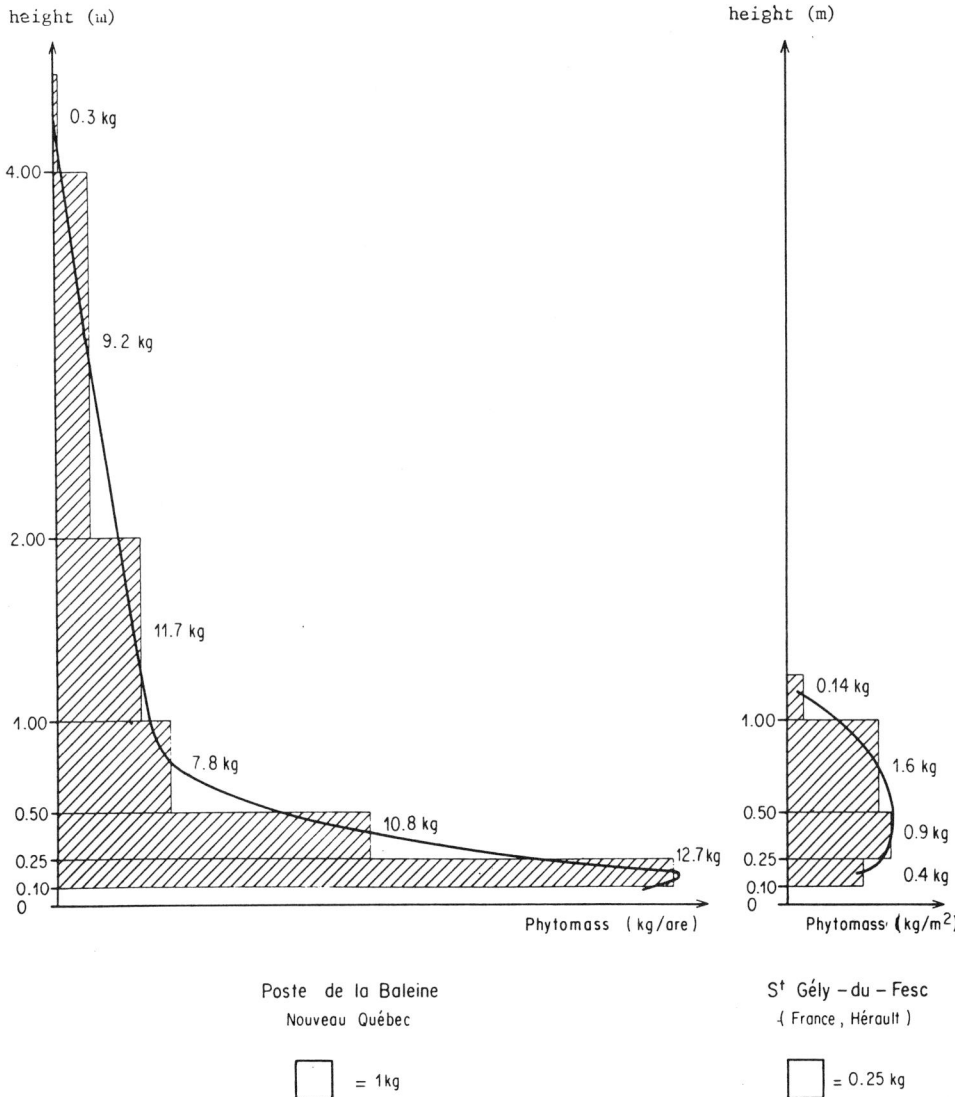

Figure 4.5 Vertical distribution of phytomass in a forest in New Quebec (left), and in a mediterranean garrigue vegetation (right)

source of water in the soil and the force that drives sap to ascend in trunks, branches, and leaves.

What then, you may ask, is the importance of this for the definition of an ecological station? The clue can be found in the law given above that states 'as high as available water ...'. The ecological station, or site, is exactly the place where the interrelations of parent material, soil, and groundwater determine the availability of water for plant growth.

In practice, of course, the main ecological factor causing vegetation or an ecological station to be homogeneous is man's influence, because he so strongly affects the species composition. And he, above all, modifies the mosaic of patches with homogeneous site conditions, splitting it up into smaller and smaller parcels, thus creating artificial landscapes. The limits of a field, a road, or a ditch are all determined by the action of man.

Man especially influences vegetation succession by, for example, harvesting, either directly or indirectly by wild animals or cattle. Thus, he causes a perturbation of the natural succession, as can be exemplified by referring to Figure 4.5 again.

In undisturbed circumstances, the vertical distribution of biomass (Figure 4.5, left) reflects the natural equilibrium between natural constraints and the performance capabilities of the local flora, as was explained earlier. In contrast, the vertical profile in Figure 4.5 on the right was obtained in 'degraded' mediterranean garrigue vegetation, dominated by *Quercus coccifera*. This species resprouts rapidly after a fire, and, also, the stems are densely crowded, thus suffocating other species. The shape of this profile is temporary. It lasts only until a competitive species, in this particular case often *Pinus halepensis*, arrives and restores the equilibrium as expressed in the left curve.

The spatial hierarchy of ecological units

The ecological units that were examined in the preceding sections correspond to mapping scales ranging respectively from 1: 100,000,000 for zones to 1: 2,000 for stations. When mapping, their natural hierarchy appears by the fact that the mapping units are spatially nested: an ecological zone contains various ecological regions which, in turn, contain various ecological districts, which contain ecological sectors, etc.

But the characterization of each unit within a specific and well-confined area requires the use of characteristics that often belong to smaller constituent units or larger comprising units than the one considered. For example, the ecological sectors of two adjacent valleys in Languedoc, namely, Saint-Martin-de-Londres and Fambetou are characterized by:

- a summer drought, which exists in the *whole* mediterranean region;
- *two* slightly different 'étages de végétation', and
- two different types of *mosaïcs* of vineyards, woods, arable crops, grasslands, etc., reflecting differences in soil types and the influence of man.

Consequently, mapping each of these attributes does not suffice for delimiting the sector. It is necessary to add weighing factors to each attribute. For example, in this case the presence of vegetations dominated by *Pinus halepensis* was regarded as being of less importance than the parent material being constituted by lacustrine deposits or the valley bottoms being covered by frequently flooded grasslands. In other cases, however, the presence of certain species may well be the most important attribute to characterize an ecological sector.

These remarks oblige to temper the hope that geographical information systems (GIS) will solve all our problems. They are certainly helpful tools for mapping ecological zones, regions, districts and stations, but the weight of the factors characterizing each of these units cannot be automatic. The ecologist must decide on the weight to give each. In some cases, the bedrock will be more important than the type of weathering, or man's influence, or the presence of a remarkable species. In our opinion, the failure of Mc Harg's planning system can also be attributed to a lack of methods for weighting the factors.

Conclusions

Before drawing conclusions, we want to remind you of a classical rule: the first thing to do when starting a mapping project is to specify the objectives, the second is to choose the mapping scale, the third to construct the legend.
In this third step, the hierarchy of factors and their respective weight must be discussed with the users of the map. Then the spatial nesting of the ecological units will emerge automatically.

The hierarchy presented in this paper is 'natural', because it proceeds from ecological laws. It also corresponds with earlier remarks that the levels of integration of biological units are successively strong and feeble: the gene is strong, the nucleous is feeble, the cell is strong, the organ is feeble, the individual is strong, the guild is feeble. The same can be said of our ecological units, namely: the ecological station is strong, the landscape is feeble, the ecological district is strong, the region is feeble, and the ecological zone is strong again. More precisely:

- The ecological station is a strong unit, the 'atom' of our science, because its homogeneity is the result of its functioning; famous ecologists even considered it as a 'quasi-organism'.

- Elements of a landscape are not very strongly integrated. Consequently, landscape ecologists still have difficulties defining precise measures of heterogeneity to unambiguously delimit the boundary between two landscapes with the help of information content, for example (Godron et Bacou, 1975).

- An ecological sector depends on strong geomorphological processes, which affect it as a whole; its limits are often clearly linked to physiographic limits.

- The ecological region is not a strong unit; its limits are vague, as they result from statistical computations with data that are particularly uncertain in mountainous regions.

- The climatic zones are a strong feature of the globe, reflected by the fact that the maps of equatorial zone, tropical zones and so on, designed by many authors, are remarkably similar.

Finally, it appears that ecologists can tell, as do the physicists: *the scale creates the phenomenon.*

Acknowledgements

I thank Frans Klijn for his help in composing this text on the basis of a French original.

References

Braun-Blanquet, J., 1925. Zür Wertung der Gesellschaftsliene in der Pflanzensoziologie. *Viertelsjahresschr. Naturf. Ges. in Zürich*, 70: 122-149.
De Candolle, A.P., 1812. *Théorie élémentaire de la botanique*. Paris.
Falinski, J., 1973. Herb layer filling by plant cormus in the *Querceto-Carpinetum*. *Phytocoenosis*, 2/2: 123-142.
Flahault, Ch., 1901. *Premier essai de nomenclature phytogéographique*, 3ème congrès intern. de Bot., 29 pp.
Forman, R.T.T. and M. Godron, 1986. *Landscape Ecology*, Wiley & Sons, New-York, 619 pp.
Godron M. and A.M. Bacou, 1975. Sur les limites 'optimales' séparant deux parties d'une biocénose hétérogène. *Ann. Univ. Abidjan*, série E (Ecologie) 8/1: 317-324.
Manil, G., 1963. Niveaux d'écosystèmes et hiérarchie de facteurs écologiques. *Bull. Soc. Ac. Roy. Belg.* 49, 6, 32: 603-623.
Moebius, K., 1877. *Die Auster und die Austerwirtschaft*. Berlin.
Schouw, J.F., 1832. *Grundzüge einer Pflanzengeographie*. Berlin.
Tricart, J., 1965. *Principes et Méthodes de la Géomorphologie*. Masson, Paris.

Spatially nested ecosystems: guidelines for classification from a hierarchical perspective 5

Frans Klijn

ABSTRACT - Ecosystems are complex systems that can be described and classified by a large number of characteristics. Also, they can be distinguished at many spatial scale levels, all related to each other. This can be understood by regarding them as being spatially nested, thus forming a mosaic at the earth's surface.
Both the complexity and the dependence on spatial scale pose problems for classification. To solve these, we argue for a deductive top-down approach based on hierarchy principles. It is based on a comprehensive hierarchical model of an ecosystem, related to spatial and temporal scale levels. From this model we derive a framework for classification and mapping at different spatial scales: a set of guiding principles.
Finally, we discuss and exemplify the applicability for environmental policy analyses.

How to proceed in the maze of overwhelming complexity?

Ecosystems are very complex systems. They are made up of a wealth of structurally related biotic and abiotic components that are also functionally related by innumerable processes: fluxes of energy and matter. Thus, ecosystems may be studied by zooming in on different aspects. For instance, we may emphasize the structural characteristics by looking at the species composition of the vegetation in relation to controlling site factors. This is the common approach of vegetation ecologists (Runhaar et al., in this book). But we may also concentrate on energy budgets (Odum, 1983) or on matter budgets (Lenz, in this book). This confronts us with the question of where to begin.

For an answer, we may derive some clues from the title of this book. These clues are 'classification' and 'environmental policy'.

For classification purposes, it seems obvious that we should look for characteristics of ecosystems that can easily be recognized or measured in the field: the characteristics that Zonneveld (in this book) calls 'diagnostic characteristics'. In practice, this would mean concentrating completely on structural characteristics, because functional characteristics cannot be recognized very easily.

On the other hand, the desired applications for environmental policy analyses seem to require an emphasis on processes. More specifically, on processes that result from environmental hazards. An analysis of acidification, for example, reveals that it influences numerous ecosystem processes. Processes that can be ordered in chains of events, such as the chain of dispersion, deposition, chemical buffering in the soil, and biotic response. In addition to following such chains, we can distinguish simultaneous processes with similar effects, such as buffering by the solution of $CaCO_3$, the weathering of silicates, the exchange of kations, and the decay of aluminium and iron (hydr)oxides (after Ulrich, 1980; Verstraten, 1982; De Vries and Breeuwsma, 1986).

If we would like to have structural characteristics for our classification, because they are easier to measure or recognize, we could relate the processes to so-called controlling factors — structural characteristics that control the processes. However, since we want to cover a large number of environmental hazards, ranging from climatic change to disturbance by noise, we would have to define all the parameters relevant for all the processes related to all environmental hazards. Thus, we might be drowned in complexity.

Still another complicating factor must be mentioned. It concerns the fact that some important environmental problems have global dimensions, such as climatic change, whereas others are confined to certain regions, such as overgrazing. Obviously it is extremely difficult, if not impossible, to develop a classification equally appropriate for problems at the global scale, at a local scale, and at all intermediate scales at the same time. Even if such a classification would be possible, we would still be confronted with the practical problem of mapping the relevant characteristics at such divergent scale levels. So, perhaps we should first decide on the spatial scale level which would then guide the classification.

Above, we illustrated how classifying ecosystems is complicated due to the fact that ecosystems are intricate systems, resulting in a large number of possible approaches. To know where to begin, a number of decisions must be made. We will summarize these in terms of questions and answers. First, is the restriction to structural characteristics the answer? Only partly, it seems. Second, is deciding on the purpose of the classification the answer? Again, apparently only partly. And third, is the spatial scale level the decisive factor? Again, we think not entirely.

In fact, we are convinced that we can only find our way out of this maze by an approach which is both holistic and deductive. In other words, we argue for a top-down approach from a theoretical basis. This will predominantly be founded on hierarchies.

The approach implies the following steps:
- postulating a simplified hierarchical model of an ecosystem comprising all relevant components in subsystems, as a means to structure the complexity;
- defining a classification framework for hierarchically nested ecosystems related to this hierarchical model;
- classifying and mapping ecosystems at different spatial scale levels in order to obtain multi-purpose ecosystem maps, related to each other both structurally and functionally;
- zooming in on the most relevant processes and parameters for each specific application, which is also related to the main ecosystem components and to certain spatial scales.

In the next section we shall go into the ecosystem model. Then, we shall explain the classification framework related to this as the 'pièce de résistance', and illustrate it with some examples of ecosystem classifications and maps produced for the Netherlands' environmental policy. This will be followed by a section in which we exemplify some of the applications Udo de Haes and Klijn introduced in the first chapter of this book.

A comprehensive hierarchical ecosystem model

A simplified ecosystem model can be based on ideas concerning interactions between the main spheres of our planet (Van der Maarel, 1976; Van der Maarel and Dauvellier, 1978). This spheric model can be transformed into a

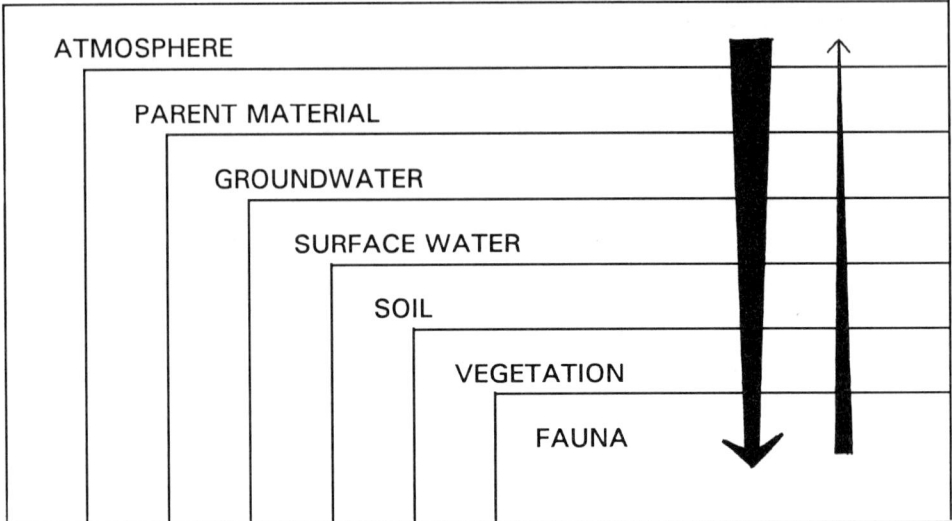

Figure 5.1 Hierarchical model of an ecosystem, showing a hierarchy of relative dependence between the major components (after Van der Maarel and Dauvellier, 1978; Bakker et al., 1981; Piket et al., 1987)

hierarchical model of an ecosystem which encompasses all the abiotic and biotic components that are ecologically relevant, irrespective of temporal or spatial scale (Bakker et al., 1981; Piket et al., 1987). An adapted version (Klijn, 1988) is presented as Figure 5.1. It can be regarded as a general, hierarchically structured model. It is based on an aggregation of the various abiotic and biotic constituents of ecosystems into a few major 'components' or 'subsystems'. These components relate to each other in a way representing a rule rather than a case.

The model may appear a somewhat deviating conception of an ecosystem because it contains so many abiotic components. Ecosystems are often considered to be confined to the relation network of biota and the directly ecologically relevant variables. In the case of a plant community such variables would be moisture availability, nutrient availability, acidity, and others. However, such an approach is too narrow for environmental policy. Since, for example, susceptibility assessment requires data on abiotic characteristics, the abiotic environment should be incorporated into the model. In addition, abiotic factors almost entirely control the distribution of ecosystems at global scales. This is a second reason to take the abiotic components fully into account in a general purpose model.

Hierarchical relations

The model is related to hierarchy theories, today again popular in ecology (O'Neill, 1988; Urban et al., 1987; Jørgensen, 1992). The hierarchical character of it is manyfold. It reflects hierarchies of volume, time of evolution and change, direction of fluxes of energy and matter, relative dependance of the various components resulting from these fluxes and, also, spatial scale. First and most importantly, it expresses that the lower components are relatively dependent on those above, as indicated by the downwardly directed arrow. However, this does not mean that one could deny the reverse influence of the lower components on the upper ones, as is also indicated, this time by an upwardly directed arrow.

All these hierarchies can be ordered in two categories, namely hierarchies of structure and hierarchies of processes. To exemplify this, we summarize a number of observations on hierarchies by Bakker et al. (1981) and others, arranging them in the two categories.

Hierarchies of structure:
- reservoirs diminish in size (parent material > water > soil > vegetation > fauna);
- patterns of the upper components are reflected in the lower ones (climate > soil; climate > vegetation; parent material > vegetation).

Hierarchies of processes:
- energy transport is commonly directed downwards;
- matter transport is generally directed downwards;
- the genesis of the lower components is determined by the upper ones (e.g. wind > dunes > sandy soils);
- the existence of lower components depends on the upper components (e.g. parent material > porosity > groundwater);
- changes in the relatively independent components have unavoidable effects on dependent components (e.g. climatic change > surface water discharge > soil erosion > vegetation).

Hierarchies of scale

These two categories of hierarchies also relate to spatial and temporal scales. We shall briefly elaborate on these scale aspects, because they are relevant for the framework for classification and mapping.

As for spatial scales, we may observe that there is a difference in the spatial scale at which ecosystem components cause patterns on the earth's surface, predominantly reflected by the distribution of biota (Leser, 1976; Walter, 1979; Bailey, 1987). In this book, Godron also addresses the question which ecosystem components determine the patterns we observe. We may recall that climate zones, for example, are a global phenomenon mainly determined by latitude. Geological processes and lithology determine the pattern of mountain ranges and valleys at large scales, thus influencing relief, hydrology, and soil formation. Soils show more fine-grained patterns, whereas vegetation superimposes an even finer pattern of various succession stages. In fact, the patterns on the earth's surface we observe are a reflection of the hierarchy of structure mentioned above.[1]

As for temporal scales, we may look at the natural rate of change of the various ecosystem components. Natural climatic change generally takes tens of thousands of years. Geological and geomorphological processes such as mountain building, weathering, or meandering of rivers need as much time, but some processes may occur in decades, and disasters within even a day. Soil characteristics may change over thousands of years or centuries, although erosion may cause soil degradation within even a few hours. Vegetation may react within a year, although natural succession generally takes some decades or even centuries. Fauna, of course, is the most rapidly responding component in ecosystems: it may simply fly away. These differences in the temporal scale of natural processes are an expression of the hierarchy of processes. This also accounts for the time lag which is so often encountered in the ecological effects of processes first implying abiotic changes, such as acidification or climatic change. These time and space aspects of the components of the hierarchical ecosystem model are visualised in Figure 5.2 in a simplified way.

---

[1] Of course, patterns can be regarded as being the result of processes. On the other hand, patterns and other structural characteristics influence the various processes. In more general 'ecological' terminology, one might state that there is an intricate relation between structural and functional characteristics. The distinction thus becomes somewhat artificial.

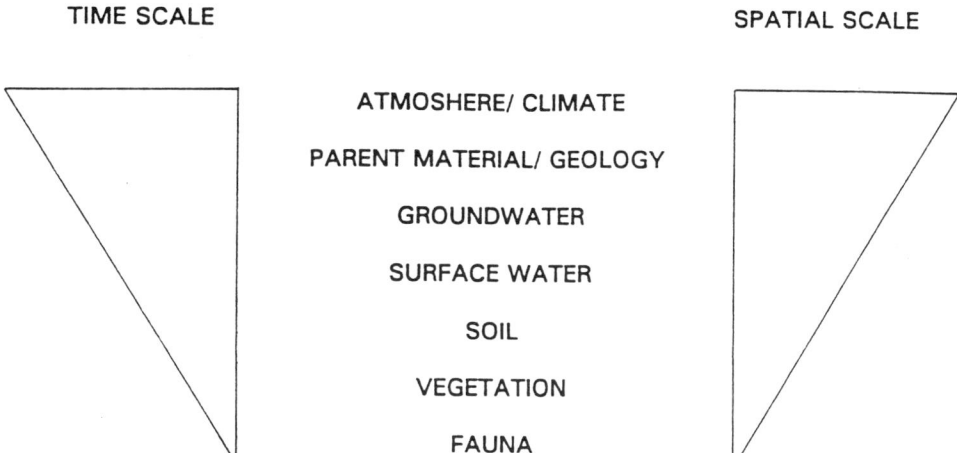

Figure 5.2 The relation between ecosystem components and spatial and temporal scales

From the hierarchical model towards a classification framework

After having postulated the hierarchical ecosystem model, we are confronted with the problem of classifying ecosystems with regard to all the relevant components, as well as for various spatial scale levels. This is, of course, practically impossible, but we may find a way out which is primarily based on the recognition of 'correlative complexes' (Kwakernaak, 1982).
The term 'correlative complex' was introduced to indicate that the whole can be known by only some of its characteristics, due to the fact that many characteristics are highly correlated. In other words, some ecosystem characteristics are diagnostic for the whole. If we select the most relevant characteristics at a certain spatial scale as classification characteristics, we can estimate quite a number of other ecosystem characteristics with sufficient accuracy. Of course, these correlative complexes result from the processes between the hierarchically arranged ecosystem components, as depicted in Figure 5.1.

The question arises how to define which are the most appropriate classification characteristics at a specific spatial scale level. We base our answer on:
- a hierarchical approach to the classification for different spatial scale levels; in this case it concerns spatially nested ecosystems;

- the selection of classification characteristics from those ecosystem components that control the pattern at the specified scale level.

The combination of this is the forementioned classification framework.

A hierarchical approach for different spatial scale levels

When classifying ecosystems, different levels of homogeneity can be distinguished for different spatial scales: from entire oceans or climatic zones to small ponds, hedgerows, or even smaller. For mapping this means that patterns can generally be subdivided in finer patterns again and again, from global to very detailed. This results in a kind of classification by subdivision.

The resulting maps can also be regarded as hierarchical, with the more detailed maps showing the internal variability of the units defined at the scale level above. This hierarchy may be understood as a hierarchy of nested ecosystems.

For the classification and mapping of such nested ecosystems the following principles are relevant:
- When zooming in, the detail is steadily increasing. Classes become more narrow and the number of boundaries increases.
- Boundaries that have been defined at a certain level must be retained at more detailed classification levels. The classification characteristics of higher classification levels overrule those at lower levels.
- The boundaries between already existing mapping units may, however, be defined more accurately at more detailed mapping scales.

Hierarchical classifications are by no means new. In fact, they can be encountered in the classification of soils, vegetation, parent material, groundwater systems, or any entity where the problem of spatial scale is encountered. In most cases, this has resulted in the definition of hierarchical classification schemes.

Sometimes, the different classification levels are explicitly intended for different spatial scale levels. Then we could speak of a spatial hierarchy. In other cases, the classification has a systematic[2] hierarchical structure, irres-

---

[2] The use of the word 'systematic' may need some explanation. It was chosen because in our opinion it best reflects the 'systematics' of organisms, as well as the 'syn-systematics', which is sometimes used in vegetation science. These concepts refer to hierarchical classifications with well-defined rules of nomenclature. In this context, the term 'taxonomical' could be used instead.

pective of spatial scale. This distinction is the same as was made by Zonneveld in chapter 2 of this book.

In 'systematic hierarchies' all classification levels belong to one single classification. For example, in the FAO-legend for soil mapping we may find soil groups, subdivided into main and secondary soil units. In the US Soil Taxonomy (Soil Survey Staff, 1975) we find orders, sub-orders, great groups, groups, etc. Duchaufour (1977) distinguishes classes, subclasses, groups, subgroups, etc. And of course, the taxonomy of species and the taxonomy used in plant sociology are well-known examples of systematic hierarchies.

The problem of relating systematic hierarchical classifications to spatial hierarchies often remains implicit, although Van der Maarel (1976) drew attention to the possibility of relating plant sociological concepts to his spheres, i.e., almost to the main components of the hierarchical model of Figure 5.1. Professional mappers, however, are used to pragmatic solutions and seem not to have any problems. In fact, they simply generalize a pattern in order to get it on the map. However, it is obvious that spatial hierarchies and systematic hierarchies do not match easily. This may be illustrated by thinking of a landscape mosaic, formed by individual forest patches with different species composition among agricultural fields with different crops and grassland communities. Combining the forests with similar dominant species, in order to obtain a higher level in the systematic hierarchy, does not yield larger patches. In a more natural landscape, valley-bottom communities with sedges could not be integrated into larger units, because they could never be united with hillside communities with Oak or Beech forests, as distinguished in a purely systematic approach. This notion made Tüxen define the concept of 'sigma-associations' (Tüxen, 1978; Theurillat, 1992) with an entirely new taxonomy. In fact, this is merely an attempt to classify complexes, without changing the point of view of plant sociology.

So, instead of desperately trying to match a systematic hierarchy with a spatial hierarchy, we had better look at the hierarchical approaches that were developed for the many integrated ecological land classifications (e.g. Bailey, 1976, 1981; Brink et al., 1965; Christian and Stewart, 1968 (SCIRO); Isachenko, 1973; Leser, 1991; Wiken and Ironside, 1977). In most cases, different nomenclature and classification characteristics were used for different spatial scale levels. In these cases, the result is not one single classification of ecosystems at different systematic levels, but rather a series of classifications for specified spatial scale levels.

In the spirit of the latter approach, we use the nomenclature given in Table 5.1. It is largely based on the nomenclature of Canada (Wiken and Ironside, 1977; Lands Directorate Environment Canada, 1981; Bailey, 1981) and the USA (Bailey, 1981; 1985; 1989; Bailey et al., 1985; Hughes and Larsen, 1988; Omernik, 1987), but for some levels the terminology has been replaced by concepts that are frequently used in the European tradition of landscape ecology, such as ecotope (Neef, 1967; Leser, 1991) and ecoseries (Wagner, 1968; Müller, 1970).

The various levels are indicated by terms beginning with eco- to emphasize the character of the classification and ending with commonly used terminology for geographical entities of different size. In the table, the approximate mapping scales and basic mapping units are also given.

Table 5.1 Nomenclature for hierarchical ecosystem classification with indicative mapping scales and basic mapping units

	INDICATIVE MAPPING SCALE	BASIC MAPPING UNIT
ECOZONE	1: > 50.000.000	> 62.500 km²
ECOPROVINCE	1:10.000.000 - 50.000.000	2.500 - 62.500 km²
ECOREGION	1: 2.000.000 - 10.000.000	100 - 2.500 km²
ECODISTRICT	1: 500.000 - 2.000.000	625 - 10.000 ha
ECOSECTION	1: 100.000 - 500.000	25 - 625 ha
ECOSERIES	1: 25.000 - 100.000	1,5 - 25 ha
ECOTOPE	1: 5.000 - 25.000	0,25 - 1,5 ha
ECO-ELEMENT	1: < 5.000	< 0,25 ha

Classification characteristics at different spatial scales

After having defined the various classification levels, we have to decide on how to classify and map ecosystems at each spatial scale level. The classification requires definition or characterization, whereas mapping requires recognition.

In this context, a number of different concepts are frequently used. For example, Vink (1975) distinguishes between differentiating characteristics and diagnostic characteristics. The differentiating characteristics are used for defining the types, and the diagnostic characteristics provide for a more elaborate description of the types. Because the diagnostic characteristics are correlated with the differentiating characteristics, they may be used in the field as an aid for recognition, i.e., for deciding on the ecosystem type.

In contrast, Zonneveld (1979; this book) only recognizes diagnostic characteristics as important properties for classification and mapping. These diagnostic characteristics are used both for defining the land units and for practical mapping. He defines them in such a way that in practice they correspond to the differentiating characteristics of Vink, but they may also cover Vink's diagnostic characteristics. We consider this at least somewhat confusing.

In addition, we recall that the diagnostic characteristics *sensu* Zonneveld are mainly used for mapping. To this end, they must be macroscopically recognizable characteristics. However, this does not necessarily relate to causal relations. For example, the species composition in natural forest stands may be relatively easy to establish, but differences are determined by the combined influence of site factors, such as topoclimate, recurrent fire, soil, or groundwater. These controlling factors, as Bailey (1987) calls them, may of course be regarded ecologically as the most important characteristics. However, they cannot all be recognized equally directly, as is obvious for topoclimate. It is much easier to establish the species composition as an indicator for climatic differences than to measure all the relevant climatical variables separately over many years.

Now, by very strictly distinguishing between classification and mapping, or, in other words, between definition and recognition, we may formulate an alternative approach to either Vink's or Zonneveld's.

For classification it seems logical to base it on characteristics that *cause* the pattern, instead of *show* it. We call these 'classification characteristics'. They are used to define ecosystem types and may be specified in the legend of a map. Thus, for classification characteristics we should use those ecosystem characteristics that in reality determine the pattern. They cause the differentiation, i.e. are the real 'differentiating characteristics'.

For practical mapping, however, we may use all characteristics that would be of help. These will be called 'mapping characteristics' or, if desired, 'diagnostic characteristics'. Mapping characteristics should of course reflect the pattern caused by the controlling factors, otherwise they are not adequately indicative. We do realize that we deviate from Zonneveld's definition of 'diagnostic characteristics' in this book, but we consider it more appropriate to distinguish classification characteristics in connection with causal relations. In fact, for the taxonomical classification of species such a distinction is also made. A species is certainly recognized by means of morphological characteristics, but is taxonomically classified by its phylogenesis. Also, in geomorphology it is common practice to classify landforms according to genesis, resulting in the distinction of, for example, glacial, fluvio-glacial, or fluvial landforms, even when the genetic processes ceased long ago.

This distinction between classification and mapping characteristics enables us to use the hierarchical model of an ecosystem as a guiding principle for selecting classification characteristics for different spatial scale levels. This is possible because it reflects a hierarchy of dependence, i.e., a hierarchy of controlling factors. Thus, we may derive a classification framework from the model, which, reduced to its essentials, implies that we should use classification characteristics that determine the pattern at the scale level concerned.

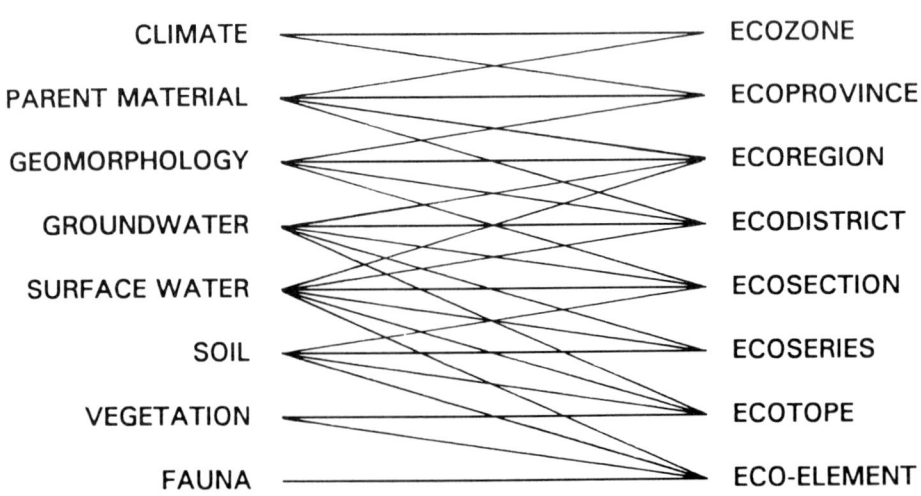

Figure 5.3 The relation between spatial scales and the ecosystem components which may yield the most adequate classification characteristics

This framework is given in Figure 5.3, in which the correspondence between determining ecosystem components and scale level is visualized. The ranking of ecosystem components in this figure is the same as in the hierarchical model, while the spatial scale levels have been indicated with the corresponding terminology. 'Parent material' has been divided into 'parent material/ geology' and 'relief/ geomorphology'. The latter can be considered as the reflection of the former in terms of landform, but cannot be regarded as a separate component because it has no volume.

We may now question how such a top-down deductive framework stands in a practical context. To this end, we shall look more closely at the characteristics that are used for classifying and mapping ecosystems or individual ecosystem components at different spatial scale levels in practice.

We then come across striking resemblences in pattern. The zonal differentiation of soil groups (FAO, 1988) correlates perfectly with vegetation zones (Walter, 1979). At the spatial scale of continents, we can observe a reflection of physiography in both soil pattern and vegetation pattern, and so on. These observations may be explained as follows. Although it may have been the intention to map soils or vegetation at global scales, the pattern which results on the map reflects the controlling factors, because vegetation and soil are only indicative for climatic differences at this spatial scale. In other words, the classification characteristics that have been defined in the legend are not the same as the characteristics which apparently cause the pattern. It would be better to call the maps respectively an ecoclimatic or bioclimatic map, and a pedoclimatic map. The legends, i.e., the classifications, should be adapted correspondingly. This would be more straightforward.

Summarizing the above, we argue that a classification of ecosystems should be performed at different spatial scales by focusing on those components that determine the pattern at a certain scale level. Thus we can use the generalized hierarchical ecosystem model as a guideline for selecting classification characteristics. On the other hand, practical mapping may be based on all characteristics that would be of help for recognition on either remotely sensed images or in the field.

Brief characterization of the various classification levels

Above, we argued for a guideline for selecting the most appropriate classification characteristics for different spatial scale levels. Now, we shall specify the various classifications in general terms.

Ecozones are related to the world-embracing climate zones. They can be distinguished on the basis, for example, of the Köppen system for regional climates (Trewartha, 1968). The pattern is reflected by the zonal pattern of soils (Dokuchaev Soil Institute, 1963; FAO, 1988) and vegetation (Walter, 1979). Hence, ecozones largely correspond with the zonobiomes of Walter (1979) or the domains and some divisions of Bailey (1989). The main ecozone types may be indicated as the arctic, the subarctic with tundra vegetation as the main diagnostic characteristic, the boreal zone with taiga coniferous forests, the temperate zone, the mediterranean, the (semi-)deserts, the dry tropical zone with savannas and the wet tropical zone with rain forests.

Ecoprovinces are determined by geological and geomorphological characteristics at very large scales as well as the climatic variations resulting from large physiographic differences. The pattern of ecoprovinces largely corresponds with the pattern of provinces as distinguished by Bailey (1989). Also, the orobiomes and some pedobiomes of Walter (1979) would be comparable.
As examples, we mention mountain ridges such as the Rocky Mountains or the Alps, or physiographic units such as the Scandinavian Shield or the Great Plains.

Ecoregions are homogeneous with respect, again, to geological and geomorphological characteristics. However, the geological subdivision is more detailed, distinguishing between different rock types in large groups. A distinction between areas dominated by sandstones, calcaric rocks, marls, and various igneous rocks, respectively, may be relevant at this scale level. Also, a further subdivision according to altitude and main groundwater flows may be made. As an example, the Netherlands' ecoregion map is shown in Figure 5.4 (Klijn, 1988).

Ecodistricts are spatial units that are homogeneous to slowly changing geological, geomorphological, groundwater and surface water characteristics. These largely correspond with soil groups as determined by parent material. At this scale, we can distinguish physiographic units such as large valley systems with Fluvisols, individual volcanoes with Andosols, deltas with Inceptisols, or salt plains with Solodic Planosols.

Typical examples from the Netherlands' ecodistrict map are Polders, Drained lakes, Calcareous coastal dunes, Lowland peats and Isolated ice-pushed ridges (Figure 5.5).

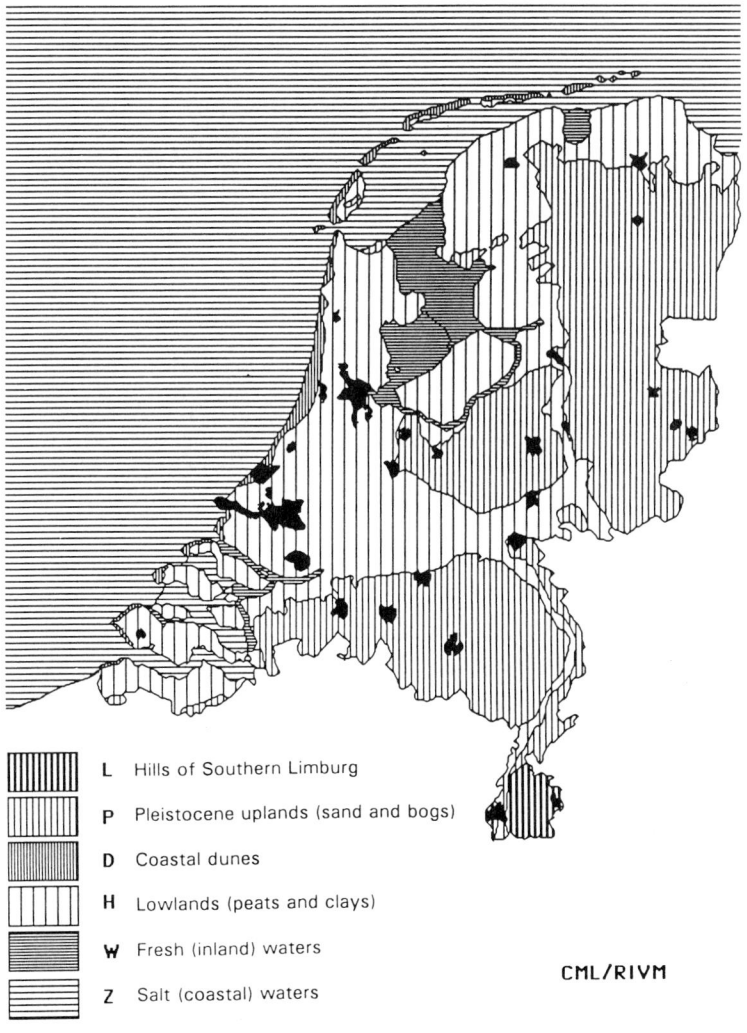

Figure 5.4 Ecoregions of the Netherlands. Map representing an ecosystem classification with geological and geomorphological characteristics as classification characteristics, intended for a mapping scale of approximately 1: 5,000,000

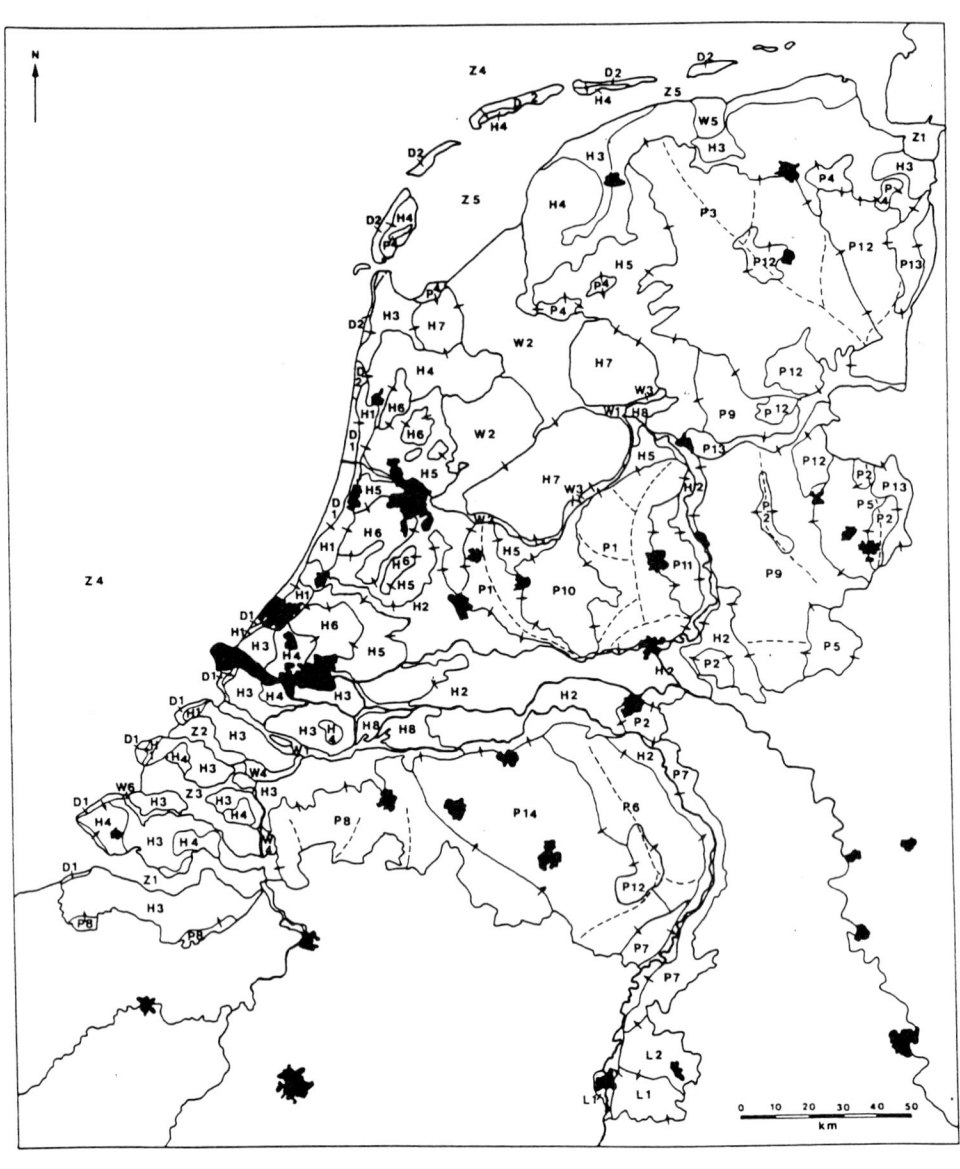

Figure 5.5 Ecodistricts of the Netherlands. Map representing ecosystems at a scale level one step below the ecoregions, which illustrates that ecodistricts are spatially nested within the ecoregions (indicated with capitals as in Figure 5.4). Geological, geomorphological, groundwater and surface water characteristics were used for the classification

Legend Figure 5.5

L1	Cretaceous hardrock landscape	H1	Beach barrier landscape
L2	Loess landscape	H2	Fluvial landscape
		H3	Sub-recent dikings
P1	Ice-pushed ridge complexes	H4	Marine clay inversion landscape
P2	Isolated ice-pushed ridges	H5	Lowland peats
P3	Till plateau	H6	Drained lakes
P4	Isolated till elevations	H7	Polders
P5	Other till areas	H8	Deltas
P6	Tectonic ridges		
P7	Old river terraces	W1	Sedimentation basins
P8	Fluvial sands landscape	W2	Fresh inner seas
P9	Eastern cover sands landscape	W3	Lakes
P10	Glacier lobe valleys	W4	Closed estuaries
P11	Sandr landscape	W5	Diked sea intrusions
P12	Bogs and bog reclamation landscape	W6	Brackish lakes
P13	Lowland brook complexes		
P14	Tectonic subsidence area	Z1	Estuaries
		Z2	Saline lakes
D1	Calcareous coastal dunes	Z3	Sea branches
D2	Non-calcareous coastal dunes	Z4	Coastal seas
		Z5	Wadden seas

Ecosections are spatial units which are homogeneous with respect to individual but large geomorphological features, such as lowland brook valleys, divides, individual slopes, slumps, mud-flows, debris cones, etc. They are distinguished on the basis of geomorphology, soil groups and groundwater regime as controlled by geomorphology: leaching or upward seepage.

Ecoseries are homogeneous to abiotic characteristics of soil, groundwater and surface water, which can be established without instruments other than a hand auger. The site characteristics used for classifying ecoseries must preferably be stable for periods of some decades. This almost automatically implies a kind of ecological soil classification, because the soil can be considered the most important site factor for plant growth in most parts of the world.
Within ecoseries, vegetation types can be encountered which differ in vegetation structure. These can be understood as succession stages, or semi-permanent stages due to land use practices, thus forming series of vegetation types related in time. Hence, an ecoseries classification and mapping is especially valuable for estimating potential or climax vegetations, or for forecasting vegetation developments (see also Claessen et al., this book).

The term *ecotope* is one of the oldest in landscape ecology (see e.g. Neef, 1967; Haase, 1973, 1989; Leser, 1991). The use of this term in practice is confined to ecosystems of a certain dimension, such as a raised bog, an agricultural field, a small forest patch or a dune slack grassland. Hence, a certain spatial scale can be attached to the ending -tope. Stevers et al. (1987) have specified the ecotope concept as: '... a spatial unit which is homogeneous as to vegetation structure, succession stage and the main abiotic site factors that are relevant for plant growth'. They have elaborated an ecotope classification for the Netherlands (see Runhaar and Udo de Haes, who treat this classification in this book).

The definitions and descriptions by Stevers et al. (1987) do not take into account the practical requirements of mapping. Their classification scheme is still primarily focused on systematic classification. Consequently, ecotopes could be as small as ditch banks or road margins. We would rather advocate incorporating a criterion of 'a certain spatial extension' of ecotopes, cf. Van der Maarel and Dauvellier (1978), i.e. a minimum size.

For the most spatially restricted ecosystems the term *eco-element* is suggested. Eco-elements may develop by vegetation processes such as string-formation in bogs, or dominance of certain species with rhizome-multiplication. Hedgerows or termite mounds may also be regarded as eco-elements. These small-scale ecosystems can be mapped as patches, but only at very detailed mapping scales. In most instances their location will be indicated on maps by means of symbols.

The link to applications

Above, we presented the outlines of a scale-dependent set of related ecosystem classifications. The classification characteristics for the various spatial scale levels are chosen in relation to a hierarchically structured model of an ecosystem.

We now may question whether the resulting classifications are useful for the desired applications. After all, classifications are supposed to be tools (Udo de Haes and Klijn, this book), and thus the proof of the pudding is in the eating. So let us try them.

First, we shall relate environmental hazards to ecosystem components and spatial scale levels in a general way. Then, we shall focus on susceptibility assessment of ecodistricts to exemplify the procedure.

Environmental themes as chains of processes in ecosystems

An analytical approach to any environmental problem requires that we concentrate on specific environmental hazards, related relevant processes and related relevant ecosystem characteristics. Unfortunately, an almost infinite number of environmental hazards can be distinguished. However, they could be grouped into a small number of central themes, as distinguished for environmental policy purposes in the Netherlands (RIVM, 1989). These themes are relatively exclusive groups of environmental hazards; although there is some overlap, they certainly interact, and the list is by no means limitative.

For reasons of convenience, however, we shall follow this distinction, which covers:
- climatic change
- acidification
- pollution (with toxic substances)
- eutrophication
- desiccation
- destruction
- fragmentation
- disturbance (by noise or odour)

The majority of these themes can be understood as chains of environmental processes which affect structural characteristics of ecosystem components. In terms of the system levels distinguished by Chorley and Kennedy (1971) they could be well studied as process-response systems that are arranged in cascades. These chains of processes follow the direction of fluxes of energy and matter, as indicated in the hierarchical model by the downward arrow. In all cases, they finally result in harm to flora and/ or fauna, including human beings.

The zooming-in on a specific environmental hazard may be guided by the relation between the initial point of attack and the resulting cascade of effects on the one hand, and the hierarchy of ecosystem components on the other. This relation is illustrated in Figure 5.6.

It seems that the point of attack of environmental hazards differs widely, from acidification, which primarily influences the atmospheric composition, to physical destruction of vegetation or fauna caused by land use changes (Klijn,

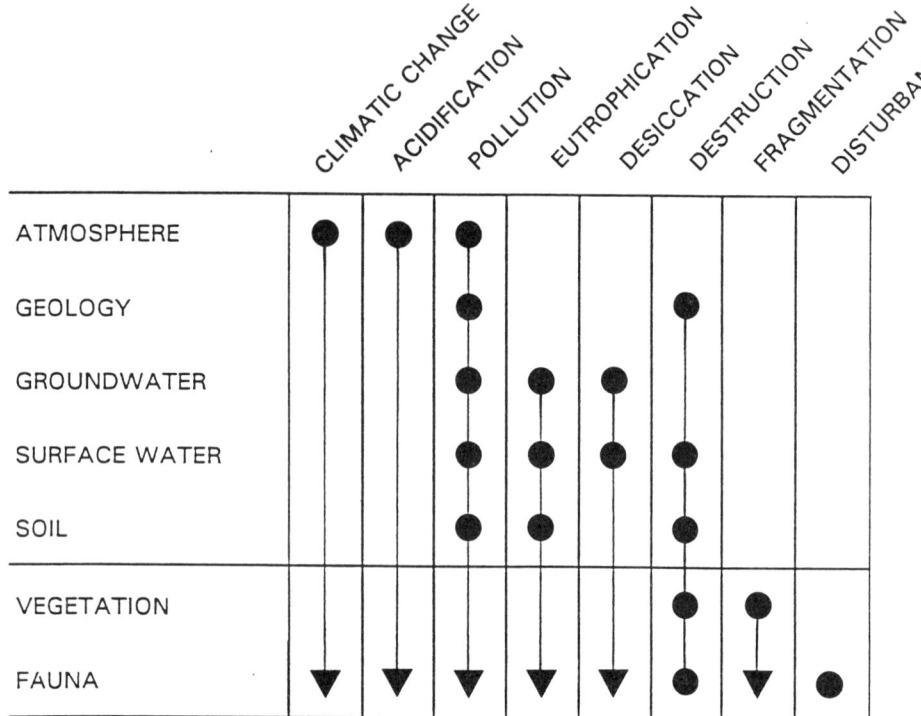

Figure 5.6 Points of attack of groups of environmental hazards, indicated by dots, and subsequent effect-cascades, indicated by arrows. Some environmental hazards may attack different components as first ones

1988; RIVM, 1989). Because of these different points of attack, the themes have different time scales. Those affecting the atmosphere often result in changes of the vegetation or fauna only over long periods, as exemplified by acidification. The effects of sulphuric acid on the environment were already being observed and attention drawn to them in the seventeenth century (Evelyn, 1661), but it was not before the seventies of this century that massive societal awareness really arose. In contrast, the destruction of vegetation by clear-cutting or the disturbance of fauna by noise, have immediate effects.

The relation between point of attack and the hierarchical model may help in deciding on the most relevant scale level to deal with a problem.[3] This follows from the fact that the determinant processes are connected with the most relevant ecosystems components in Figure 5.6, whereas the ecosystem components in their turn are connected with the classification levels as depicted in Figure 5.3.

Analyses at spatial scale levels above the highest point of attack are thus useless. A scale level below the one corresponding to this point of attack may be most appropriate if the most important conditional processes take place in the corresponding component, or if the most significant effects are expected in that component.

Some examples

We shall illustrate the procedure of analytically zooming-in by focusing on some themes that run through abiotic environmental processes. We shall leave out total destruction, which is so obvious an intervention that quantifying the impact is fairly easy. Also, we shall not cover fragmentation or disturbance, because these processes affect the fauna in a relatively direct way through chorological processes. This means that the effect-chain is short and that no previous abiotic processes are involved. For the moment, we also leave out climatic change.

Thus, acidification, pollution with toxic substances, eutrophication, and desiccation remain.

For these themes we shall briefly treat the procedure of susceptibility assessment, which was carried out with the ecodistrict classification and map for the Netherlands. It implies:
- identifying the relevant processes related to the themes;
- identifying the relevant parameters in the various components that control the processes;
- determining the relations between these parameters and their individual contribution to the controlling of the processes, in order to estimate the total carrying capacity of the system;

---

[3] Of course, the extension of an intervention is also important for the decision of the most relevant spatial scale. This depends especially on the surface area at which a human activity is carried out, or the number of emission points related to a human activity.

- and finally, integrating this information into a degree of susceptibility, thus yielding a table to convert the legend units of the ecosystem classification into susceptibility classes.

As the first step, each environmental theme must be unraveled into a number of relevant processes. Such an unraveling into individual processes which predominantly determine the susceptibility of ecosystems will result in a list of processes to be taken into account. This list is specified in Table 5.2.

Table 5.2 The main abiotic processes determining the susceptibility of ecosystems for a selection of environmental themes

ACIDIFICATION
- $CaCO_3$ - dissolving
- Silicate weathering
- Cation exchange
- Aluminum silicate degradation (setting free of Al)
- Iron hydroxide degradation
- Supply with cations through upward seepage
- Denitrification

EUTROPHICATION
- Leaching
- Denitrification
- P-fixation (adsorption)
- P-precipitation (Ca/Mg PO_4)

POLLUTION
- Adsorption/desorption
- Precipitation
- Leaching

DESICCATION
- Drying out
- Acidification
- Mineralization
- Subsidence

Table 5.3 Abiotic ecosystem characteristics that largely control the processes discerned in the foregoing table

	Acidification	Eutrophication	Pollution	Desiccation
PARENT MATERIAL				
Primary CaCO3 content	■	■	■	
Content of weatherable silicates		■		
Texture (clay, silt)	■	■	■	■
SOIL				
Texture (clay content, esp. Fe and Al)	■	■	■	■
Fe- and Al-(hydr)oxides	■	■		
Depth of decalcification	■	■	■	■
Organic matter content	■	■	■	■
Type of organic matter		■		■
Groundwater level/fluctuations (redox conditions)		■	■	■
GROUNDWATER/SURFACE WATER				
Groundwater level/fluctuations	■	■		■
Direction and rate of groundwater flow	■	■	■	■
Quality of upward seepage (cations, salinity)	■	■	■	■
Quality of surface water (cations, salinity)	■	■	■	■

As the second step, we must identify the structural ecosystem characteristics that control these processes. When we do so for the four environmental themes selected above, we find that for different themes, different sets of parameters are needed. Partly, however, the parameter sets overlap, which allows the selection of a limited set of parameters with multi-purpose character. In Table 5.3, the parameter sets for the four selected themes have been specified. It illustrates the overlap.

As the third step, we must quantify the various parameters for the different legend units, as well as their contribution to the overall counteracting capacity. In this case, we are concerned with ecodistricts. It is obvious that such a quantification does not follow directly from the ecodistrict map, because the classification characteristics are, in fact, of a more general nature. Fortunately, the existence of the 'correlative complexes' mentioned above enables a quite accurate estimate of the value and internal spatial variability within the classification units of the characteristics that have not been measured separately. Such an estimate may be based on elaborate empirical research, existing data bases, or expert judgement. The latter method, combined with scrutinizing existing maps on single parameters (Van Duijvenbooden and Breeuwsma, 1987), has been used to assess the susceptibility of ecodistricts.

After some simple calculations, four susceptibility classes were distinguished for each theme. A reclassification of the mapping units was then carried out. Finally, the resulting susceptibility maps were discussed in an attempt to match the results of the semi-quantitative approach with the judgement of a number of experts. Examples of such susceptibility maps for eutrophication and pollution are shown in Figures 5.7 to 5.10.

Meanwhile, more sophisticated quantitative approaches have been developed for some of the environmental problems concerned here (e.g Kuylenstierna and Chadwick, 1991). These allow a quantitative underpinning of the susceptibility classes in terms of critical loads. The procedure as such, however, does not differ from the one described here.

For susceptibility assessments in general, it is of the utmost importance to define precisely the environmental hazard, and possibly even the substance, as well as to specify the threatened components or ecosystem characteristics. Thus, within the environmental theme 'eutrophication', a distinction has to be made between N and P, because of their different behaviour and chemical fate. Or, as a second example, for the theme of pollution we have to distinguish between organic toxicants and heavy metals, which behave entirely differently. For the latter, we must also distinguish between accumulation in the soil or leaching to the groundwater, which both form environmental risks. Such specifications result in differences in ecosystem susceptibility, as demonstrated by comparing the patterns of Figures 5.7 and 5.8, or 5.9 and 5.10.

Of course, susceptibility assessment is only one of a number of possible applications for environmental policy. It belongs to the static approach as distinguished by Udo de Haes and Klijn in the first chapter of this book. We shall not go into other applications now. For an example in the field of environmental impact assessment, EIA, we refer to Claessen et al., elsewhere in this book. They treat the application of ecotopes and ecoseries classifications for the Netherlands in predictive modelling for water management policy.

Epilogue

Since ecosystems are very intricate systems that can be classified by means of many different characteristics, it is difficult to decide on where to start. We argued that only a holistic and deductive approach, i.e., top-down from a theoretical basis, would show us out of the maze.
This approach was founded on a simple and general conception of an ecosystem. To this end, the ecosystem was depicted by means of a generalized hierarchical model with the main ecosystem components arranged according to relative dependence and a number of other hierarchies.

Instead of attempting to develop one systematic hierarchical classification scheme for all scale levels, we should be aware of the fact that a spatial hierarchy requires a different approach. In our opinion, the classification of spatially nested ecosystems can only be satisfactorily performed by developing a series of related ecosystem classifications.
Each classification should be specific for a pre-fixed spatial scale level, and should be related to the scale levels immediately above and below. The relation with the classifications at adjacent scale levels should be based on 'dependence' relations between the ecosystem components.

We argued that the classifications should be based on controlling factors, e.g., climatic and geological characteristics for large spatial units, and characteristics of soil and vegetation for the smallest spatial units. Thus, we could use the hierachical ecosystem model as a guideline for selecting the most appropriate classification characteristics, because it is related to spatial scales: each ecosystem component influences the pattern on the earth's surface at a different scale level.
In Figure 5.3 this guideline, or classification framework, was depicted by relating the hierarchy of ecosystem components to spatial units, i.e., ecosystems at various spatial scale levels. For the different spatial scale levels we proposed a simple nomenclature, based largely on the terminology used in Canada and the United States.

Since the so-called controlling factors determine the pattern on the earth's surface, they should be used to define the classes or legend units. From these the best 'classification characteristics' can be retrieved.

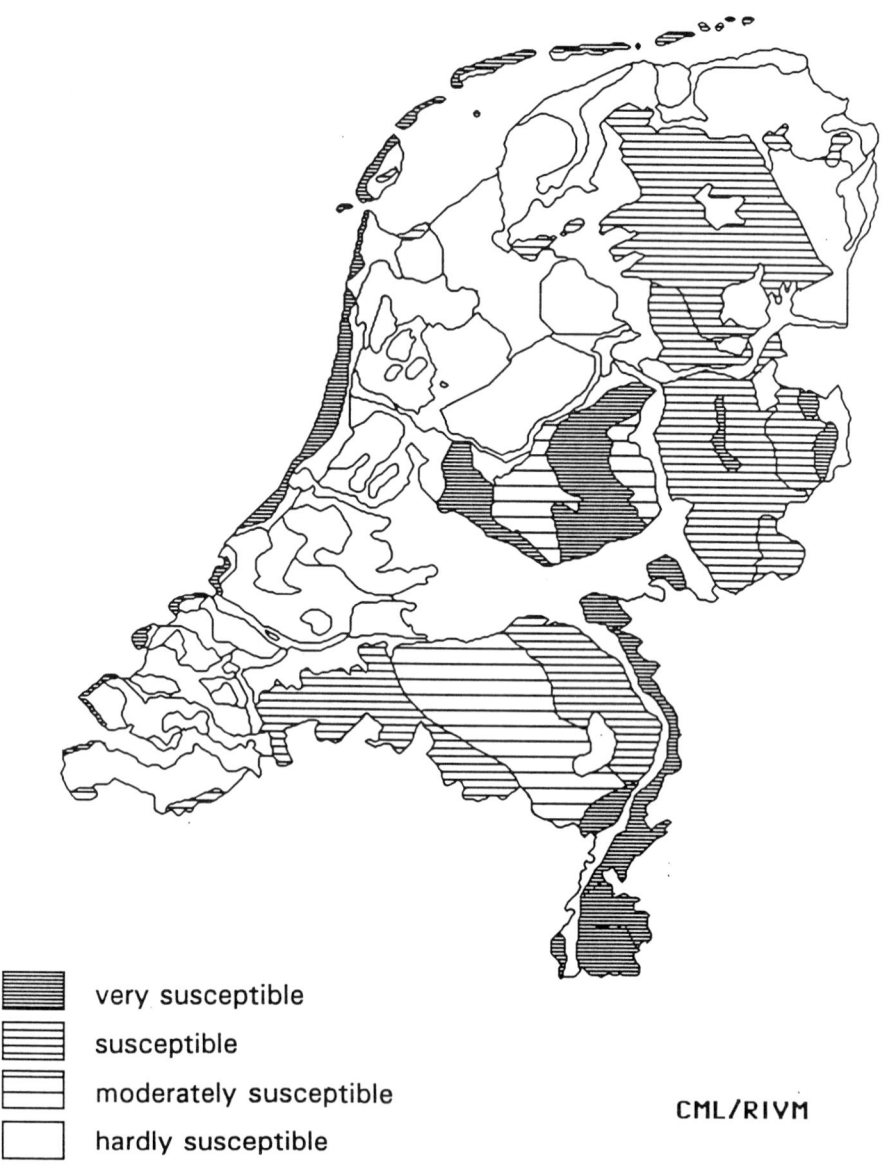

Figure 5.7 Eutrophication: susceptibility of ecodistricts to leaching of nitrate to the groundwater

SPATIALLY NESTED ECOSYSTEMS: GUIDELINES FOR CLASSIFICATION   111

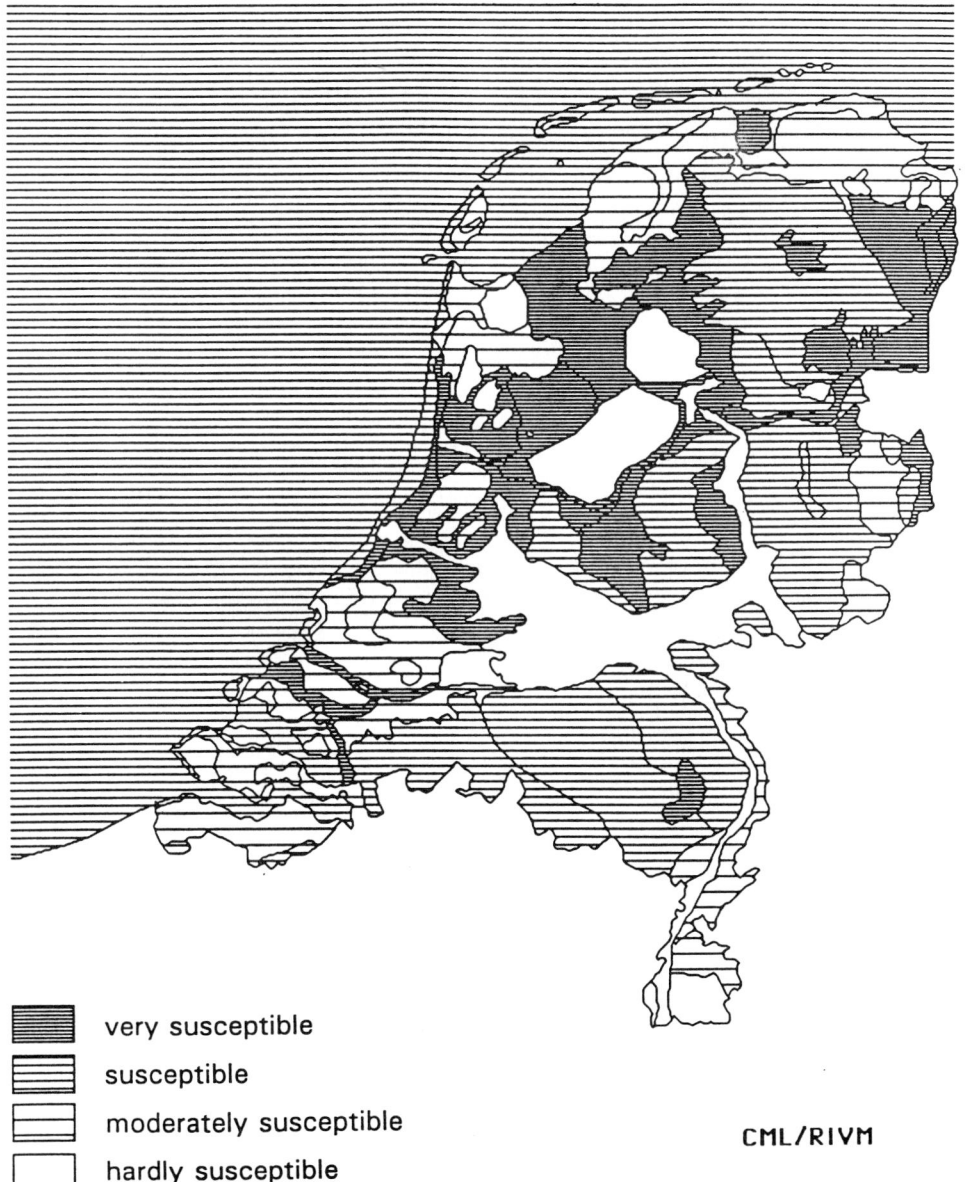

Figure 5.8 Eutrophication: susceptibility of ecodistricts to saturation of soils or eutrophication of surface waters by phosphorus

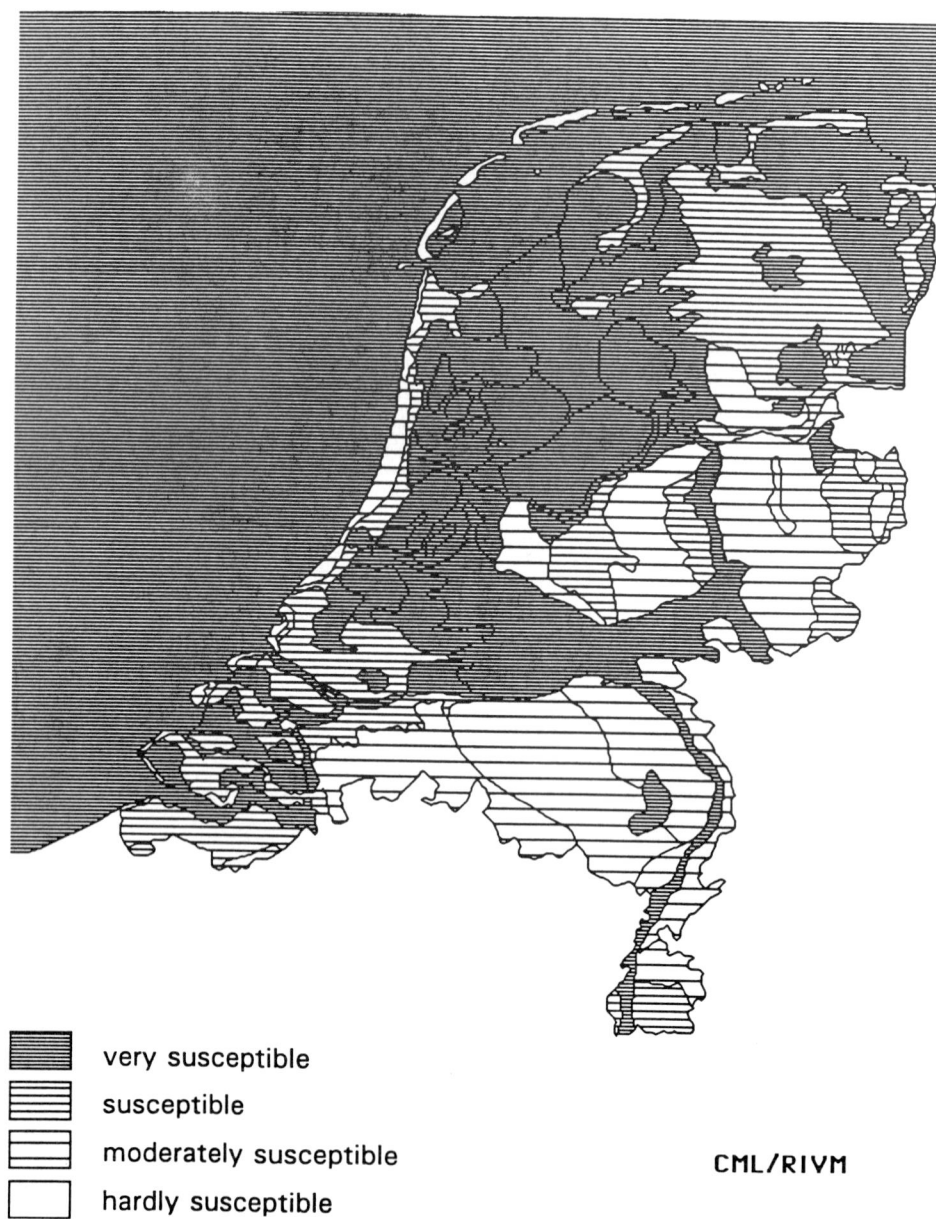

Figure 5.9 Pollution: susceptibility of ecodistricts to accumulation of heavy metals in the topsoil

SPATIALLY NESTED ECOSYSTEMS: GUIDELINES FOR CLASSIFICATION 113

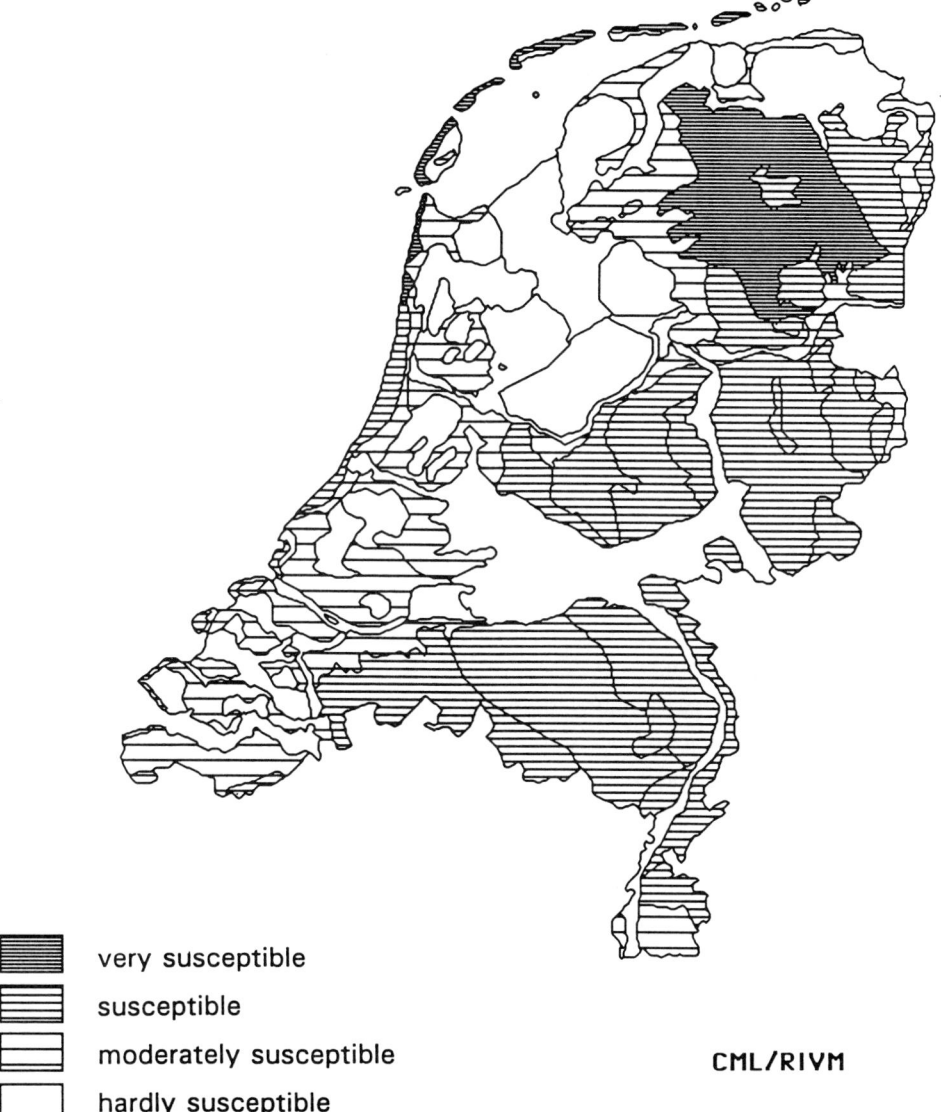

Figure 5.10 Pollution: susceptibility of ecodistricts to leaching of heavy metals to the groundwater

114   F. KLIJN

In contrast, for practical mapping all characteristics may be used which help in recognition. These could be called 'mapping characteristics' or 'diagnostic characteristics'.

The top-down approach, which was followed to achieve comprehensive classification, did not start with the requirements of the desired applications, because this would have implied too great an emphasis on processes which cannot easily be mapped.
However, the hierarchical model can be related to environmental hazards easily, because these hazards can be depicted as effect-chains: they attack at a certain ecosystem component, and subsequently affect the dependent components, always finally influencing the biota. The point of attack and resulting processes are often related to abiotic components of the environment.

The relation between environmental hazard and point of attack in terms of ecosystem component on the one hand, and the relation between ecosystem component and spatial scale on the other, consequently also imply a relation between environmental hazard and spatial scale: environmental hazards act at different spatial scales.

Applications illustrate that it is indeed possible to match a general framework for ecosystems classification with an analytical approach to environmental hazards. The latter concerns zooming-in on specific characteristics of the legend units and spatial units that have not been used for the classification.
This possible matching is predominantly based on the existence of 'correlative complexes'. The parameters that are required for the applications can be estimated with sufficient accuracy due to the existence of correlations between controlling factors and dependent characteristics.

References

Bailey, R.G., 1976. *Ecoregions of the United States*. USDA Forest Service, Intermountain Region, Ogden, Utah.
Bailey, R.G., 1981. Integrated approaches to classifying land as ecosystems. In: Laban, P. (ed.). *Proceedings of the workshop on land evaluation for forestry*. ILRI publ. 28: 95-109, Wageningen.
Bailey, R.G., 1985. The factor of scale in ecosystem mapping. *Environmental Management* 9/4: 271-276.
Bailey, R.G., 1987. Suggested hierarchy of criteria for multi-scale ecosystem mapping. *Landscape and Urban Planning* 14: 313-319.

Bailey, R.G., 1989. Explanatory Supplement to Ecoregions Map of the Continents. *Env. Cons.* 16/4: 307-310.
Bailey, R.G., S.C. Zoltai and E.B. Wiken, 1985. Ecological regionalization in Canada and the United States. *Geoforum* 16/3: 265-275.
Bakker, T.W.M., J.A. Klijn and F.J. Van Zadelhoff, 1981. *Nederlandse kustduinen. Landschapsecologie.* Pudoc, Wageningen, 144 pp.
Brink, A.B.A., J.A. Mabbutt, R. Webster and P.H.T. Beckett, 1965. *Report of the Working Group on land classification and data storage.* Military Engrg. Exp. Establ. Rep. no. 940. Christchurch, England, 97 pp.
Chorley R.J. and B.A. Kennedy, 1971. *Physical Geography. A Systems Approach.* Prentice-Hall International Inc., London, 370 pp.
Christian, C.S. and G.A. Stewart, 1968. Methodology of integrated surveys. In: *Aerial surveys and integrated studies.* Proc. Toulouse Conf. 1964, UNESCO, Paris, pp. 233-280.
De Vries, W. and A. Breeuwsma, 1986. Relative importance of natural and anthropogenic proton sources in soils in the Netherlands. *Water Air and Soil Pollution* 28: 173-184.
Dokuchaev Soil Institute, 1963. *Soil-geographical zoning of the USSR.* Jerusalem: Israel Prog. Sci. Transl.
Duchaufour, P., 1977. *Pedology. Pedogenesis and classification.* English edition (translated by T.R. Patton), George Allen & Unwin, London, 448 pp.
Evelyn (1661) *Fumifugium or; the Inconvenience of the Aer and Smoake of London Dissipated; together with some Remedies Humbly proposed.* Paper to King Charles II, London.
FAO, 1988. *FAO/ Unesco Soil Map of the World, Revised Legend.* World Resources Report 60, FAO, Rome. Reprinted as Technical Paper 20, ISRIC, Wageningen, 1989.
Haase, G., 1973. Zur Ausgliederung von Raumeinheiten der chorischen und regionischen Dimension. *Petermann's Geogr. Mitt.* 117: 81-90.
Haase, G., 1989. Medium scale landscape classification in the German Democratic Republic. *Landscape Ecology* 3/1: 29-41.
Hughes, R.M., and D.P. Larsen, 1988. Ecoregions: an approach to surface water protection. *Journal WPCF* 60/4: 486-493
Isachenko, A.G., 1973. *Principles of landscape science and physical geographic regionalization.* J.S. Massey (ed.), Melbourne, Australia, 311 pp.
Jørgensen, S.E., 1992. *Integration of Ecosystem Theories: A Pattern.* Kluwer Academic Publishers, Dordrecht etc.
Klijn, F., 1988. *Milieubeheergebieden. Deel A: Indeling van Nederland in ecoregio's en ecodistricten. Deel B: Gevoeligheid van de ecodistricten voor verzuring, vermesting, verontreiniging en verdroging.* CML reports no. 37/ RIVM report no. 758702001, 183 pp.
Kuylenstierna and Chadwick, 1991. The Sensitivity of Soil and Ecosystems to Acidic Depositions. In: N.H. Batjes and E.M. Bridges (eds.), *Proceedings of the International Workshop on Mapping of Soil and Terrain Vulnerability to Specified Chemical Compounds in Europe at a Scale of 1: 5 M.* ISRIC, Wageningen.
Kwakernaak, C., 1982. *Landscape Ecology of a Prealpine Area.* Publ. Fys. Geogr. Bodemk. Lab. 33, University of Amsterdam.
Lands Directorate Environment Canada, 1981. Various publications.
Leser, H., 1976. *Landschaftsökologie.* Verlag Eugen Ulmer, Stuttgart.
Leser, H., 1991. *Landschaftsökologie.* 3rd (revised) Ed. Verlag Eugen Ulmer, Stuttgart.

Müller, S., 1970. Öko-Serien der baden-württembergischen forstlichen Standortskartierung am Beispiel der Kalkverwitterungslehme. *Mitt. der Deutschen Bodenkundlichen Gesellschaft*, 10: 43-46.
Neef, E., 1967. *Die theoretische Grundlagen der Landschaftslehre.* Verlag H. Haack, Gotha-Leipzig.
Odum, H.T., 1983. *Systems Ecology, an introduction.* John Wiley & Sons, New York.
Omernik, J.M., 1987. Ecoregions of the conterminous United States. *Annals of the Association of American Geographers* 77: 118.
O'Neill, R.V., 1988. Hierarchy Theory and Global Change. In: T. Rosswall, R.G. Woodmansee and P.G. Risser (eds.), *Scales and Global Change*. John Wiley & Sons Ltd, Chichester/London.
Piket, J.J.C. et al., 1987. *Atlas van Nederland. Part 16: Landscape*. Staatsuitgeverij, The Hague, 23 pp.
RIVM (ed. F. Langeweg), 1989. *Concern for tomorrow. National environmental survey 1985-2010.* RIVM, Bilthoven, 344 pp.
Soil Survey Staff, 1975. *US Soil Taxonomy*. US Department of Agriculture, Washington DC.
Stevers, R.A.M., J. Runhaar and C.L.G. Groen, 1987. *Het CML-ecotopensysteem. Uitwerking voor Noord-, West- en Zuidwest-Nederland.* CML reports no. 34, Leiden, 110 pp.
Theurillat, J.P., 1992. L'analyse du paysage végétal en symphytocoenologie: ses niveaux et leurs domaines spatiaux. *Bull. Ecol.* 23/1-2: 83-92.
Trewartha, G.T., 1968. *An Introduction to Climate*. 4th ed. McGraw Hill, New York.
Tüxen, R., 1978. Versuch zur Sigma-Syntaxonomie mitteleuropäischer Flusstal-Gesellschaften. *Ber. Int. Symp. Int. Vereinigung Vegetationsk.* 1978: 273-286.
Ulrich, B., 1980. Production and consumption of hydrogen ions in the ecosphere. In: T.C. Hutchinson and M. Havas, *Effects of acid precipitation on terrestrial ecosystems*. Plenum Press, New York/ London, pp. 255-282.
Urban, D.L., R.V. O'Neill and H.H. Shugart Jr., 1987. Landscape Ecology. A hierarchical perspective can help scientists understand spatial patterns. *Bioscience* 37/2: 119-127.
Van der Maarel, E. 1976. On the establishment of plant community boundaries. *Ber. Deutsch. Bot. Ges.* 89: 415-443.
Van der Maarel, E. and P.L. Dauvellier, 1978. *Naar een Globaal Ecologisch Model voor de ruimtelijke ontwikkeling van Nederland.* Staatsuitgeverij, The Hague.
Van Duijvenbooden, W. and A. Breeuwsma (eds.), 1987. *Kwetsbaarheid van het grondwater.* RIVM report 840387003, 69 pp.
Verstraten, J.M., 1982. *De bodem als bufferend systeem tegen verzuring.* University of Amsterdam, 16 pp.
Vink, A.P.A., 1975. *Land Use in Advancing Agriculture.* Springer Verlag, Berlin etc., 394 pp.
Wagner, A., 1968. Oekoserie, Oekoseriengruppe und Standortstypengruppe. Neue Begriffe in der forstlichen Standortsaufnahme. *Allgemeine Forstzeitschrift* 42: 731-732.
Walter, H., 1979. *Vegetation of the earth and ecological systems of the geo-biosphere.* 2nd ed., Springer Verlag, New York Inc.
Wiken, E.B. and G. Ironside, 1977. The development of ecological (biophysical) land classification in Canada. *Landscape Planning* 4: 273-275.
Zonneveld, I.S., 1979. *Land Evaluation and Land(scape) Science*. ITC Textbook on Photo-Interpretation, Volume VII, Chapter VII. 2nd edition, ITC, Enschede, the Netherlands.

Ecosystem classification by budgets of material; the example of forest ecosystems classified as proton budget types 6

Roman J.M. Lenz

Nothing is as practical as a good theory (David Ricardo, 1772 - 1823)

ABSTRACT - A conceptual framework for a highly flexible approach to classification is proposed, based on a hierarchy of different levels of organization of matter and on the necessity of integration and scaling. First, it is argued that we should regard three organizational levels at the same time. Second, it is demonstrated that the larger the man-induced disturbances, the more adequate a classification seems by focusing on processes rather than structural characteristics. Related to the hierarchy of organizational levels, trans-level systems may be recognized, to be understood as reaction chains affecting different organizational levels. Finally, this approach is exemplified by a classification of forest ecosystems based on material balance and proton budgets in forests.

Introduction and theoretical framework

The huge amount of ecological data already available and the observations and experiences in ecosystem management necessitate investing more in integration (see also Odum, 1977) by ecologists. Classification is one way of integrating such diverse knowledge. We all know that a complex object can be approached from different points of view, each having its own advantages. For classification this means that various classifications and methods must be taken into consideration, and we should not challenge the development of the classification or the methodology.

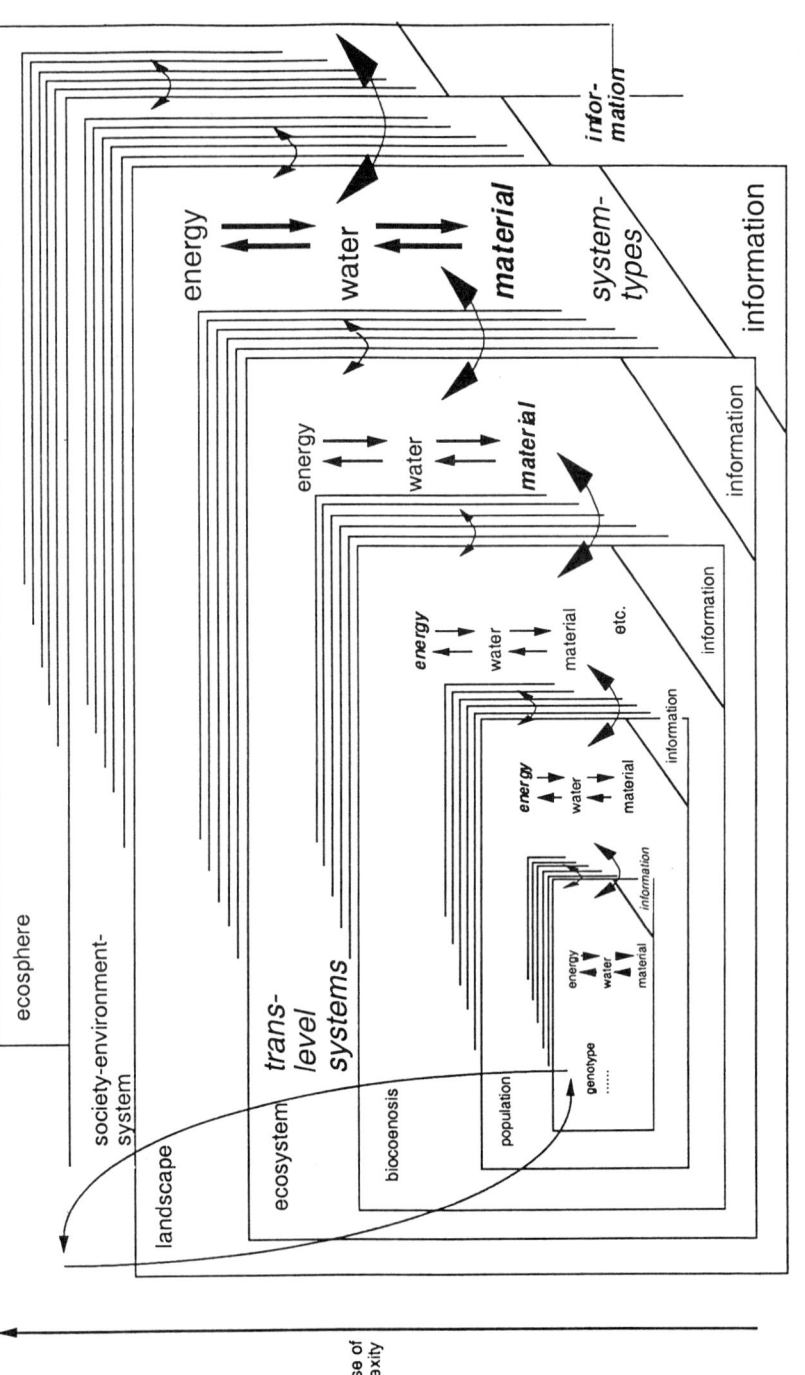

Figure 6.1 Levels of the organization of matter: budgets and relationships

Therefore, classification concepts for the moment will be concepts under transition, with their advantages for specific purposes, and adequate only in relation to the present state of knowledge and available data. Before going into the classification of ecosystems, we should position the ecosystem in relation to other systems of lower or higher order. For ecologists, the organizational levels of Haber (1978, 1993), who divided the ecosystem described by Miller (1975) into ecosystems, landscapes and (human) society-environment-systems, appear to be a good basis for such a positioning.

A hierarchy of organizational levels

The idea of a hierarchy of organizational levels is in widespread use today. Such a hierarchy is shown in Figure 6.1, additionally specifying the relevant budgets at the various organizational levels, and their relationships (Lenz, 1991; Lenz and Haber, in prep.). The general description of organizational levels (Egler, 1970; Miller, 1975) is untouched, although exceptions due to the dependence of natural and anthropogenic systems on specific environmental factors have to be taken into account. The environment can be specified in terms of site factors and, more generally, environmental factors. System types evolve by adapting to natural or anthropogenic environmental factors, resulting, for instance, in forest ecosystems or agro-ecosystems. Thus, the continuity of a typical set of environmental factors and their dynamics is the dominating explanatory mechanism of the genesis of the system type. Studying a system type, the frame conditions given at the higher level and the compartmental interrelations at the lower level are somehow the external (environment) and internal objects. Therefore, preferably at least three organizational levels have to be regarded.

Many environmental problems, however, can only be fully understood by analysing them at many different organizational levels, e.g. the problem of the release of fluorocarbohydrates by man into the atmosphere, reacting with the stratospheric ozone layer, and thus changing the irradiation for all organizational levels.
Hence, if there are trans-level phenomena, such as reaction chains due to chemicals or radiation, passing across several organizational levels and connecting them in such a way as to be considered a system of its own, then the hierarchy of organizational levels and the role of single levels may become merely background information. Such so-called trans-level systems have to be

thought of as exceptions to systems according to organizational levels, and, therefore, they do not have a priori common characteristics with the systems of the hierarchy of organizational levels. Even a new hierarchy of dominance created by the cross-level matter flow can be recognized, resulting in specific structures like shoot-root ratio, root ramification pattern, or others (Ulrich, 1993, pers. com.).

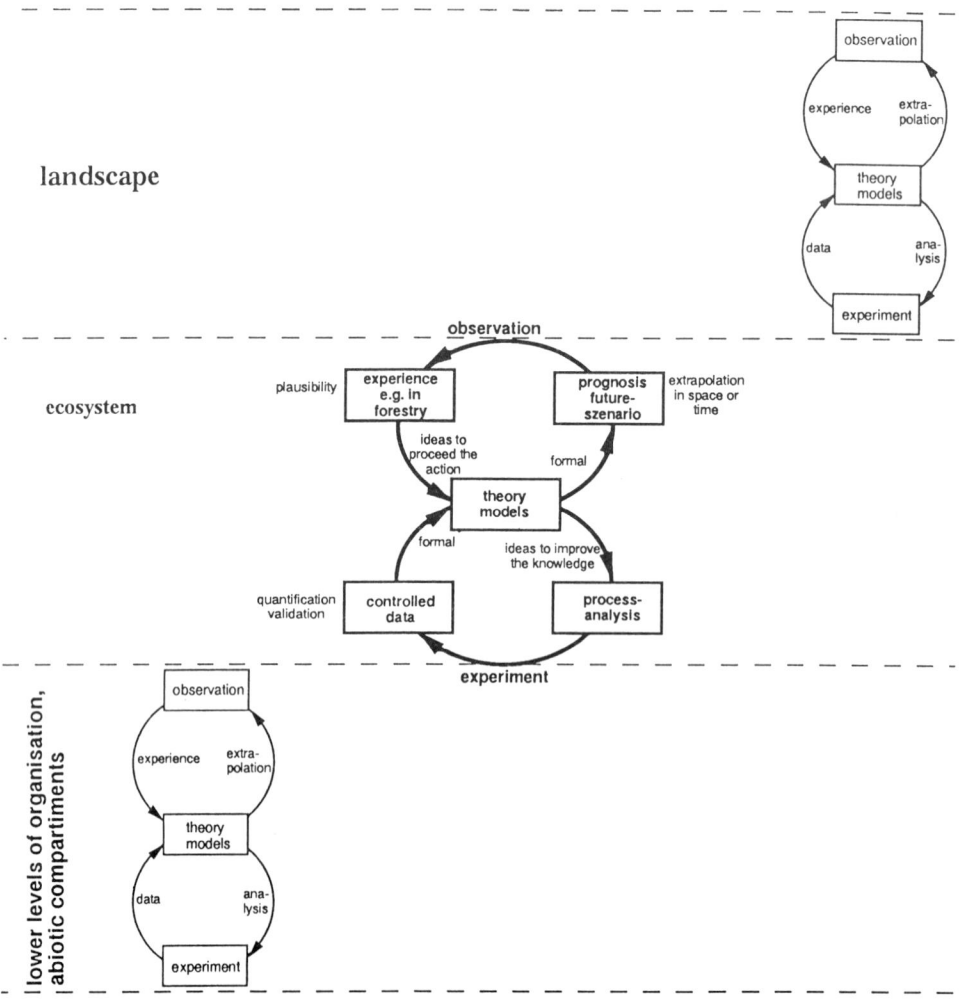

Figure 6.2 Research approaches in relation to organizational levels with the ecosystem as focal object

ECOSYSTEM CLASSIFICATION BY BUDGETS OF MATERIAL 121

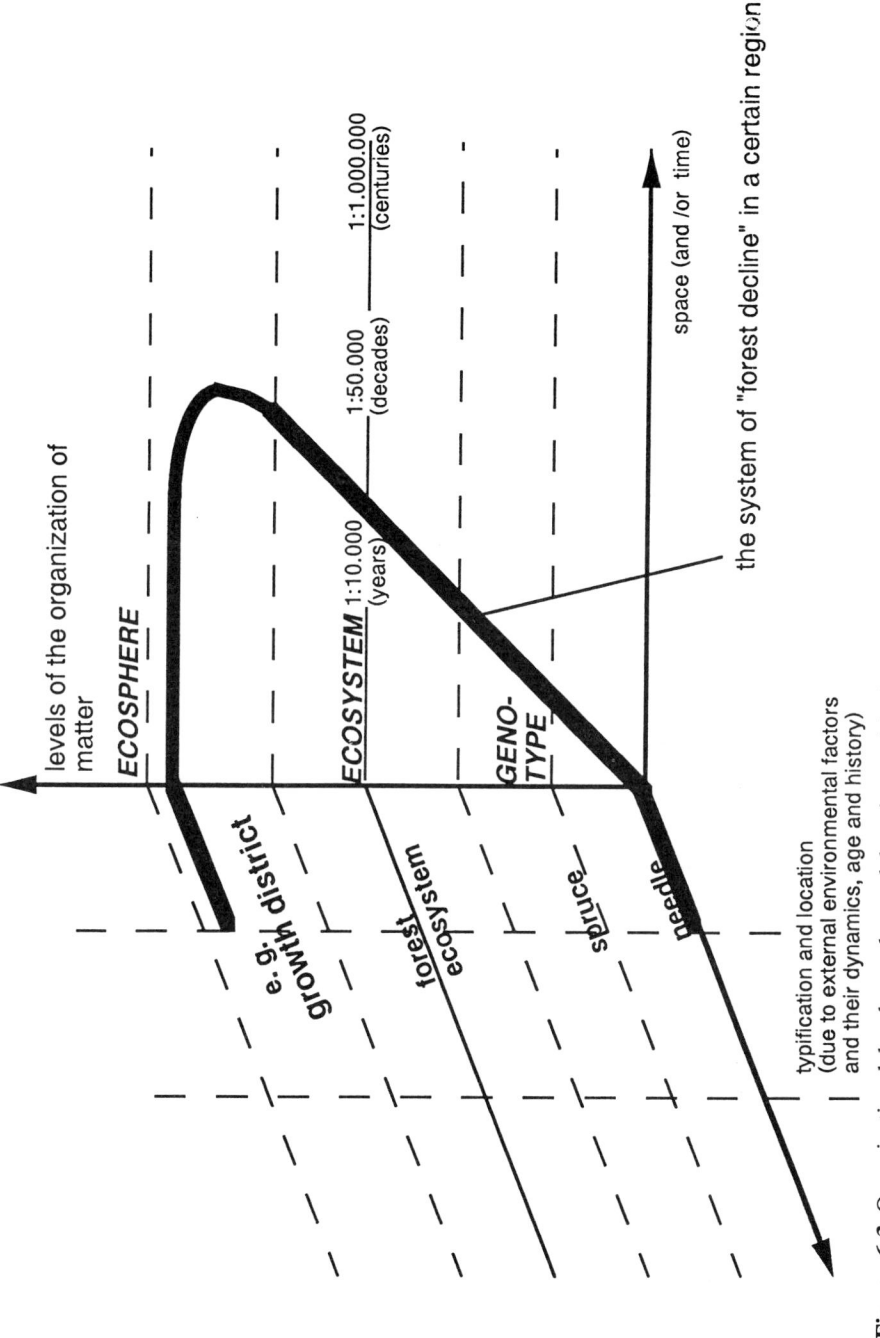

Figure 6.3 Organizational levels, scales and locations with their specific set of environmental factors, as determinants of an ecological system

Levels of observation related to organizational levels: the staircase approach

Especially in relation to trans-level phenomena affecting different organizational levels, different research approaches have to be introduced, showing that descriptive-empirical (which also means more holistic) and explanatory-quantitative (which also means more reductionistic) approaches are complementary. In Figure 6.2 the focal object of consideration is the ecosystem, embedded in frame-conditions of a landscape and composed of biotic and abiotic compartments. Under steady-state conditions, the ecosystem types are related to both environment and compartments, but they can be better described by observations, i.e. holistic comparisons, in some cases, or by experimental work, i.e. reductionistic analyses, in others.

Therefore, to include all relevant information in classification and description of systems at any organizational level, a 'staircase' or 'scaling' of research steps appears to be the most useful approach. This combines comparative and quantitative research and is related to the various organizational levels (see Miller, 1975; Haber, 1993; O'Neill et al., 1986, etc.) and, also, takes into account that there are continuous transitions between observations and experiments, and between structures and processes. In this sense, structures are the result of processes, yet new or strongly modified structures also modify processes, thus generating 'higher' structures and patterns (Lenz and Schall, 1989; Lenz, 1991; Ulrich, 1993, pers. com.).

Figure 6.3 shows an attempt to illustrate the relation between organizational levels for a study of forest decline in a certain region. Cause-effect relationships always determine the extension of a study. In addition, the specific location, its history and other factors have to be taken into account. Forest decline, for example, can only be understood by focusing on deteriorations of the balance of material in forests. This implies an analysis at several levels below the ecosystem level. Figure 6.4 illustrates the specific staircase for a study of forest decline, specifying the cause-effect relations at a number of related organizational levels (Lenz and Haber, 1992).

Level of Organisation	Reaction time Sec.-Min.	Min.-Days	Days-Years	Decade	Century	Damage Symptoms	Main Changes in Processes
Cell (μm^2) *Cell physiology*	transport biochemistry, antioxidants	vascular tissue, chloroplasts				Needle chlorosis	phloem collapse, starch stowage, impairment of transport enzymes by Mg^-, K^- and other deficiencies
Organ (cm^2) *Eco-physiology*		Assimilation Respiration Magnesium allocation	Biomass compartments				Mg deficiency causing reduced photosynthesis chiefly taking place in youngest needles
Genotype (m^2) *Aut-Ecology*			Balances: Carbon Water Magnesium	Biomass compartmentation		Needle necrosis and abscission	Mg deficiency reducing wood increment in favour of crown growth. Needle development controlled by nitrogen supply
Species/ Population (ha) *Dem- and Synecology*				Stand density Stand biomass	Aut-Ecological needs	Tree mortality	Leaf area index controlled by nitrogen supply if all other factors are adequately supplied
Ecosystem/ Landscape (km^2) *Ecosystem and Landscape Ecology*				Temperature variation Basic cation supply Water supply	Temperature Precipitation Radiation	Stand mortality Retrogression	Excess of nitrogen and of protons causing impoverishment of basic cations, increase of toxic Al concentrations, systemic water stress and nutrient imbalances

Figure 6.4 A hierarchical scheme of structures and processes for spruce (*Picea abies*) and spruce-dominated ecosystems to simulate damage symptoms related to key processes (after Lenz and Schall, 1991)

Structural and functional characteristics: processes as systems' determinants

Ecosystem classification often starts with a description of the visible structures, such as organisms and their habitats, vegetation and its composition, land-use types, etc. But in the long term, we should try to include knowledge of processes, because 'dynamics govern structure', as Van der Maarel (p. 15, 1980) put it. Especially today, the dynamics of environmental factors are often disturbed by man. Therefore, focusing only on the structural characteristics will sometimes lead to misinterpretations. Structures may be relict, which means that they originated under entirely different environmental conditions. Environmental planning should also give advice based on the detection of deviations from a more natural steady state of the object under consideration. This implies that functioning and processes are becoming more and more important for classification.

A process-orientation enables classifying types of reactions to specific environmental impacts. In general, structural characteristics as indicators of environmental changes will only be valuable for more obvious impacts and, therefore, are inappropriate for early warning of current environmental changes. We thus maintain that classifications of ecosystems must be open and flexible to allow combinations of structures and functioning on the one hand, with primarily man-induced disturbances on the other hand.

The most appropriate classification of ecosystems for early warning of environmental changes can be achieved by:

(i) observation of structure and functioning,
(ii) analysis of the relevant processes,
(iii) combining (i) and (ii) in a systems' description (theory, model); but adding
(iv) that observations and experiments should be combined within one level, as well as between neighbouring levels.

Below, we shall go into an example of a classification according to these four guidelines.

Forest ecosystems as proton budget types[1]

As was explained elsewhere (cf. Haber et al., 1991; Lenz and Haber, 1990; Lenz and Stary, 1991), essential processes of the nutrient and proton budget in forest ecosystems can be estimated and interpreted on a regional scale. The results reflect the ongoing destabilization and disturbance of the nutritional fundamentals of productivity in our forests, even if they are not always quantified. Hence, further process-oriented case studies are necessary.

In addition, our enormous environmental problems, especially on a regional and global scale, and the high esteem with which forests as semi-natural or natural ecosystems are regarded in our society, makes it essential that integrative methods be used to optimize programmes for the revitalization of forests. To accomplish this, the system we must understand and manage is the organizational level of an ecosystem, framed by a landscape and internally regulated by organisms and abiotic components.

The intention of the following chapters is to systematically extrapolate the central processes and sensitivities from ecosystems onto a regional scale. We shall take the 'accident' of acidification as an example for testing the classification of ecological systems. Acidification seems to be a good example of a trans-level system, with specific perturbations and stress on various levels of organization and type-specific characteristics (cf. Lenz, 1991). Thus we are also contributing to an integrated assessment a missing link in the acid rain debate (Streets, 1989).

Methodological concept

The nutrient and proton fluxes in meso-climatically comparable regions play a central role in the prediction and evaluation of the reactions of ecosystems with similar species composition. This approach refers to the concept of material budgets and critical loads (Van Breemen et al., 1984; Hauhs, 1989; 1990; Hauhs et al., 1989; Nilsson and Grennfelt, 1988; Paces, 1986; Ulrich, 1987, 1989a,b) primarily applied to European forests.

The concept of critical loads means, that in the long term there should be no protons which cannot be buffered by the weathering of primary silicates in the soil (Nilsson and Grennfelt, 1988). Critical processes concerning the proton

---

[1] The following sections are a revised version of a paper published in the journal Modeling Geo-Biospere Processes (Lenz, 1992)

budget are defined as rates of depletion of the pool of base cations in the ecosystem.

As long as there is buffering due to a depletion of the pool of base cations, this process is the driving force for the ecosystem running into a new stage, at which the critical load can only be compensated by the rate of weathering. So, for the medium-term, it is necessary to include the buffering capacity of the base cation pool, which can be estimated by the base saturation of a soil column of given depth (Lenz and Haber, 1990; Lenz, 1991). If the pool in the soil is more or less empty, the exceedence of critical loads can be described by rate types, which are based on the actual weathering rate in comparison to the deposition rate.

These definitions and simplifications of reality are related to the scale of observation. An adequate compromise must be sought, to achieve area-related recommendations for monitoring and management.

Proton budgets as classification characteristic

In the following, a compartment model will be used to illustrate the key processes of internal and external proton production and consumption. The compartment model is based on evaluations of Ulrich (1987, 1989a,b) and on more detailed descriptions of the compartments by Matzner (1988). The model is a simplified model of a forest ecosystem with special emphasis on the deviation of balance of material, resulting in proton production and/or consumption.

This model includes most of the described proton fluxes due to natural and anthropogenic processes. The critical processes can generally be differentiated into internal and external rates (Van Breemen et al., 1984), related by a feedback loop. Internal processes can be regarded as biologically determined processes, which are adapted to their environment. External refers to acid loads as a kind of input to the biologically regulated part of the ecosystem. If there is a sufficient pool of bases, as well as a high base cation release via weathering, then nitrogen supply and mesoclimatically and biocoenotically determined productivity limit the internal proton production rates. The ranges of proton fluxes were derived from literature (see Figure 6.5).

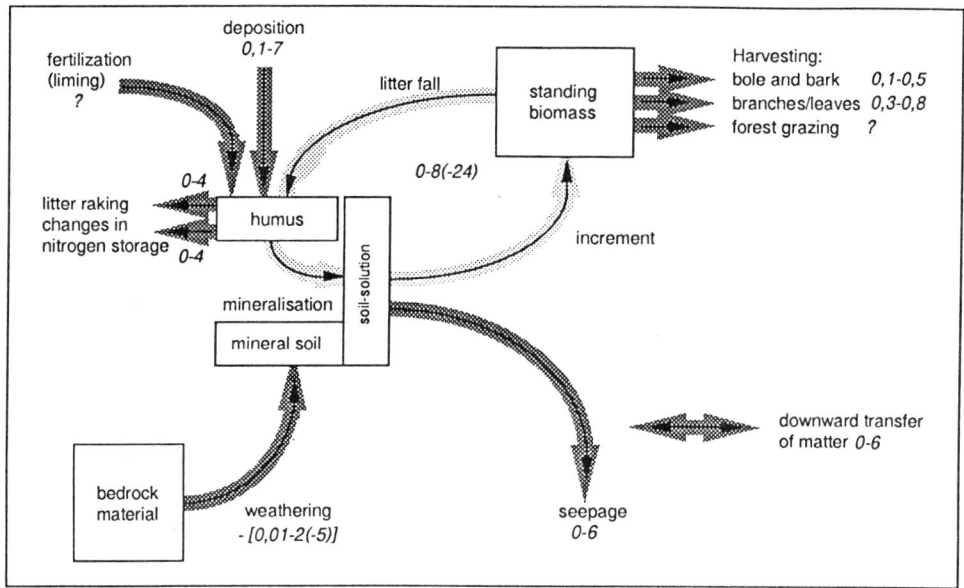

Figure 6.5 Internal (light arrows) and external (dark arrows) proton production and consumption processes in forest ecosystems (with special emphasis on utilization; kmol ha^{-1} yr^1). Sources: Berden et al., 1987; Bredemeier and Ulrich, 1989; Van Breemen et al., 1984; Fölster, 1985; Glatzel, 1989; Hauhs, 1989; Hauhs et al., 1989; Matzner, 1988; Nilsson and Grennfelt, 1988; Paces, 1986; Reuss and Johnson, 1986; Schulze et al., 1989; Sverdrup and Warfvinge, 1987a,b; Ulrich, 1987, 1989a,b; Ulrich and Meyer, 1987

Depending on the dominance of the specific external or internal processes of proton production or consumption, seven proton budget types — five principal ones and two subtypes — can be distinquished (Table 6.1, Figure 6.6). The typification in Figure 6.6 is only conceptual, which means that the relevant compartments and dominant processes are marked by the relative thickness of the arrows.

The proton budget types

In primary succession, proton production in early stages is largely determined by the amount of weathering (i.e. rate of the soil formation is particulary important). Therefore, the establishment rate of the two important pools (the soil and the organic matter) in mesoclimatically similar regions (which are determining the potential natural productivity) is normally regulated by weathering (type 1 in Fig 6.6).

No.1 Geogenetic rate type

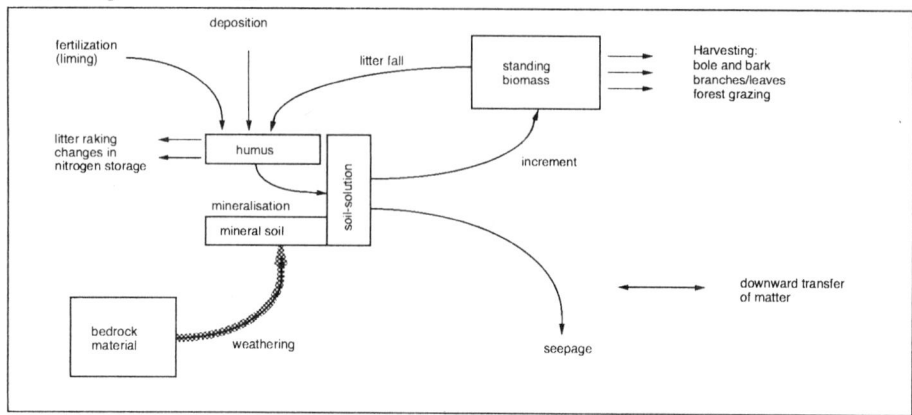

No.2a Biogenetic-geogenetic steady-state type (seasonal climate)

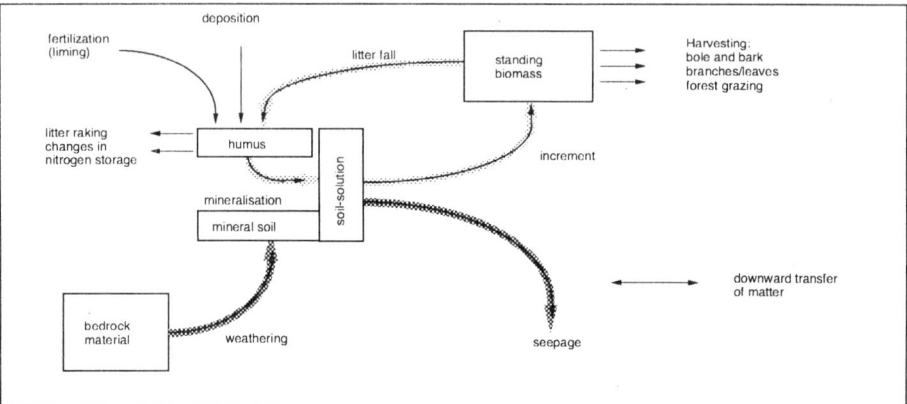

No.2b Biocoenotical equilibrium type

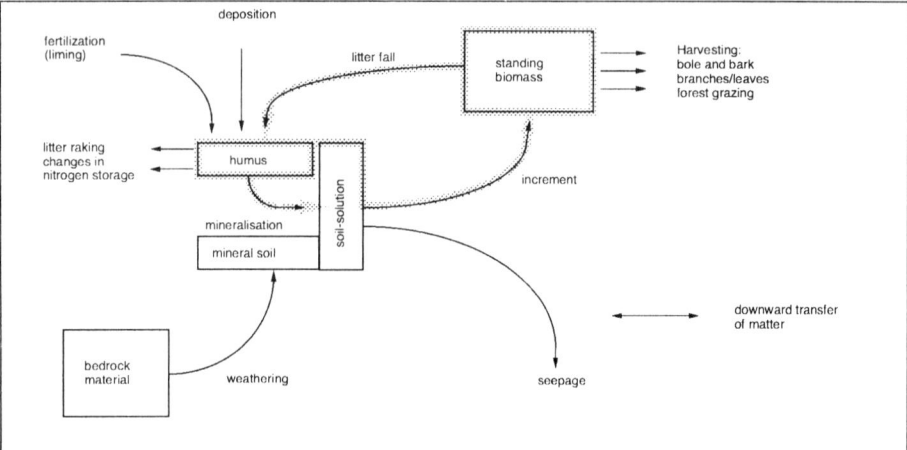

Figure 6.6 nos. 1, 2a, 2b System types of Figure 6.5; the dotted arrows are the key processes

Nr.	Name	Key-factors
1	Geogenetic rate type	Initial phase in a primary succession
2a	Biogenetic-geogenetic steady state type	Seasonal climate and climax vegetation
2b	Biocoenotical equilibrium type	Wet tropical climate and climax vegetation
3a	Pool (of bases) type	High cation exchange capacity and base saturation in the soil
3b	Flow through type	Depleted soils under humid conditions
4	Orogenous type	Extreme matter input / output relations along a slope
5	Artificial high yield type	Fertilization and harvesting are dominating the site factors

Table 6.1 Forest ecosystem types, based on material balance and proton budget

Under natural conditions, an incremental form of type 1 develops, which builds an increasingly available pool of nutrients. It is assumed that nitrogen is incorporated in the nutrient cycle proportionally to the increase of the nutrient pool of the ecosystem from the atmosphere by N-fixation. This assumption is based on the situation that organisms and their growth are limited by nitrogen and phosphorous, and that they have to optimize the use of these nutrients under the constraints of the frame given by other environmental conditions (climate, soil type). Several groups of organisms contribute to this optimization, and they are often limited in their activities by withdrawals of carbon, nitrogen, phosphorous and base cations — naturally, by storage or anthropogenically, through bioremoval.

As long as the rates of supply from the atmosphere and the bedrock material are not exceeding the needs of the biocoenoses, a steady state will be established (type 2a). Under ideal conditions this phase can reach a 'biocoenotical equilibrium', which can become more or less independent from the geogenous initial conditions.

No.3a Pool (of bases) type

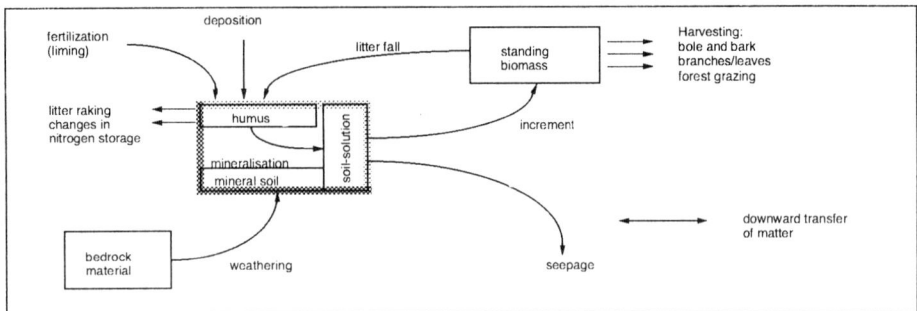

No.3b Flow through type

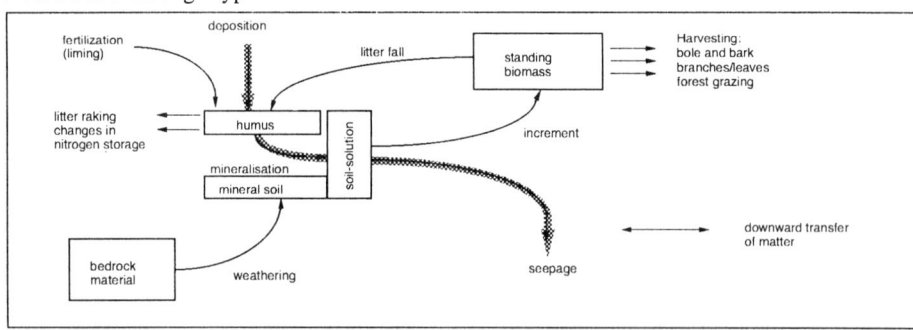

No.4 Orogenous type (matter transfer along a slope)

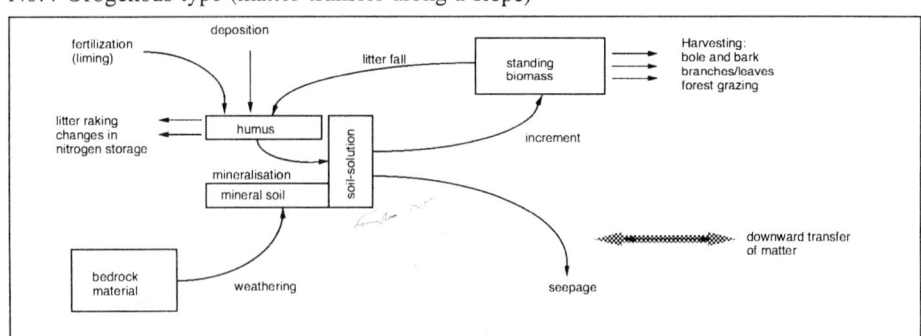

Figure 6.6 nos. 3a, 3b, 4 System types of Figure 6.5; the dotted arrows are the key processes

No.5 (Artificial) high yield type

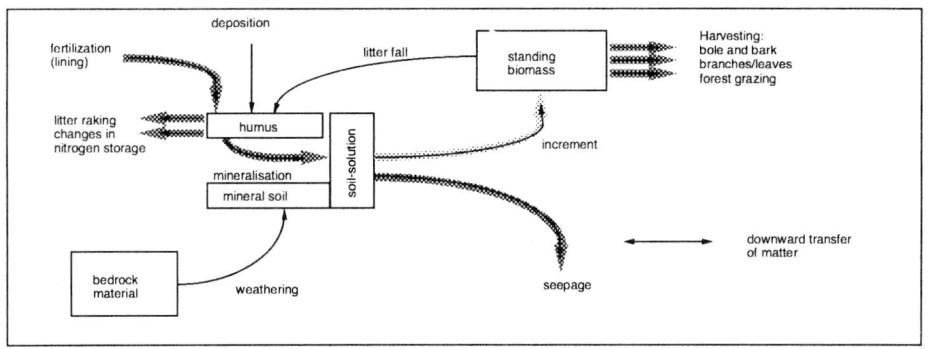

Figure 6.6 no. 5 System type of Figure 6.5; the dotted arrows are the key processes

In regions with a seasonal climate this final and ideal phase cannot be reached without pools in the soil. Thus, totally internal cycles are established only in the wet tropics (type 2b). In our nemoral zono-biomes in Central Europe, the capacity of pools in the humus layer and mineral soil are determinants of optimal productivity and biocoenotical equilibrium phases.

The bigger the pools which occur within meso-climatically similar conditions, the greater the elasticity of the systems of type 2 (steady state). Thus, the capacity of the pool of nutrients in the soil determines the sensitivities and possibilities of utilization of these system types (type 3a).
If there is a depletion of the pools (e.g., by water surplus and seasonal mineralisation pushes) and the actual load does not permit any regeneration, a through-flow of acids will follow, carrying with them base cations. These flow-through types (type 3b) enforce an adaptation of the biocoenosis to a low nutrient availability. The vegetation is compelled to build up the most closed internal mineral cycle possible, as is typical for vegetations of dunes, or for the larch or *Calamagrostis* vegetations in alpine regions and acidic grasslands. This can be interpreted as a strategy to obtain 'optimal'-productivity under such conditions. This is also based on the assumption that organisms try to improve their habitat by a site-specific optimal closing of material cycles. More demanding species are normally eliminated under either dry or wet, or other extreme conditions.

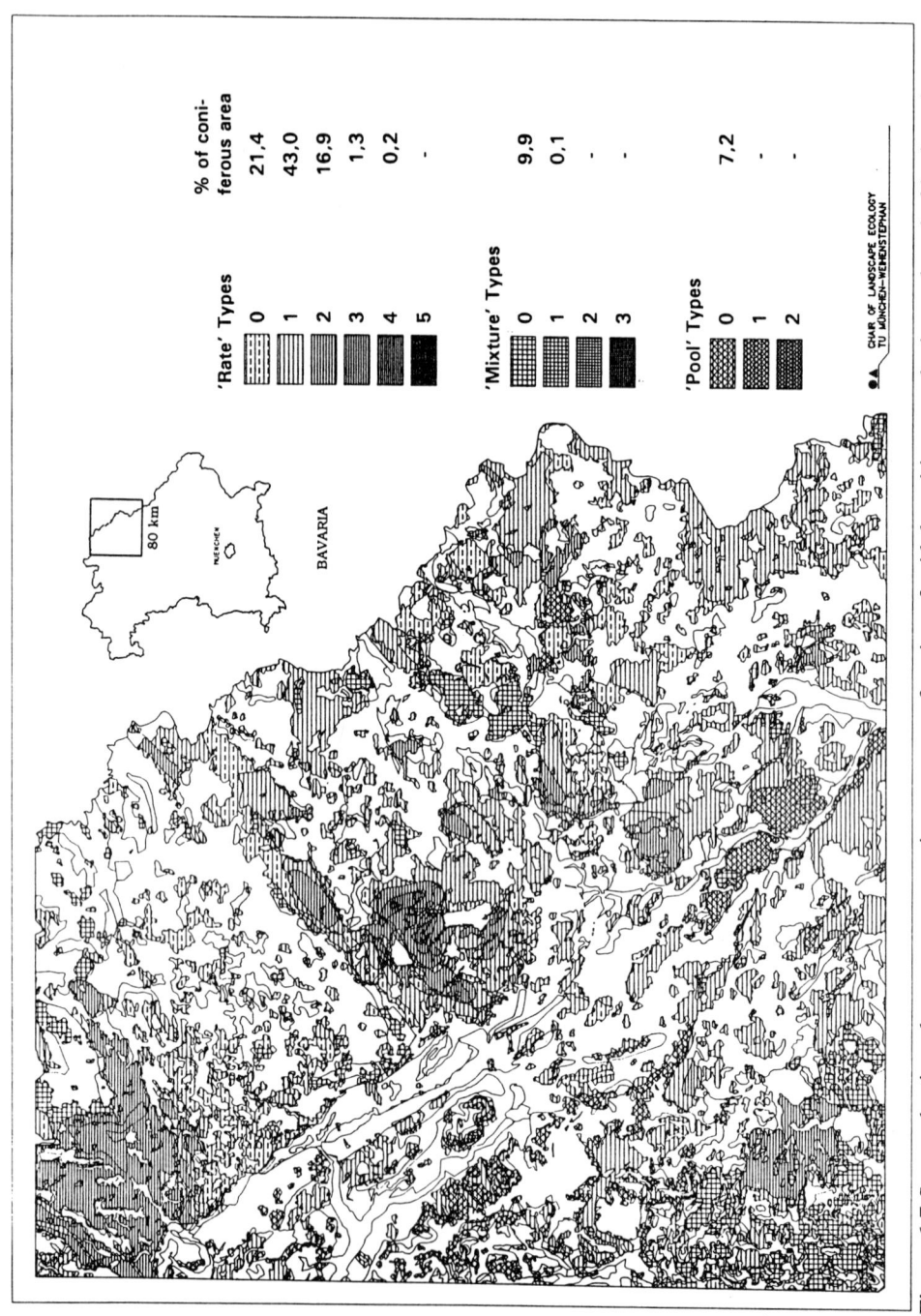

Figure 6.7 Spruce dominated ecosystem - reaction types as a function of acid deposition, weathering and pool of basic cations

Similar processes can be observed under specific climatic or orogenous conditions that cause drought or important downward transfer of matter, e.g., on a ridge in mountainous regions (type 4). Such 'orogenous depletion types' have their counterparts in lowlands and wet meadows. There, a naturally highly productive biocoenosis is established, which depends on the input of nutrients from the surroundings or from flooding (see Haber, 1990; or Haber in this book, chapter 3).

Many types of systems can be transformed into so-called high yield types by human interference, mainly in the form of additional inputs of nutrients. If these high yield types are removed or replaced, an artificial kind of steady state will be established, which may be connected to leaching losses, if the uptake of nutrients does not adapt to the rates of the input (see type 5).

The types described here are in a sense extremes, types 2a (biogenetic-geogenetic steady state type), 3a (pool type) and 3b (flow-through type) for spruce ecosystems in the Harz (Hauhs, 1989), in Northeastern-Bavaria (Figure 6.7) and Europe (Figure 6.8). On a local scale, several mixtures of these types occur, for which some processes may compensate, at least for the medium term. These partially compensational feedback processes should be quantified to identify the local, dominating processes. Only then can the best measures for internal pool-related restoration be determined (Lenz and Haber, 1992).

Concluding remarks

Using these system types for mapping and evaluation of forests on a regional scale, it should be possible to describe more systematically the indicators to be monitored and to decide on the measures of management.
On a regional and especially global scale, the measures to be carried out are evident. These are (i) drastic reduction of the acidic and nitrogen deposition, combined carefully with (ii) site-specific compensational and restorational measures, such as liming and eventually fertilization.
But for regional to local scales, the required measures also depend on the goal of biocoenosis composition, i.e. the 'target ecosystem'. The biocoenotically optimal reclamation can be developed only on a long-term basis, requiring the re-introduction of system-specific, deep-rooting tree species as one of the most

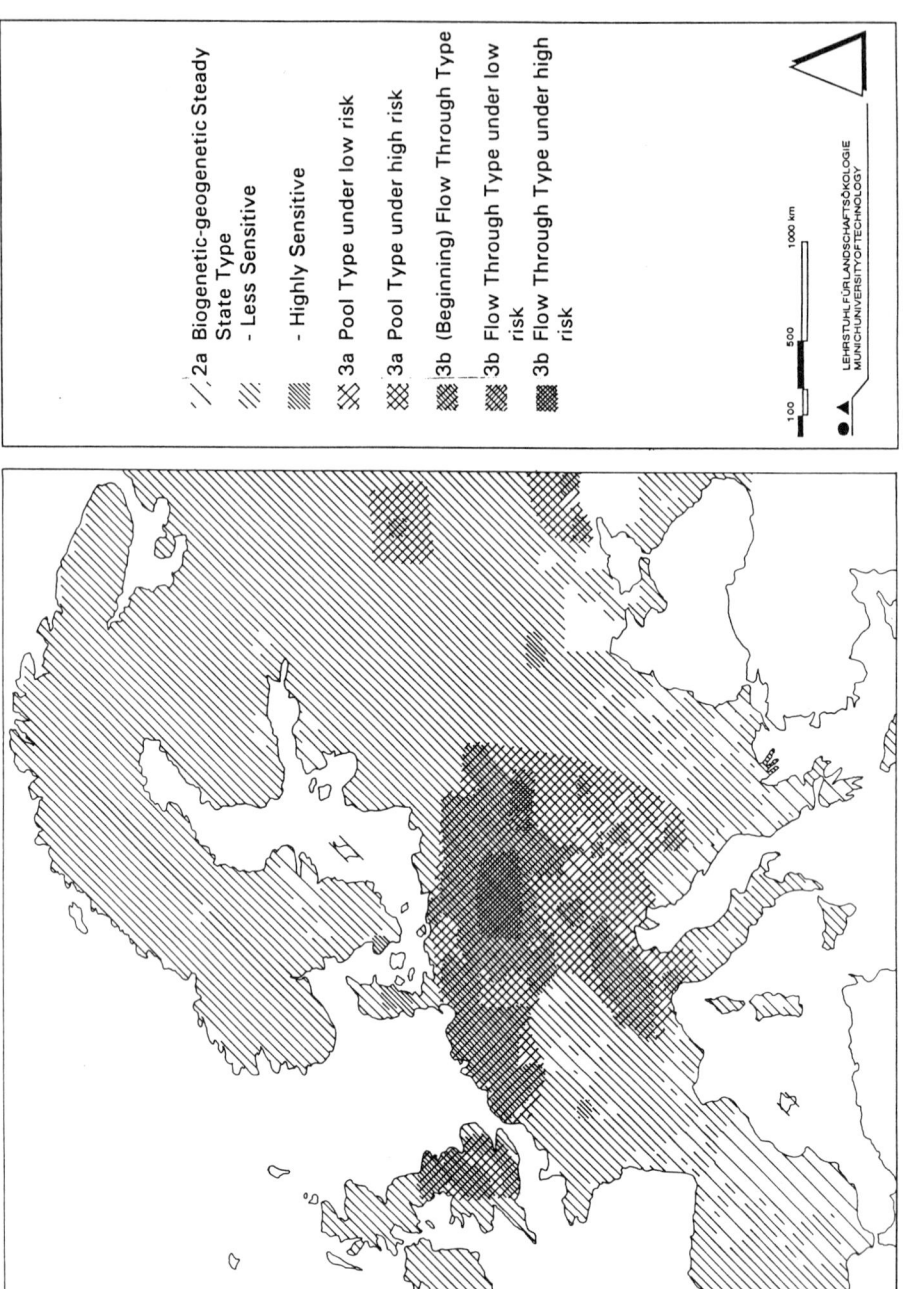

Figure 6.8 Spatial distribution of the main forest ecosystem reaction types in Europe, based on the material balance

important aspects of silviculture. At present, most of the plantations in Northeastern-Bavaria use (commercial) tree species, which are unable to close the nutrient cycles under present conditions. In any case, these target species and target biocoenoses have no real chance of survival, if there is no compensation for the former and actual depletion of nutrients caused by acid deposition and forest utilization (Lenz and Stary, 1991).
In general, we have to recognize that the human impact, either from the atmosphere or by fertilization and harvesting, must not exceed the carrying capacity. This carrying capacity may be defined as the possibility for biological regulation in such quantity and intensity that further disturbances will be minimized.

It has to be stated that the ecosystem types described in this paper are simplifications due to the lack of data from more representative case-studies. Only the combination of theoretical modelling and site-specific measurements will lead to a real improvement of the quantifications of the budgets.

Acknowledgements

I would like to thank Drs. Frans Klijn for reviewing and discussing this contribution, and Prof. Dr. Wolfgang Haber, Dr. Ralph Hantschel, and Dipl. Ing. Peter Schall for discussing the issue and collaborating in this field of research.

References

Alcamo, J., M. Amann, J.P. Hettelingh, M. Holmberg, L. Hordijk, J. Kämäri, L. Kauppi, P. Kauppi, G. Kornai and A. Mäkela, 1987. Acidification in Europe: Simulation Model for Evaluating Control Strategies. *Ambio* 16/5: 232-245.
Berden, M., J. Nilsson, K. Rosen and G. Tyler, 1987. *Soil acidification - extent, causes and consequences. An evaluation of literature information and current research.* Nat. Swed. Env. Protection Board, Rep. 3292, Solna, Sweden, 164 pp.
Bredemeier, M. and B. Ulrich, 1989. Depositionsbedingte und ökosysteminterne Anteile der Säurebelastung von Waldböden. *AFZ* 11: 256-260.
Egler, F.E., 1970. *The way of Science. A philosophy of Ecology for the Layman.* Hafner Publishing Company, New York, 145 pp.
Fölster, H., 1985. Proton Consumption Rates in Holocene and Present-day Weathering of Acid Forest Soils. In: J.I. Drever (ed.), *The Chemistry of Weathering.* NATO ASI Series C: Mathematical and Physical Sciences, Vol. 149. Reidel, Dordrecht, pp. 197-209.

Glatzel, G. 1989. Internal proton generation in forest ecosystems as influenced by historic landuse and modern forestry. In: B. Ulrich (ed.), *Internationaler Kongreß Waldschadensforschung: Wissenstand und Perspektiven*, 2-6 Okt. 1989 in Friedrichshafen am Bodensee. Vorträge Bd. 2, pp. 335-349.

Haber, W., 1978. Fragestellung und Grundbegriffe der Ökologie. In: K. Buchwald and W. Engelhardt (eds.), *Handbuch für Planung, Gestaltung und Schutz der Umwelt*. Bd. 1. BLV, München, pp. 80-89.

Haber, W., 1990. Basic Concepts of Landscape Ecology and Their Application in Land Management. *Physiol. Ecol. Japan*, 27 (Special Issue): 131-146.

Haber, W., 1993. Ökologische Grundlagen des Umweltschutzes. In: K. Buchwald and W. Engelhardt (eds.), *Umweltschutz - Grundlagen und Praxis*. Economica Verlag, Bonn, 98 pp.

Haber, W., R. Lenz, P. Schall, R. Bachhuber, W.-D.Grossmann, K. Tobias and H.F. Kerner, 1991. *Prüfung von Hypothesen zum Waldsterben mit Einsatz dynamischer Feedbackmodelle und flächenbezogener Bilanzierungsrechnung für vier Schwerpunktforschungsräume der Bundesrepublik Deutschland*. Ber. d. Forschungszentrums Waldökosysteme Göttingen, Reihe B Band 20, 132+53 pp.

Hauhs, M., 1989. Decline of Norway Spruce at the Harz Mountains - A hypothesis based on acid deposition and site sensitivity. *International Congress on Forest Decline Research*, Friedrichshafen, 1-6.10.1989, Poster Abstracts Vol. 2, Poster 254: 535-536.

Hauhs, M., 1990. Lange Brahmke: An Ecosystem Study of a Forested Watershed. In: D.C. Adriano (ed.), *Advances in Environmental Science*, Vol. 5 (Case Studies). Springer, New York, pp. 275-305.

Hauhs, M., K. Rost-Siebert, G. Raben, T. Paces and B. Vigerust, 1989. Summary of European Data. In: J.L. Malanchuk and J. Nilsson (eds.), *The Role of Nitrogen in the Acidification of Soils and Surface Waters*. Nordic Council of Ministers, Miljorapport 1989:10. Gotab, Stockholm, pp. 5-1 - 5-37.

Lenz, R., 1991. *Charakteristika und Belastungen von Waldökosystemen NO-Bayerns - eine landschaftsökologische Bewertung auf stoffhaushaltlicher Grundlage*. Ber. Forschungszentrum Waldökosysteme Göttingen, Reihe A: 200 pp.

Lenz, R., 1992. Forest Ecosystems as Proton Budget Types - A Landscape-Ecological Evaluation, Illustrated by NE-Bavaria and Central Europe. *Modeling Geo-Biosphere Processes* 1: 115-129.

Lenz, R. and W. Haber, 1990. Longterm assessment of spruce vitality in the Fichtelgebirge (West Germany) under ongoing acid deposition. *Vegetatio* 89: 121-135.

Lenz, R. and W. Haber, 1992. Approaches for the restoration of forest ecosystems in northeastern Bavaria. *Ecol. Modelling* 63: 299-317.

Lenz, R. and W. Haber W. (in prep). *Classification theory of ecological systems - are the generalizations only the exceptions?*

Lenz, R. and P. Schall, 1989. *Darstellung waldschadensrelevanter Ökosystembeziehungen als Grundlage von dynamischen Modellen und Hypothesensimulationen am Beispiel der Stickstoffhypothese*. Verh. Ges. Ökol. 17. Goltze, Göttingen, pp. 633-641.

Lenz, R. and P. Schall, 1991. Belastungen in fichtendominierten Waldökosystemen. Risikokarten zu Schlüsselprozessen der neuartigen Waldschäden. *AFZ* 46: 756-761.

Lenz, R. and R. Stary, 1991. Concepts and Methods for Compensation of Acid Deposition, Melioration and Restoration of Acidified Forest Areas in Northeastern Bavaria (Fed. Rep. of Germany). In: O. Ravera (ed.), *Terrestrial and aquatic ecosystems: perturbation and recovery*. Ellis Horwood, 422-433.

Matzner, E., 1988. *Der Stoffumsatz zweier Waldökosysteme*. Ber. Forschungszentrum Waldökosysteme/Waldsterben, Göttingen, Reihe A, Bd. 40, 217 pp.

Miller, G.T., 1975. *Living in the Environment. Concepts, Problems, and Alternatives*. Wadsworth Publishing Company, Belmont, California.

Nilsson, J. and P. Grennfelt, 1988. *Critical Loads for Sulphur and Nitrogen*. Miljörapport 1988: 15. Gotab, Stockholm, 418 pp.

Odum, E.P., 1977. The emergence of ecology as a new integrative discipline. *Science* 195, No. 4284: 1289-1293.

O'Neill, R.V., D.L. De Angelis, J.B. Waide and T.F.H. Allen, 1986. *A Hierarchical Concept of Ecosystems*. Princeton Univ. Press, Princeton.

Paces, T., 1986. Weathering and Mass Balance in Small Drainage Basins: Environmental Applications in the Bohemian Massif (Central Europe). *Sci. Geol. Bull* 39/2:131-150.

Reuss, J.O. and D.W. Johnson, 1986. *Acid Deposition and the Acidification of Soils and Waters*. Ecological Studies Vol. 59. Springer, New York, 120 pp.

Schulze E.-D., O.L. Lange and R. Oren, 1989a. *Air pollution and forest decline. A study of spruce (Picea abies) on acid soils*. Ecological Studies Vol. 77. Springer, New York, 475 pp.

Streets, D.G., 1989. Integrated Assessment: Missing Link in the Acid Rain Debate? *Environmental Management* 13/4: 393-399.

Sverdrup, H. and P. Warfvinge, 1987a. *The kinetics of mineral weathering*. Technical Report, Dept. of Chemical Engineering II, Lund Institute of Technology.

Sverdrup, H. and P. Warfvinge, 1987b. *The kinetics of cation production of the weathering of feldspars* (Original in Swedisch). Technical Report, Dept. of Chemical Engineering II, Lund Institute of Technology, 1987, 33 pp.

Ulrich, B., 1987. Stability, elasticity and resilience of terrestrial ecosystems with respect to matter balance. In: E.-D. Schulze and H. Zwölfer (eds.), *Potentials and limitations of ecosystem analysis*. Ecological Studies Vol. 61. Springer, New York, pp. 11-49.

Ulrich, B., 1989a. Material balance characteristics of forest ecosystems subjected to acid deposition. *International Congress on Forest Decline Research*, Friedrichshafen, 1-6.10.1989, Poster Abstracts Vol. 2, Poster 339, pp. 705-706.

Ulrich, B., 1989b. Effects of Acid Deposition on Forest Ecosystems in Europe. *Adv. Environm. Sci.* 2: 189-272.

Ulrich, B., 1993. Personal communication.

Ulrich, B. and H. Meyer, 1987. *Chemischer Zustand der Waldböden Deutschlands zwischen 1920 und 1960, Ursachen und Tendenzen seiner Veränderung*. Ber. Forschungszentrum Waldökosysteme, Reihe B, Bd. 6, 133 pp.

Van Breemen, N., C.T. Driscoll and J. Mulder, 1984. Acidic deposition and internal proton sources in acidification of soils and waters. *Nature* 307: 599-604.

Van der Maarel, E., 1980. Towards an ecological theory of nature management. *Verh. Ges. Ökologie* 8: 13-24.

The use of site factors as classification characteristics for ecotopes

7

Han (J.) Runhaar and Helias A. Udo de Haes

ABSTRACT - This chapter describes a classification of small-scale ecosystems or ecotopes using abiotic and biotic site factors as classification characteristics. The plant species composition of the types distinguished can be used for mapping purposes. To this end, ecological species groups have been defined that are characteristic for the different ecosystem types.

The classification is meant for use in environmental impact assessment. So far, efforts have focused on the classification of terrestrial ecosystems and their description in terms of floristic composition. However, our goal is to classify all ecosystems occurring in the Netherlands and, also, to include fauna in the description. The use of the classification in environmental impact assessment is discussed, and a comparison is made with vegetation classifications.

Our conclusion is that the use of site factors for the classification of ecosystems is of more than merely practical importance, because it facilitates the integration of knowledge about the abiotic and biotic components of ecosystems.

Introduction

When he introduced the term ecosystem, Tansley (1935) emphasized that the physical environment formed a part of it: 'Though the organisms may claim our primary interest, when we are trying to think fundamentally we cannot separate them from their environment, with which they form one physical system'. He introduced the term as an alternative for community, a concept he disliked because it suggests that plant and animal life can be studied separately from their physical environment.

The ecosystem concept of Tansley has had a large influence in ecology. Nevertheless, integrated studies in which equal attention is given to biotic and abiotic aspects of ecosystems are rare, as Leser (1991) concludes. This is reflected in ecosystem classifications, which are often restricted to the biotic components of the ecosystem, such as vegetation or other parts of a biocenosis, although the underlying relations with the abiotic environment are sometimes used as 'guiding principle' in the sense of Zonneveld (chapter 2 of this book).
Integration is probably furthest developed in applications, such as land evaluation. The basic unit in land evaluation, the land unit, is defined on the basis of both biotic and abiotic characteristics (Zonneveld, 1989). However, it is described by means of classifications of separate components of ecosystems, such as soil or vegetation.

In this chapter we present an ecosystem classification in which both biotic and abiotic aspects are used for the classification and description of the types (Figure 7.1). In this *ecotope* classification, both abiotic and biotic factors are used as classification characteristics if they determine the species composition. The species composition is used to identify the ecosystem types in the field. For this purpose, ecological species groups have been defined that list the species indicative for the ecosystem type.

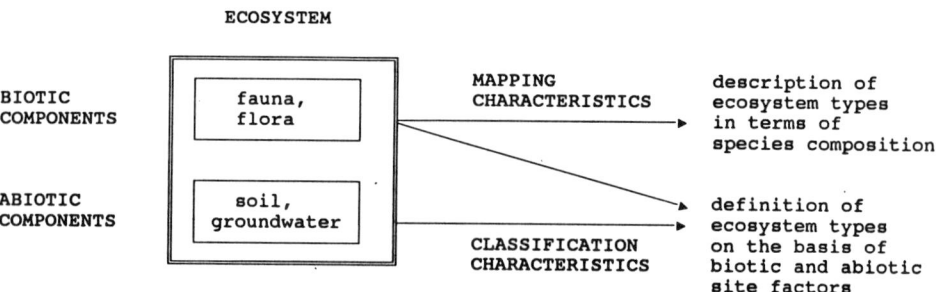

Figure 7.1 Use of abiotic and biotic ecosystem characteristics in the ecotope classification

This classification was developed for practical applications, i.e., as a tool to predict changes resulting from human interference in the context of environmental impact assessment. The use of ecologically relevant site factors as classification characteristics enables us to express changes in the (a)biotic environment caused by human activities, in terms of changed site conditions and corresponding changes in ecosystem types (Figure 7.2).

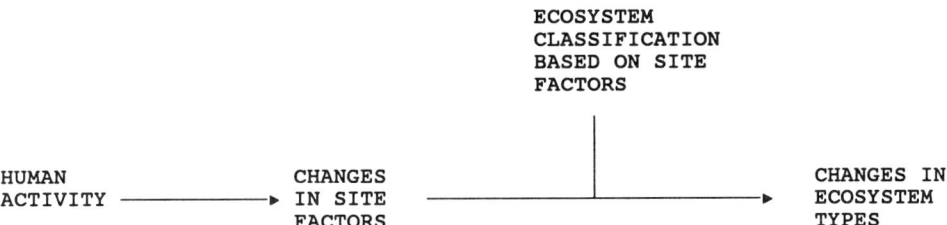

Figure 7.2 Use of an ecosystem classification based on abiotic factors in environmental impact assessment to predict changes in ecosystem types

Vegetation is the biotic component that is most directly affected by changes in the environment caused by human activities. Therefore, we started with the classification of terrestrial ecosystems in which macrophytes often form the dominant biotic component. Also, we limited the description of the biotic component of the ecosystem types to the vegetation (Stevers et al., 1987; Runhaar et al., 1987). More recently, we made a start with the description of terrestrial ecosystems in terms of soil fauna (Sinnige et al., 1991; 1992), as well as with the classification of aquatic ecosystems using waterplants and aquatic macro-evertebrates for the description of the types (Verdonschot et al., 1992).

This chapter first deals with the classification of terrestrial ecosystems and the description of these ecosystems in terms of floristic composition, i.e., the part that is furthest developed. Then we briefly discuss the use of the classification in environmental impact assessment, and we compare the classification with vegetation classifications. Finally, we shall discuss adding fauna to the description of the terrestrial ecosystem types, as well as the extension of the classification to aquatic ecosystems. We shall end with some conclusions on the use of site factors as classification characteristics for ecosystems.

Ecotope as basic unit

An ecosystem is a concept without dimension. An ecosystem can be as small as a decaying leave on the forest floor or as large as the earth. Both comply with the requirements that, first, they consist of biotic and abiotic components, and, second, that there are sufficient interactions between the components to justify distinguishing them as separate systems.

For a classification, however, we should be able to refer to tangible objects of a certain size. As the basic spatial unit we chose the ecotope. Troll (1950, 1986, 1970) defines an *ecotope* as the smallest part of a landscape that is homogeneous in edaphic factors and, to a certain degree, in biotic factors. His ecotope is equivalent to the *biogeocenosis* of Sukachev and Dylis (1968), to the *site* as used in Anglo-American literature[1] (Troll, 1968) and to the *'station écologique'* of Long (1974). In essence, an ecotope is an ecosystem, but an ecosystem of a certain size and homogeneity. Leser (1991) characterizes the difference between ecotope and ecosystem as that between a spatial and a functional concept (Figure 7.3).

The definition of Troll still leaves much room for differences in interpretation as to the scale of the 'smallest landscape unit'. For terrestrial situations dominated by vegetation, the definition of an ecotope has therefore been specified as follows (Stevers et al. 1987):

'A spatial unit that is homogenous in vegetation structure, stage of succession and in the dominant abiotic factors that determine the species composition of the vegetation'.

This resembles the definition of the *site type* of Hills (1976): 'a small area of land which is homogeneous with respect to all features effective in conditioning the development of natural vegetation on a local area'. However, a difference is that vegetation structure and stage of succession are not included in Hills's definition.

---

[1] The term ecotope is sometimes used in Anglo-American literature, but mostly with a different meaning. For example, Tansley (1939) and Daubenmire (1968) use it to indicate the abiotic part of the ecosystem, thus using it as an equivalent for the 'geotope' of Leser (Figure 7.3) or the 'physiotope' of Neef (1986). Whittaker et al. (1973) use the term ecotope in an abstract meaning referring to the multidimensional hyperspace which represents the full range of external circumstances to which species are subjected.

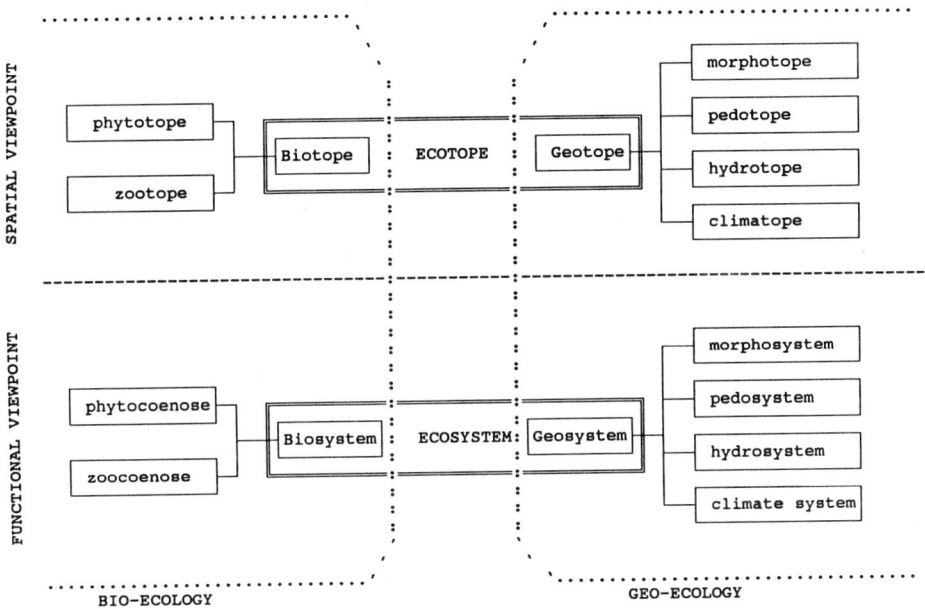

Figure 7.3 Ecotope and ecosystem as two complementary concepts used in respectively a spatial (top) and a functional (bottom) context. Adapted from Leser (1991)

Classification characteristics and definition of ecotope types

For the classification of ecotopes we use abiotic and biotic factors that determine the species composition of the ecosystem. In this case, since we have limited ourselves to the vegetation, we are interested in the species composition of the vegetation only.

In theory, many different types of factors can be used (Figure 7.4). We might, for example, use *physiological factors* that directly influence the plants: availability of light, carbon, oxygen, water, nitrogen, phosphorus, etc. Although these factors could be used profitably for modelling simple agricultural ecosystems, they are not very practical for the classification of natural ecosystems, because they are too variable in space and time and too difficult to measure. Instead, we have chosen to use what we call *operational site factors*: properties of the ecotope that largely determine the physiological conditions in which the plants live. Since there is much agreement in plant ecological literature (see e.g. Ellenberg, 1979; Etherington, 1982; Klapp, 1965; and Landolt, 1977) that moisture regime, nutrient availability, acidity,

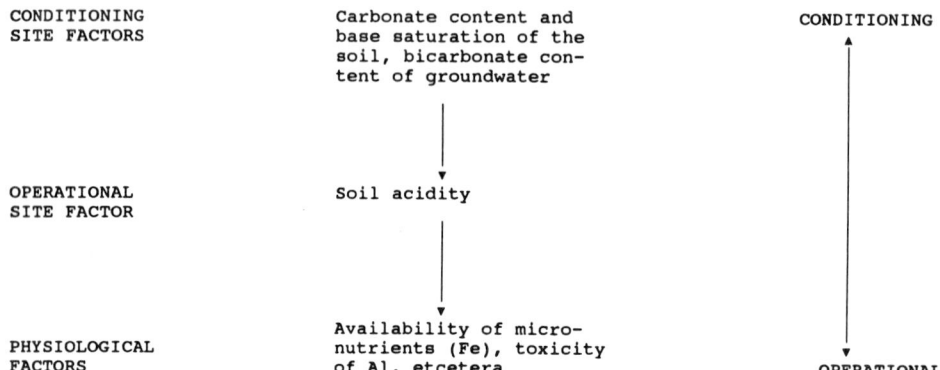

Figure 7.4 Relation between operational site factors used for the classification of ecotopes and physiological factors (bottom) on the one hand, and conditioning site factors (top) on the other, exemplified for acidity

salinity, and 'dynamics' are the most relevant abiotic site factors for plant life, we have chosen these factors as classification characteristics. Operational site factors,[2] in turn, are influenced by *conditioning site factors*, such as soil composition and hydrology. These can be used for ecosystem classifications at a higher scale level, for example, for the classification of ecoseries (Klijn et al., 1992; see also Klijn or Claessen et al., in chapters 5 and 10 of this book). However, for an ecotope classification such conditioning factors are not practical: the relation between conditioning factors and species composition is too indirect. An example is upward seepage, a factor which is an important conditioning factor for moisture regime, nutrient availability, and acidity (Grootjans, 1985). In some areas with poor, sandy soils, upward seepage with bicarbonate-rich groundwater is a prerequisite for mesotrophic conditions. In other areas mesotrophic conditions can be caused by a high buffering capacity of the soil or the influence of bicarbonate-rich surface water (Van Wirdum, 1991). Therefore, upward seepage is not a very suitable classification charac-

---

[2] We use the terms operational and conditioning in a relative sense to distinguish between more operational and more conditional site factors (see Figure 7.4). Thus, we deviate from Van Wirdum (1986) who limits the use of the term operational to physiological factors and calls all properties of the environment that influence these physiological factors 'conditional'.

teristic for a nationwide ecotope classification, because its influence on the species composition varies per region.

Vegetation structure has been chosen as the main biotic classification characteristic, although it is not a site factor. However, it can be considered as indicative for the operational factors time and vegetation management (see the appendix of this chapter).
Climate is an important factor that influences vegetation both directly, e.g. through frost, and indirectly, e.g. through soil development. However, it has not been used as a classification characteristic because in the Netherlands climatic differences are small. However, if the ecotope classification were to be used for a larger region, climate would be the first to be added as a classification characteristic.
The order in which species establish themselves and chorological relations with the surrounding environment are also important factors influencing the species composition. However, they are not very suitable as classification characteristics, as we shall discuss later.

We may, analogous to the state factor equation of Jenny (1941) for soils, formulate the following equation concerning the floristic species composition of an ecosystem:

$$S_{\text{pecies composition vegetation}} = f\,(cl, sal, moist, nutr, ac, dyn, man, est, chor, time)$$

in which:

cl	= climate	dyn	= dynamics
sal	= salinity	man	= vegetation management/grazing
moist	= moisture regime	est	= order of establishment
nutr	= nutrient availability	chor	= chorological relations
ac	= acidity	time	= time

For each classification characteristic several classes have been distinguished (Table 7.1 and appendix). Ecotope types are then constituted by combinations of these classes, such as ecotope type G22, which stands for 'Grassland on wet, weakly acid to neutral sites with low nutrient availability'.
Not all the theoretically possible combinations of classes have been distinguished as ecotope types. Some combinations are ecologically irrelevant. For example, in ecosystems that are very rich in nutrients, the influence of acidity

Table 7.1 Characteristics and classes used for the classification of terrestrial ecotopes

CHARACTERISTIC	CLASSES
moisture regime	wet, moist, dry
nutrient availability	low, moderate, and high
acidity	acid, weakly acid to neutral, alkaline
salinity	fresh, brackish, saline
dynamics	sand drift, perturbation
vegetation structure	pioneer, grassland, tall herbaceous, dwarf shrub, shrub, forest

is much less pronounced than in nutrient-poor ecosystems. Acidity is then hardly reflected by the species composition. Therefore, acidity has not been used as a classification characteristic in very nutrient-rich ecosystems. In addition, many combinations of classes do not occur in the Netherlands. In total, about 100 ecologically relevant combinations of biotic and abiotic site factors have been distinguished (Table 7.2).

Ecological species groups

To describe the ecotope types in terms of species composition we use *ecological species groups*. These comprise plant species characteristic for a certain ecotope type. Since many species occur in more than one ecotope type, the species have often been assigned to more than one species group. The original species groups were defined in collaboration with the National Herbarium of the Netherlands (Runhaar et al., 1987). Later additions were made by Dirkse and Kruijsen (1993), who added mosses and liverworts, and Van Raam and Maier (1993) who added Characeae. As an example, Table 7.3 lists one of the ecological species groups.

Table 7.2 Types distinguished in the ecotope classification for terrestrial ecosystems. Subtypes are not shown

	fresh						brackish	saline
	low nutrient availability			moderate nutrient availability		nutrient rich		
	acid	weakly acid	alkaline	acid to weakly acid / acidity not determined[*]	alkaline[*]			
Wet	P21 pioneer G21 grassland S21 shrub B21 forest	P22 pioneer G22 grassland S22 shrub B22 forest	P23 pioneer G23 grassland S23 shrub	P27 pioneer G27 grassland R27 tall herbs S27 shrub B27 forest		P28 pioneer G28 grassland R28 tall herbs S28 shrub B28 forest	bP20 pioneer bG20 grassland	zP20 pioneer zG20 grassland
Moist	P41 pioneer G41 grassland S41 shrub B41 forest	P42 pioneer G42 grassland S42 shrub B42 forest	P43 pioneer G43 grassland S43 shrub B43 forest	P47 pioneer G47 grassland R47 tall herbs S47 shrub B47 forest	P46 pioneer G46 grassland S46 shrub B46 forest	P48 pioneer G48 grassland R48 tall herbs S48 shrub B48 forest	bP40 pioneer bG40 grassland	
Dry	P61 pioneer G61 grassland S61 shrub B61 forest	P62 pioneer G62 grassland S62 shrub B62 forest	P63 pioneer G63 grassland S63 shrub B63 forest	P67 pioneer G67 grassland R67 tall herbs S67 shrub B67 forest		P68 pioneer G68 grassland R68 tall herbs S68 shrub B68 forest		

[*]) So far only ecotopes on moist sites with moderate nutrient availability have been subdivided into acid to weakly acid and alkaline types.

Table 7.3 The ecological species group of ecotope type P21, comprising species characteristic for pioneer vegetations on wet, nutrient-poor, acid sites. Only the most common mosses and liverworts are specified. In the second column, other ecotope types in which the species occurs are indicated. For codes see Table 7.2, V and W indicate aquatic ecotope types. Sources: Runhaar et al. (1987), mosses and liverworts according to Dirkse and Kruijsen (1993)

Species characteristic for ecotope type P21	Other ecotope types in which the species commonly occurs:
Drosera intermedia	
Drosera rotundifolia	G21, G22
Juncus bulbosus	P22, V11, V12
Lycopodium inundatum	
Rhynchospora alba	
Rhynchospora fusca	
Mosses:	
Calypogeia fissa	G21, G22, G41, H41
Cephalozia bicuspidata	P41, G41, H22, H41
Dicranella cerviculata	P41, G21, G41
Drepanocladus fluitans	G21, G22, H21, W11
Gymnocolea inflata	G21, G41, H41
Sphagnum compactum	G21, G41
Sphagnum molle	G21, G41

The initial assignment of species to ecological species groups is based primarily on national and international literature regarding, for example, indicator values of plant species for abiotic factors (Klapp, 1965; Ellenberg, 1979; Clausman et al., 1987). This information was often contradictory and incomplete; not all the species occurring in the Netherlands were mentioned, and there was little or no information on the ecological tolerance of species. Therefore, expert judgement was used to complete the initial assignment of species to ecological species groups.

As a second step the consistency of the groups was tested using approximately 50,000 relevés from all over the Netherlands. The relevés were used to check

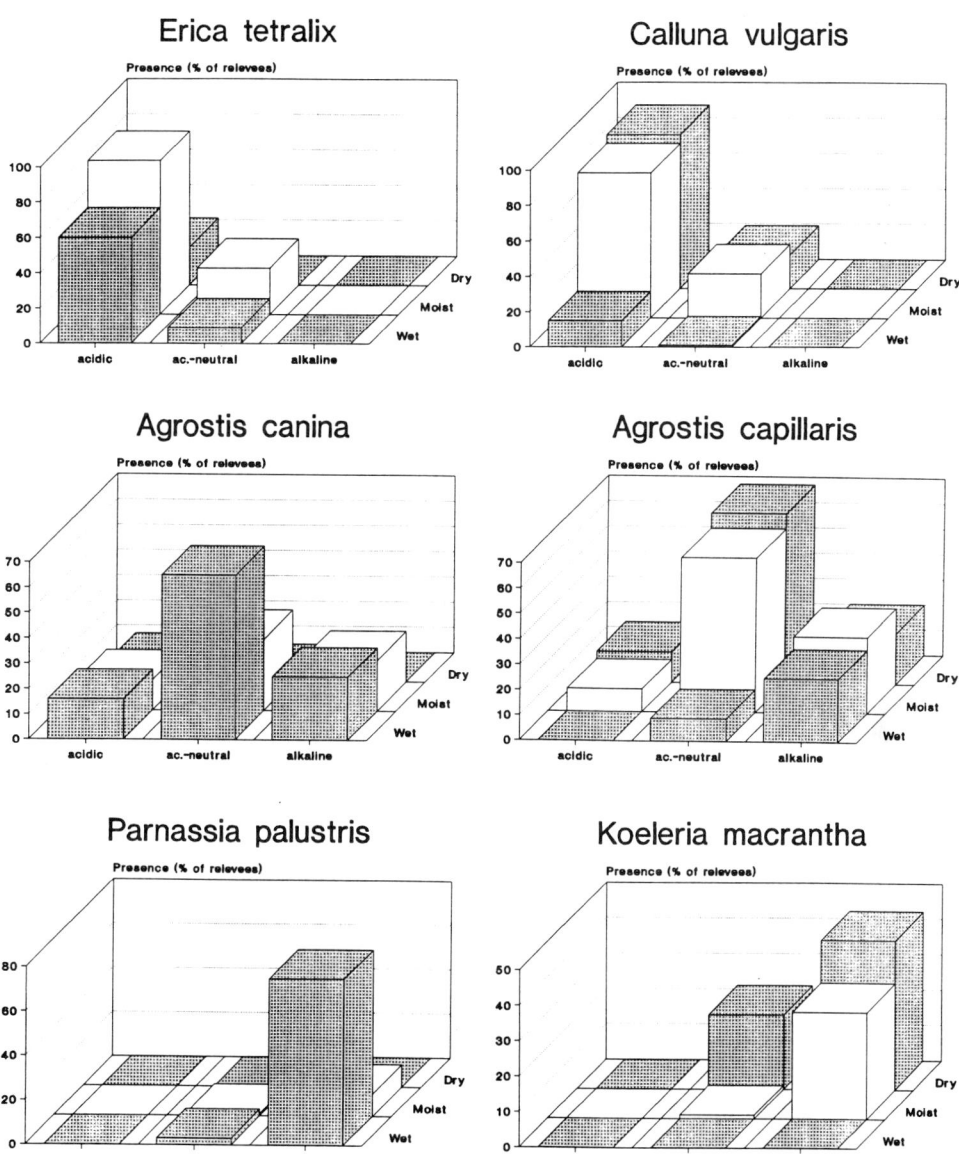

Figure 7.5 Frequency distribution of species in relation to moisture regime and acidity. Only the distribution of species in herbaceous vegetations on nutrient-poor, non-saline sites is shown. The axes for nutrient availability, salinity, and vegetation structure are left out

Table 7.4 Frequency distribution of species in relevés belonging to ecotope types of herbaceous vegetation on wet to dry, acid to alkaline, nutrient-poor sites. The ecotope type has been determined by means of the species composition. This kind of information has been used to check the internal consistency of the species groups. %fr=frequency (% of the sites); %ab=abundance (mean cover on the sites)

Moisture regime	WET			MOIST			DRY		
Acidity	acid	acid-neutral	alkaline	acid	acid-neutral	alkaline	acid	acid-neutral	alkaline
Ecotope types	P21,G21	P22,G22	P23,G23	P41,G41	P42,G42	P43,G43	P61,G61	P62,G62	P63,G63
Nr. of relevees	N= 225	N= 254	N= 4	N= 222	N= 80	N= 10	N= 203	N= 203	N= 105
Frequency/abund.	%fr %ab	%fr %ab	%fr %ab	%fr %ab	%fr %ab	%fr %ab	%fr %ab	%fr %ab	%fr %ab
Aceras anthrop.	. .	. .	. .	. .	. .	10. 2.	. .	. .	. .
Agrostis cap.	16. 4.	9. 6.	25. 1.	9. 3.	61. 12.	30. 1.	12. 3.	67. 11.	20. 3.
Agrostis canina	0. 3.	65. 11.	25. 3.	12. 3.	28. 5.	20. 2.	7. 3.	3. 7.	6. 2.
Agrostis vinealis	. .	. .	. .	2. 4.	5. 8.	. .	9. 7.	26. 10.	. .
Aira car.	. .	. .	. .	. .	1. 0.	. .	. .	2. 1.	. .
Aira praecox	. .	0. 2.	. .	. .	9. 2.	. .	4. 1.	41. 2.	51. 1.
Anagallis minima	. .	. .	. .	. .	3. 2.	. .	. .	. .	. .
Anagallis tenella	. .	2. 5.	25. 19.	. .	. .	. .	. .	1. 0.	5. 0.
Anchusa arvensis	. .	. .	. .	. .	. .	. .	. .	. .	5. 4.
Anchusa off.	4. 10.	. .	. .	. .	. .	1. 1.	. .	. .	3. 2.
Andromeda pol.	8. 1.	44. 3.	. .	7. 3.	71. 9.	40. 1.	4. 2.	27. 11.	3. 2.
Anthoxanthum od.	. .	. .	. .	. .	. .	10. 5.	. .	. .	2. 5.
Anthyllis vuln.	. .	. .	. .	. .	. .	. .	. .	2. 1.	. .
Apahanis inexp.	. .	. .	25. 19.	. .	. .	10. 3.	. .	. .	. .
Arabis h-h	. .	. .	. .	. .	. .	. .	. .	. .	0. 0.
Arnica montana	30. 2.	31. 3.	. .	2. 1.	3. 1.	. .	2. 0.	. .	2. 0.
Apsparagus o-o	. .	. .	. .	. .	. .	30. 4.	. .	. .	. .
Aulacomnium pal.	. .	. .	. .	. .	4. 1.	50. 2.	. .	. .	. .
Avenula pratensis	. .	. .	. .	1. .	1. 1.	10. 2.	. .	2. 1.	7. 4.
Avenula pubescens	. .	. .	. .	. .	. .	70. 41.	. .	. .	. .
Botrychium lunaria	. .	. .	. .	. .	. .	70. 2.	. .	. .	. .
Brachypodium pin.	. .	. .	. .	. .	. .	10. 0.	. .	. .	. .
Brza media	. .	2. 2.	. .	. .	11. 1.	10. 5.	. .	1. 1.	. .
Bromus erectus	. .	. .	. .	. .	. .	. .	. .	. .	. .
Bromus tectorum	. .	. .	. .	. .	. .	. .	. .	. .	. .
xCalammophila b.	. .	. .	. .	. .	3. 1.	. .	. .	1. 3.	5. 12.
Calamagrostis can.	4. 10.	32. 10.	100. 19.	8. 5.	15. 24.	10. 3.	4. 7.	23. 24.	43. 4.
Calamagrostis ep.	. .	4. 19.	. .	. .	. .	. .	. .	. .	. .

whether species attributed to a certain species group actually occur in combination with other species of the same group. To this end, the draft species groups were used to assign relevés to ecotope types. Next, the frequency distribution of the species in relevés assigned to different ecotope types was examined (Table 7.4 and Figure 7.5).

Since the ecotope types were determined on the basis of species composition, this type of information can only be used to a limited extent. It cannot be used to test the indicative value of species or species groups, as this would result in circular reasoning. The latter can only be based on direct measurements of the abiotic factors involved. However, this kind of information can be used to test the consistency of the species groups, and it gives an indication about species of which little or no information was found in the literature. In addition, the number of ecotope types in which a species occurs gives valuable information on the ecological tolerance of species.

The method we used is similar to the reciprocal averaging method that Latour et al. (chapter 9 of this book) used to calculate the indicator values of species in the Netherlands, starting with the indicator values of Ellenberg. It may, however, be clear from the foregoing that we have our doubts concerning the use of this type of data as the only source of information to determine the indicative value of plant species.

The ecological species groups indicate which species can be expected in an ecotope belonging to a certain type. This means that they can also be used in an opposite way, i.e. as a mapping characteristic by using the species composition of the vegetation to establish the ecotope type.[3] This is a much faster way of determining which ecotope type is encountered than by measuring all the relevant site factors. Several computer programmes have been developed for this purpose, using vegetation descriptions (relevés) or floristic data (species lists) as input. In several studies (Canters et al., 1991; Witte and Van der Meijden, 1990; Plate, 1990) this method was used to establish the distribution of ecotope types in the Netherlands (Figure 7.6). Groen et al. (chapter 13 of this book) treat this procedure in more detail.

---

[3] To this end, it is assumed that the species composition is in equilibrium with the prevailing site conditions. If the site conditions have recently changed, the species composition may no longer be indicative for the ecotope type.

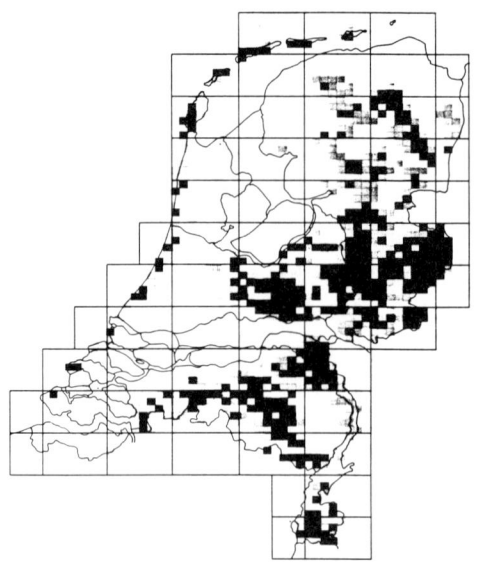

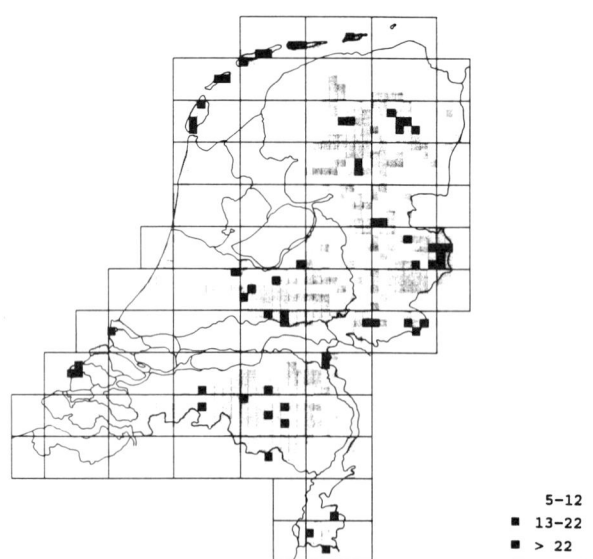

Figure 7.6 Number of species belonging to the ecological species groups characteristic for herbaceous vegetations on wet, weakly acid, nutrient-poor sites, before and after 1950, indicating the former and present occurrence of the corresponding ecosystem types. These ecosystem types have declined strongly in area and species richness since they are susceptible to lowering of the groundwater table, acidification, and eutrophication. In reality the decline is even larger, because since 1950 the impact of these hazards only increased (source: Van der Meijden et al., 1989)

Testing the classification

The classification of ecotopes in combination with the description of types by ecological species groups can be regarded as a set of hypotheses concerning the relation between species composition and site factors (Figure 7.7).

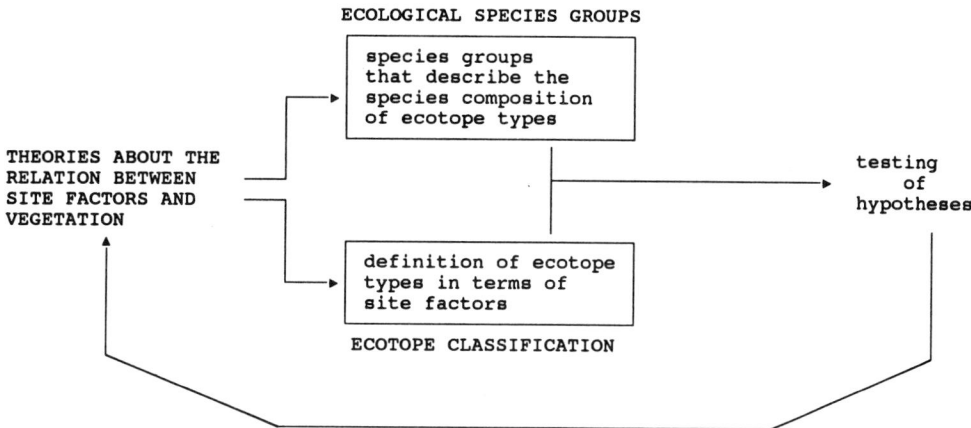

Figure 7.7 The ecotope classification and the corresponding species groups as a set of hypotheses

These hypotheses can be tested by comparing the species composition at certain locations with the measured values of the site factors used for the classification. Differences between expected and measured values of site factors, or between expected and found species composition, necessitate restating the hypotheses by adjusting either the species groups or the site factor classes. This means that the development of the classification is an *iterative* process, in which the cycle of Figure 7.7 is repeated several times, each cycle resulting in improved and further quantified relations or, in Poore's (1962) words, in a *successive approximation*. This can best be explained with an example.

An example: the classification to moisture regime

For the classification characteristic *moisture regime* three classes were defined: 'wet', 'moist', and 'dry'. Wet ecotope types are characterized by high groundwater levels, resulting in prolonged periods of low oxygen availability in the soil. Dry ecotope types are characterized by prolonged periods of low moisture availability (pF >4.2). In moist ecotope types water and oxygen are generally available throughout the growing season.

The moisture classes were first described qualitatively as above. Then, we described them in terms of species composition by assigning plant species to ecological species groups of wet, moist, and dry ecotope types. This assignment was based on the species' indicative value for moisture regime, for which we used the relations specified in Table 7.5.

Expert judgement and relevés were used as additional sources of information for assigning species to species groups.

Table 7.5 Relation between moisture classes and indicator values according to Ellenberg (1979), Klapp (1965), Clausman et al. (1987), and Londo (1975), used for the initial assignment of plant species to ecological species groups

	DRY	MOIST	WET
Ellenberg	1-4	4-6	7-10
Klapp	1-3	3-6	7-9
Clausman et al.	0-2	2-7	7-10
Londo	-	f, a, (f)	W, F

The next step was to check whether the ecotope types, as determined on the basis of the species composition of the vegetation, corresponded with measured differences in moisture regime, and to find out which abiotic parameters could best be used to describe differences in moisture regime. To this end, 202 relevés were made in 19 nature conservation areas in the Netherlands, where the groundwater regime had been monitored for at least five years (Runhaar, 1989a). Measurements with piezometers at or near the relevé sites were used to characterize the groundwater regime.

We assumed that high groundwater levels in spring, resulting in low oxygen availability in the beginning of the growing season, would be most differentiating between wet and moist ecotopes. We also assumed that in sandy soils low groundwater levels later in the growing season would be the most discriminating factor for the difference between moist and dry ecotopes.

Analysis of the data, however, revealed that spring groundwater levels, i.e., the average groundwater level in March-April, did not only correspond best with differences between sites classified as wet and moist, but also between those classified as moist and dry. This suggests that in natural vegetations the moisture supply during the first part of the growing season is more important for the competition between species than the moisture supply towards the end of the growing season. The latter is known to be important in agricultural ecosystems, because it largely determines crop production. However, it seems to have little influence on the competition between species, and consequently on the species composition in natural ecosystems.

The data enabled us to establish a quantified relation between moisture classes on the one hand, and soil texture and groundwater levels on the other (Figure 7.8 and appendix). The relation with other site factors has been tested in a similar way.

Use in environmental impact assessment

When assessing the impact of human intervention on vegetation, ideally we first predict changes in the site factors. The ecotope classification and the corresponding species groups can then be used to determine the resulting changes in ecotope type and species composition (Figure 7.2). This approach has been used in a recent study on the effects of water management measures on natural vegetations in the Netherlands. In this study, the biotic response is based on formally predicted changes in abiotic site conditions (Van der Linden et al., 1992). On the basis of the empirical relations between operational site factors and ecological species groups we could assess how ecotopes belonging to a certain type will respond to hydrological changes (see also Claessen et al. in chapter 10 of this book).

However, such a deterministic approach is not always possible. The classes used in the definition of ecotope types are not always defined unequivocally in terms of abiotic characteristics (see, for example, the appendix for the definition of nutrient availability classes). Even when they are defined quantitatively, it is not always possible to predict changes in abiotic conditions with enough accuracy.
Another approach can then be followed, in which some effects are predicted deterministically, while others are predicted on the basis of correlations. Such

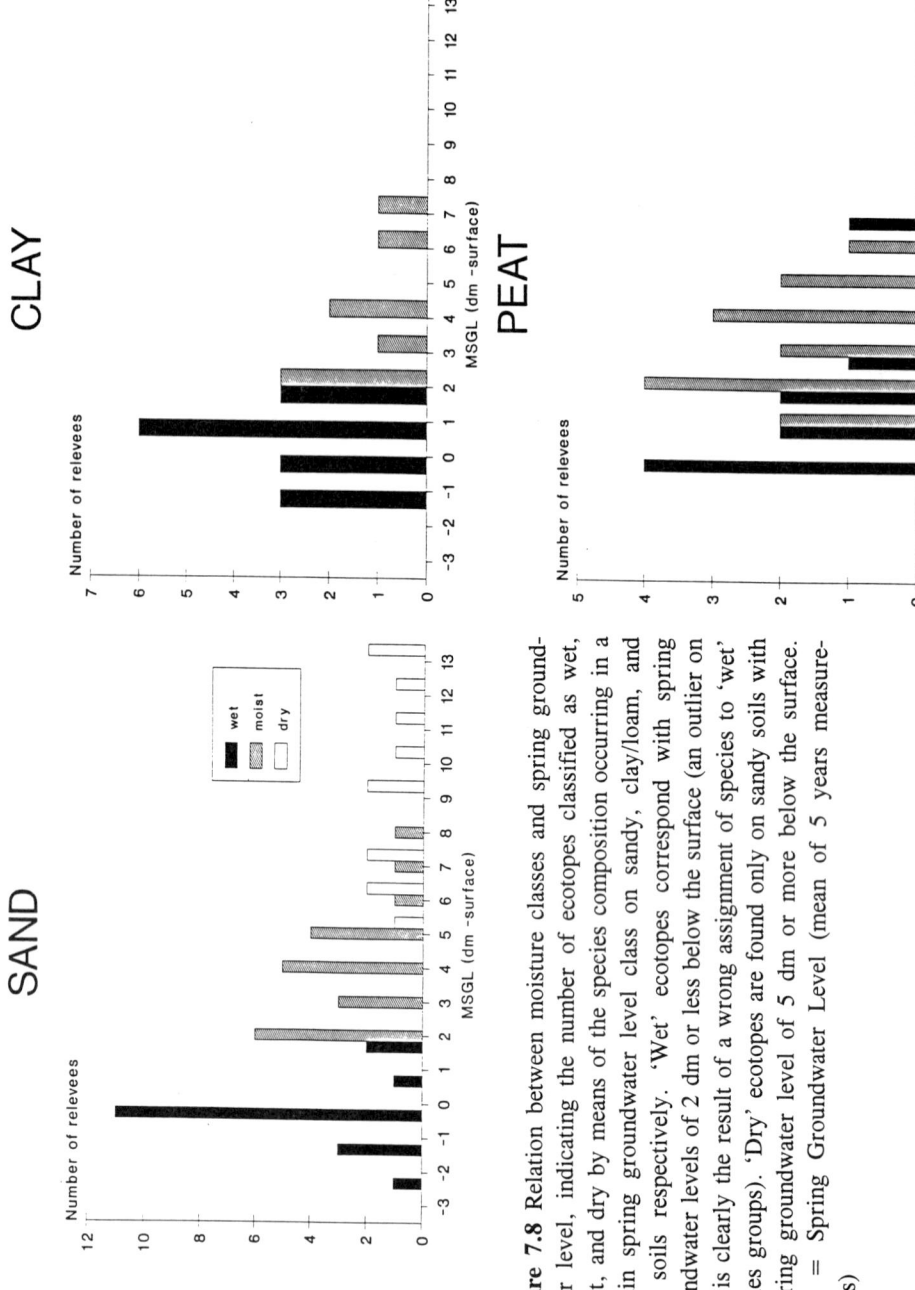

Figure 7.8 Relation between moisture classes and spring groundwater level, indicating the number of ecotopes classified as wet, moist, and dry by means of the species composition occurring in a certain spring groundwater level class on sandy, clay/loam, and peat soils respectively. 'Wet' ecotopes correspond with spring groundwater levels of 2 dm or less below the surface (an outlier on peat is clearly the result of a wrong assignment of species to 'wet' species groups). 'Dry' ecotopes are found only on sandy soils with a spring groundwater level of 5 dm or more below the surface. SGL = Spring Groundwater Level (mean of 5 years measurements)

an approach was used for the environmental impact assessment study on the island of Texel (Runhaar, 1989b), where the effects of coastal erosion and changes in dune management were predicted. In this study, the effects of a lowering of the groundwater table due to coastal erosion were predicted deterministically with a groundwater model and quantitative relations between groundwater levels and ecotope types. But, to predict the effects of increased sand drift we used a correlative approach by making a comparison with similar coastal areas that had been subjected to coastal erosion in the past. In such a way effects can be predicted when data or knowledge is insufficient for a completely deterministic approach.

So far, the ecotope classification was used 1) in a regional study to assess the effects of groundwater retrieval in dune and polder areas in the west of the Netherlands (Udo de Haes, 1987); 2) in a regional study on the effects of coastal erosion on the dune area of Texel (Stevers et al., 1984; Runhaar, 1989b); and 3) in national studies on the effects of surface water and ground-

Table 7.6 Use of the ecotope classification in environmental impact studies

STUDY	ACTIVITY	STUDY AREA	IMPACTS	SCALE
IODZH 1980	groundwater extraction and infiltration of water	dune- and polder area of South-Holland	changes in groundwater level, infiltration, soil perturbation, vegetation management	1:5,000
KUSTEX 1983	coastal erosion, groundwater extraction	dune area of Texel	erosion, sand drift, changes in groundwater level, vegetation management	1:5,000
PAWN 1990	surface water management	the Netherlands	changes in groundwater level and inlet of water	1:500,000
MER-DIV 1992	public water supply by groundwater extraction	the Netherlands	changes in groundwater level and seepage	1:500,000

water management on natural vegetations in the Netherlands (Claessen and Witte, 1991; Witte et al., 1992; Claessen et al., chapter 10 of this book).

Table 7.6 gives an indication of the scale of the studies and the types of impacts studied. It shows that the ecotope classification can be used for areas of different size and for the prediction of different types of impacts.

Comparison with vegetation classifications

Conceptually, the ecotope classification differs fundamentally from vegetation classifications, since the object of classification is different: ecosystems, including abiotic components, vegetation, and fauna. In vegetation classifications only one component of the ecosystem is classified: the vegetation.

In practice, the differences are less pronounced than one might expect from this conceptual difference. This can partly be explained by the fact that until recently the ecotope classification focused on vegetation, and that only those abiotic site factors were used as classification characteristics, which are reflected in the plant species composition.
Moreover, 'many vegetation scientists will readily agree that the basic units they are classifying are really ecosystems' (Daubenmire, 1968, p. 29), and relations with site factors often implicitly or explicitly play an important role in the definition of vegetation types. Consequently, the distinction between vegetation classifications and ecosystem classifications is not always clear.
The vegetation classification of Tansley (1939) for the British Isles, and the ecosystem classification of Doing (1974) for dune vegetations in the Netherlands, are examples of classifications in which both floristic and edaphic factors are used to define the types. In this respect they are intermediate between our ecotope classification and a purely phytosociological approach.

Another difference with vegetation classifications is found in the method used to define the types (Figure 7.9). In the ecotope classification, we use a limited set of classification characteristics and pre-defined classes. Any ecotope can be placed in this framework. The addition of new data does not lead to changes in the classification, although new insights may.
In contrast, most vegetation classifications delimit the types less strictly, because they are established by clustering relevés based on similarities in species composition. The types are usually described in terms of an average or

'ideal' species composition of clusters (Westhoff and Van der Maarel, 1973, p.630). Because the vegetation types can be regarded as nodes in a multi-dimensional continuum, relevés that are intermediate between types cannot always be classified. Also, the addition of new data may lead to changes in the classification, because the number and size of clusters depend on the data set used.

This difference can best be described as the difference between classification according to a deductive and an inductive approach (cf. Udo de Haes and Klijn in the first chapter of this book).

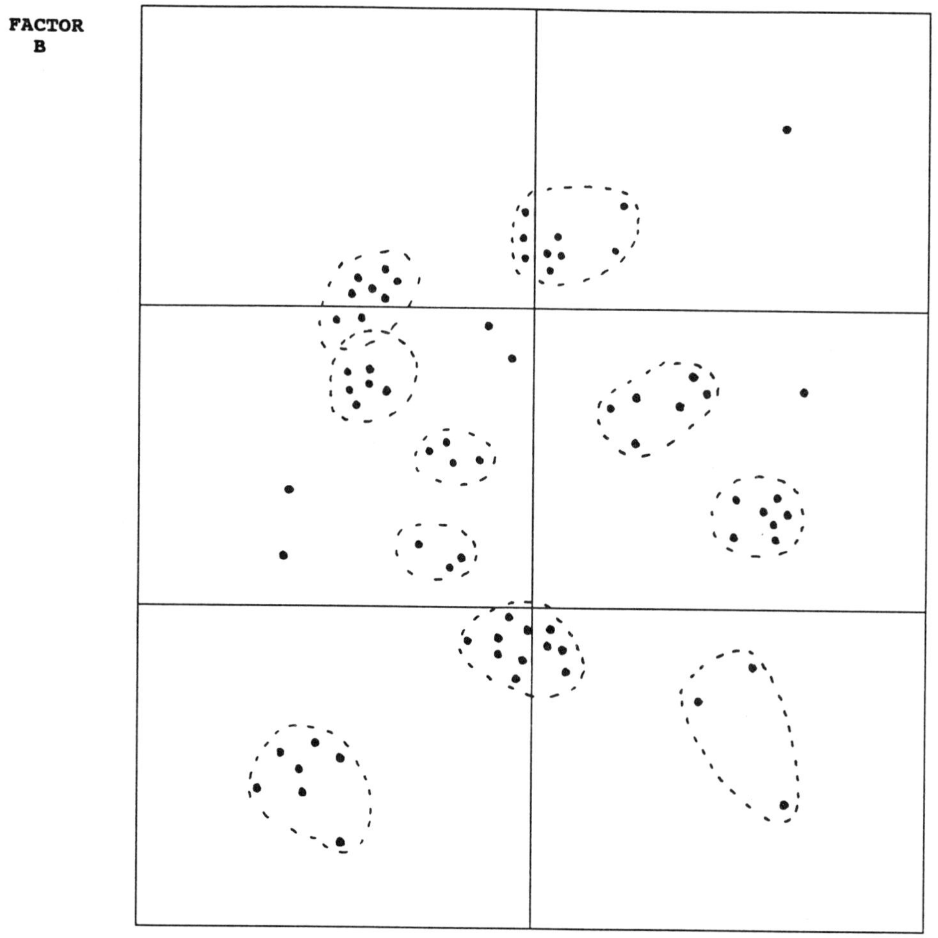

Figure 7.9 Comparison of an ecotope classification with fixed classes related to abiotic factors, in this case factors a and b, with a vegetation classification based on clustering relevés with similar species composition

As to the two approaches to classification we agree with Whittaker (1973), who states that: 'No one approach to classification can claim exclusive merit. Different approaches, based on different characteristics of communities, may reveal different relationships of interest'. As shown in the previous section, an ecosystem classification on the basis of operational site factors can be very useful for environmental impact assessment, because it directly links species composition to site factors. However, vegetation classifications appear better suited for descriptive purposes, because differences in species composition can be described in more detail.

Increasing the detail of the ecotope classification

An objection that is sometimes raised against the ecotope classification, is that the units are quite heterogeneous in floristic composition and are difficult to interpret in terms of vegetation types (Grootjans et al., 1987). Indeed, the classification is less detailed than phytosociological classifications. For the Netherlands, only approximately 100 terrestrial ecotope types are being distinguished, against approximately 200 associations in the phytosociological approach (Westhoff and Den Held, 1969).

The comparability with phytosociological classification can be enhanced by increasing the detail of the ecotope classification. By using more site factors and by using more classes for each site factor we can increase the floristic homogeneity of the ecotope types. It is, however, unlikely that we will ever be able to explain the species composition of vegetations completely in terms of site factors.

In the classification of ecotopes we started with a limited set of factors, which determine the main variation in the vegetation. To explain the remaining variation we have to use an increasingly complex combination of factors. We may then think of factors such as:

- *vertical heterogeneity:* In most sites there is some vertical heterogeneity connected with stratification of the substrate, for example, a gradient in pH caused by leaching of the topsoil. A site with pronounced stratification will differ in species composition from a more homogeneous site, even if the average value of the site factor in the rooted topsoil is the same.

- *horizontal heterogeneity:* An ecotope is assumed to be spatially homogeneous with respect to abiotic factors. In reality, however, there is always some internal spatial variability. For some ecosystems this heterogeneity is even characteristic, as could be said, for example, for the hummocks and hollows in a well-developed bog.

- *temporal variability:* In most cases the classification of ecotopes is based on the average value of the site factors. Temporal variability is disregarded. However, there are large differences in species composition between ecotopes characterized by stable conditions and ecotopes characterized by large fluctuations in site factors.

- *history:* The history of a site determines which species were present at a certain location, either as living plants or in a seed bank. This can be decisive for the competition between species.

- *chorological relations:* The surroundings determine which plant and animal species can invade an ecotope, the grazing pressure, etc.

These are complex factors that cannot be incorporated in the present ecotope classification very easily. However, most of these factors are directly or indirectly related to the soil types and the geographic region in which an ecotope occurs.

For example, the Netherlands' coastal dune areas are characterized by very dynamic conditions caused by salt spray, sand drift, groundwater fluctuations, and intensive grazing by rabbits. In contrast, ombrotrophic peat bogs are characterized by very stable conditions, with very small groundwater fluctuations, little input of salts and nutrients, little grazing, etc.

By using geographic regions and soil type as additional characteristics, we get units that are more homogeneous with respect to biotic and abiotic processes, and consequently in species composition. As an example, the ecotope type G21, 'Low herbaceous vegetations on wet, nutrient-poor, acid sites', can thus be further subdivided (Table 7.7). Indeed, the resulting units can be interpreted in terms of vegetation types more easily. Based on another example, Van Beusekom (1993) also found a better correspondence by differentiating according to the geographical region.

Table 7.7 A subdivision of ecotope type G21 (Low herbaceous vegetations on wet, nutrient-poor, acidic soil) according to ecoregion and dominant soil type. The corresponding vegetation types *sensu* the Braun-Blanquet system are indicated, as well as the differentiating species. For example *Carex trinervis* can be found in G21 only in the coastal dunes.
D = Coastal dune region; P = Pleistocene (sandy) region; H = Holocene (clay/ lowland peat) region (after Klijn, chapter 5 of this book)

ECORE-GION	SOIL TYPE	ASSOCIATION	DIFFERENTIATING SPECIES
D	non-calcaric sand (hydrovague soils)	*Empetro-Ericetum*	*Oxycoccus macrocarpos, Carex trinervis*
P	non-calcaric sand (hydropodzol)	*Ericetum tetralicis*	*Sphagnum compactum, Gentiana pneumonanthe, Juncus squarrosus*
	oligotrophic peat (raw peat soil)	*Erico-Sphagnetum magellanici, Empetro-Sphagnetum rubellii*	*Eriophorum vaginatum, Andromeda polifolia*
H	mesotrophic peat with rainwater lens (raw peat soil)	*Sphagnetum palustri-papillosi*	*Pallavicinia lyellii, Dryopteris cristata, Vaccinium vitis-idaea*

Incorporating fauna in the description of ecotope types

So far, we have focused on vegetation. However, it is our goal to extend the classification to all ecosystems at the ecotope scale level, whether or not they are dominated by vascular plants. A problem is, of course, which fauna species to use for the description of the ecotope types. For most plant species, ecotopes form the appropriate spatial scale level to explain the distribution pattern. Fauna species, however, may have large home ranges, often comprising many ecotopes (Opdam, 1984). Therefore, the use of fauna species in the description of ecotope types can best be limited to species with a low mobility.

Sinnige et al. (1991; 1992) made a start with the description of terrestrial ecotopes in terms of soil macrofauna. They concluded that many of the factors used for the present classification focusing on vegetation are also relevant for soil macrofauna.
They proposed using the litter composition, which controls the development of humus form, as an additional classification characteristic. The litter composi-

tion correlated with site factors already used for the classification, such as moisture regime, nutrient availablity, and acidity for a large part. However, litter composition is primarily determined by the dominant plant species. Especially in artificial ecosystems, such as afforestations, the litter produced may vary considerably from the litter expected under the given abiotic site conditions.

Whether or not soil texture should also be used as an additional classification characteristic is still uncertain. It is generally assumed that soil texture is a very important factor for soil fauna. However, many differences that are attributed to differences in soil texture can also be explained by differences in moisture regime and nutrient availability, which result from differences in soil texture.

In aquatic ecosystems, macrophytes often play a minor role. In many waters, macrophytes are even absent or represented by only a few species. A purely phytosociological approach to the classification of aquatic ecosystems leads to unsatisfactory results with many 'associations' characterized by very few plant species, sometimes by only one.

Therefore, the aquatic ecotope classification developed by Verdonschot et al. (1992) was based on both macrophytes and macro-evertebrates. The classification characteristics largely correspond with those for terrestrial ecotopes (Table 7.8), although there are differences in the classes distinguished.

Table 7.8 Characteristics and classes used for the classification of aquatic ecosystems (Verdonschot et al., 1992)

CHARACTERISTIC	CLASSES
depth	shallow, deep
nutrient availability	low, moderate, high
acidity	acid, weakly acid, neutral to alkaline
salinity	fresh, moderately brackish, strongly brackish, saline
dynamics	stagnant small, stagnant large (wave activity), slow streaming, swift streaming

For aquatic ecosystems, the differences with other classifications are much less pronounced than for terrestrial ecosystems, where the difference between the ecotope classification and phytosociological classifications is evident. In fact, for the aquatic ecotope classification we used prior classifications of Verdonschot (1990) and Torenbeek (1988).

It seems that in aquatic ecology ecosystems are more commonly defined directly in terms of abiotic factors (cf. Hawkes, 1975, on the classification of rivers; or Higler, 1988). Depth, stream velocity, acidity, salinity, and nutrient availability/trophic state are frequently used classification characteristics.

The difference between the aquatic ecotope classification and other aquatic ecosystem classifications is not a conceptual one, but rather stems from different approaches. The ecotope classification is a typical example of a deductive approach, in which types are defined on the basis of existing knowledge about the relation between site factors and species composition. In contrast, most other aquatic ecosystem classifications are the result of an inductive approach, in which clusters are first distinguished on the basis of similarities in species composition, with subsequent re-interpretation in terms of site factors.

Conclusion and discussion

An ecosystem classification with abiotic and biotic site factors as classification characteristics can be very useful for environmental impact assessment, both on regional and national scale levels. Since the types are defined in terms of site factors, changes in the abiotic environment can be directly translated into changes in ecosystem types.

An additional advantage is that the classification, once established, is easily understood by non-biologists. Despite the fact that developing an ecotope classification such as the one discussed requires detailed knowledge on plant and/or animal species, applying such a classification does not. This is not only an advantage for presenting the results of study to policymakers and politicians, but it also means that the classification can be easily used by other specialists involved in environmental impact assessment studies, such as hydrologists (for example, Kloosterman, 1993) or pedologists.

Apart from these practical considerations, an ecosystem classification such as the one presented contributes to integrating knowledge of specialists who work on different taxonomic groups — flora and various fauna groups — as well as on different components of the ecoystem — ecologists, pedologists, hydrologists, etc.

Although our explanation of the variation in species composition by means of site factors may appear reductionistic, our goal is truly holistic: the classification and description of ecosystems on the basis of all relevant properties, both biotic and abiotic. So far, only a modest start in this integration of knowledge has been achieved. The classification and description of the terrestrial ecotope types is still largely focused on one biotic aspect, i.e., vegetation, whereas functional aspects, such as nutrient cycling or foodweb structure, have as yet hardly been taken into account.

As for classifying ecosystems, it is interesting to see that in aquatic ecology site factors are more commonly used than in terrestrial ecology, in which — at least in Europe — the phytosociological approach is dominant.

This difference in tradition probably follows from the fact that in aquatic ecosystems site factors can be easily measured, whereas establishing the species composition is often a time-consuming job because of the many taxonomic groups involved. It is, therefore, often more practical to define ecosystem types in terms of abiotic factors.

In terrestrial systems it is just the other way around. It is easy to describe ecosystems in terms of plant species composition, while the site factors that determine the species composition are often difficult to measure because of spatial and temporal variability. This probably explains why the definition and description of terrestrial ecosystems is often limited to the plant species composition only.

Classifications on the basis of species composition as the only criterion, such as phytosociological classifications, can be useful for descriptive purposes. However, they give little insight to the relation with other components of the ecosystem. Therefore, we think that more attention should be given to the factors that control the observed differences in species composition.

In this respect we fully agree with Strahler (1975, p.243), who states: 'A fundamental principle of scientific classification is that the setting up of classes is better done according to the causes of the class differences than according to the effects that differences produce'.

Acknowledgements

We thank Professor Ies Zonneveld for his valuable comments on earlier drafts of this contribution, and for the stimulating discussion on vegetation versus ecosystem as objects of study.

References

Boeker, P., 1954. Bodenreaktion, Nährstoffversorgung und Erträge von Pflanzengesellschaften des Rheinlandes. *Bodenkunde und Pflanzenernährung* 66: 54-64.
Canters, K.J., C.P. Den Herder, A.A. De Veer, P.W.M. Veelenturf and R. De Waal 1991. Landscape-ecological mapping of the Netherlands. *Landscape Ecology* 5: 145-162.
Claessen, F.A.M. and J.P.M. Witte 1991. National water management strategies for conservation and recovery of terrestrial ecosystems. In: O. Ravera (ed.), *Terrestrial and aquatic ecosystems. Perturbation and recovery.* Ellis Horwood, New York.
Clausman, P.H.M.A., A.J. Den Held, L.M. Jalink and J. Runhaar 1987. *Het vegetatieonderzoek van de provincie Zuid-holland, deelrapport 2: Milieu-indicatie van vegetaties (Toewijs)*. Prov. Zuid-Holland, The Hague.
Clymo, R.S., 1962. An experimental approach to part of the calcicole problem. *J. Ecol.* 50: 707-731.
Daubenmire, R., 1968. *Plant communities. A textbook of plant synecology.* Harper & Row, New York.
Dirkse, G.M. and B.W.J.M. Kruijsen 1993. Classification into ecological groups of the mosses and liverworts of the Netherlands. *Gorteria* 10: 1-29. (In Dutch, with English summary)
Doing, H., 1974. *Landschapsecologie van de duinstreek tussen Wassenaar en IJmuiden.* Mededelingen Landbouwhogeschool Wageningen 74-12.
Ellenberg, H., 1979. *Zeigerwerte der Gefässpflanzen Mitteleuropas.* Scripta Geobotanica IX, Göttingen, 2e Auflage.
Etherington, J.R., 1982. *Environment and plant ecology.* John Wiley & Sons.
Grime, J.P. and J.G. Hodgson 1968. An investigation of the ecological significance of lime-chlorosis by means of large-scale comparative experiments. In: Rorison I.H. (ed.). *Ecological aspects of the mineral nutrition of plants.* Blackwell, Oxford.
Grootjans, A.P., 1985. *Changes of groundwater regime in wet meadows.* Diss., University of Groningen, the Netherlands.
Grootjans, A.P., Everts F.H. and N.P.J. de Vries 1987. Ecosysteemtypologie: bruikbaar, maar met mate. *Landschap* 4: 151-153.
Hawkes, H.A., 1975. River zonation and classification. In: Whitton B.A. (ed), *River Ecology.* Studies in Ecology 2: 312-374. Univ. Calif. Press.
Higler, L.W.G., 1988. *A worldwide surface water classification system.* Unesco, Parijs.

Hills, G.A., 1976. An integrated iterative holistic approach to ecosystem classification. In: J. Thie and G. Ironside (eds.), *Ecological (biophysical) land classification in Canada*. Proceedings of the first meeting of the Canada Committee on ecological (biophysical) land classification, pp. 73-97.

Jenny, H., 1941. *Factors of soil formation. A system of quantitative Pedology*. McGraw-Hill, New York & London.

Klapp, E., 1965. *Grünlandvegetationen und Standort*. Verlag Paul Parey, Berlin & Hamburg.

Klijn, F., A. ter Harmsel and C.L.G. Groen 1992. *Naar een ecoserieclassificatie ten behoeve van het ecohydrologische voorspellingsmodel DEMNAT-2. Onderzoek effecten grondwaterwinning*. CML report 85, Leiden/ RIVM, Bilthoven.

Kloosterman, F.H., 1993. *De landelijke hydrologische systeemanalyse. Deelrapport 2, Deelgebied Midden-Nederland*. Report OS 93-41, TNO-IGG, Delft.

Knauer, N., 1972. Beitrag zur Standortscharakteristik verschiedener Grünland-Pflanzengesellschaften. *Vegetatio* 25: 289-309.

Landolt, E., 1977. *Ökologische Zeigerwerte zur Schweizer Flora*. Veröff. des Geobot. Inst. Eidg. Techn. Hochschule st. Rübel, Zürich, nr. 64.

Leser, H., 1991. *Landschaftsökologie*. Eugen Ulmer, Stuttgart, 3rd edition.

Londo, G., 1975. *Nederlandse lijst van hydro-, freato- en afreatofyten*. RIN, Leersum.

Long, G., 1974. *Diagnostic phyto-écologique et aménagement du territoire. Tome I: Principes généraux et méthodes*. Collection d'écologie, 4. Maison & Cie, Paris.

Neef, E., 1968. Der Physiotop als zentralbegriff der komplexen physischen Geografie. *Petermanns Geogr. Mitt.* (1968): 15-23.

Opdam, P., 1985. *Delineating ecotopes as holistic landscape units: some methodological problems*. Annual Report 1984 Rijksinstituut voor Natuurbeheer (RIN), the Netherlands.

Plate, C.L., 1990. From flora statistics to an ecotope classification. *Statistical Journal of the United Nations ECE* 7: 85-99.

Poore, M.E.D., 1962. The method of successive approximation in descriptive ecology. *Advances in ecological research* 1: 35-68

Runhaar, J. 1989a. Relation between moisture-indication of the vegetation and groundwater levels. *Landschap* 6: 129-146. (Dutch with English summary)

Runhaar, J. 1989b. Coastal defence analysis Texel. In: F. Van der Meulen, P.D. Jungerius and J.H. Visser (eds.), *Perspectives in coastal dune management* SPB Academic Publishings, The Hague, pp. 197-205.

Runhaar, J., C.L.G. Groen, R. Van der Meijden and R.A.M. Stevers 1987. A new division in ecological groups in the flora of the Netherlands. *Gorteria* 13: 277-359. (Dutch, with English summary)

Sinnige, C.A.M., W.L.M. Tamis and F. Klijn 1991. *Aanzet tot een ecotopenclassificatie toegespitst op de bodemfauna*. CML report 75, Leiden. (Dutch with English summary)

Sinnige, C.A.M., W.L.M. Tamis and F. Klijn 1992. *Indeling van bodemfauna in ecologische soortengroepen*. CML report 80, Leiden. (Dutch with English summary)

Sparling, J.H., 1967. The occurrence of Schoenus nigricans L. in blanket bogs. *J. Ecol.* 55: 1-31.

Stevers, R.A.M., J. Runhaar, K.J. Canters and H.A. Udo de Haes 1984. *Beleidsanalyse Kustverdediging Texel: De effecten van kustverdedigingsalternatieven op het natuurlijk milieu.* Centrum voor Milieukunde, Leiden.

Stevers, R.A.M., J. Runhaar, H.A. Udo de Haes and C.L.G. Groen 1987. The CML Ecotope system: a national typology of ecosystems, focussed on the vegetation. *Landschap* 4: 135-150. (Dutch, with English summary)

Strahler, A.N., 1975. *Physical geography.* 4th edition. Wiley & Sons, New York.

Sukachev, V. and N. Dylis 1968. *Fundamentals of forest biogeocoenology.* Oliver & Boyd, Edinburgh / London. (English translation of 'Osnovy lesnoi biogeotsenologii', published in 1964 by 'Nauka' Publishing Office, Moscow.)

Tansley, A.G., 1935. The use and abuse of vegetational concepts and terms. *Ecology* 16: 284-307.

Tansley, A.G., 1939. *The British Isles and their Vegetation.* Cambridge University Press.

Torenbeek, R., 1988. *Hydrobiologie en waterhuishouding: Een beleidsvoorbereidende studie.* RIN-rapport 88/55. Rijksinstituut voor Natuurbeheer, Leersum.

Troll, C., 1950. Die geographische Landschaft und ihre Erforschung. *Studium Generale* 3: 163-181.

Troll, C., 1968. Landschaftsökologie. In: R. Tüxen (ed.). *Pflanzensoziologie und Landschaftsökologie.* Ber. Symp. Int. Vergl. Vegetationskunde, Stolzenau/Weser, 7: 1-21.

Troll, C., 1970. Landschaftsökologie und Biogeocoenologie. Eine terminologische Studie. *Revue Roum. Géol., Geophys. et Geograf., Série de Géographie* 14: 9-18.

Udo de Haes, H.A., 1987. Study on drinking water supply of South Holland and the role of ecology in policy analysis. *Impact Assessment Bulletin* 5: 55-71.

Van Beusekom, C.F., 1993. Kennis van ecotopen en sturing van standplaatsfactoren: onmisbaar voor effectief natuurbeheer. In: M. Cals, M. De Graaf and J. Roelofs (eds.), *Effectgerichte maatregelen tegen verzuring en eutrofiëring.* Katholieke Universiteit Nijmegen.

Van der Linden, M., J. Runhaar and M. Van 't Zelfde 1992. *Effecten van ingrepen in de waterhuishouding op vegetaties van natte en vochtige standplaatsen. Onderzoek effecten grondwaterwinning deelrapport 7.* RIVM, Bilthoven.

Van der Meijden, R., C.L. Plate and E.J. Weeda 1989. *Atlas van de Nederlandse flora. Deel 3: Minder zeldzame en algemene soorten.* RHHB Leiden/CBS Voorburg.

Van Raam, J.C. and E.X. Maier 1993. Overzicht van de Nederlandse kranswieren. *Gorteria* 18: 111-116.

Van Wirdum, G., 1986. *Water related impacts on nature protection sites.* Proceedings of the technical meeting 43 of the TNO committee on hydrological research. The Hague, 1986.

Van Wirdum, G., 1991. *Vegetation and hydrology of floating rich fens.* Thesis, Univ. of Amsterdam. Datawyse, Maastricht.

Verdonschot, P.F.M., J. Runhaar, W.F. Van der Hoek, C.F.M. de Bok and B.P.M. Specken 1992. *Aanzet tot een ecologische indeling van oppervlaktewateren in Nederland.* RIN-rapport 92/1, CML-report 78. IBN-DLO, Leersum.

Verdonschot, P.F.M., 1990. *Ecological characterisation of surface waters in the province of Overijssel (the Netherlands).* Thesis, Wageningen.

Westhoff, V. and A.J. Den Held 1969. *Plantengemeenschappen in Nederland*. Thieme, Zutphen.
Westhoff, V. and E. Van der Maarel 1973. The Braun-Blanquet approach. In: R.H. Whittaker (ed.), *Handbook of vegetation science V: Ordination and classification of communities*. Junk Publishers, The Hague, pp. 617-704.
Wheeler, B.D., S.C. Shaw and R.E.D. Cook 1992. Phytometric assessment of the fertility of undrained rich-fen soils. *Journal of Applied Ecology* 29/2: 466-475.
Whittaker, R.H. 1973. Approaches to classifying vegetation. In: R.H. Whittaker (ed.), *Handbook of vegetation science V: Ordination and classification of communities*. Junk Publishers, The Hague, pp. 323-342.
Witte, J.P.M. and R. Van der Meijden 1990. *Natte en vochtige ecosystemen*. Wetenschappelijke Mededeling KNNV nr. 200. KNNV, Utrecht.
Witte, J.P.M., F. Klijn, F.A.M. Claessen, C.L.G. Groen and R. Van der Meijden 1992. A model to predict and assess the impacts of hydrological changes on terrrestrial ecosystems in The Netherlands, and its use in a climate scenario. *Wetlands Ecology and Management* 2: 69-83.
Zonneveld, I.S 1989. The land unit. A fundamental concept in landscape ecology, and its applications. *Landscape Ecology* 3: 67-86.

Appendix

Definition of characteristics and description of classes

Moisture regime

The term 'moisture regime' is used to characterize the oxygen-water relations in the soil. Three classes are distinguished: wet, moist, and dry. On the basis of a field survey (see Figure 7.4) these classes have been defined in terms of combinations of soil texture and spring groundwater level (average groundwater level in March-April):

wet	sites with permanent or periodic low oxygen contents of the soil due to a high groundwater level; mean spring groundwater level 20 cm below surface or less.
moist	sites with, on average, sufficient oxygen and water availability. Mean spring groundwater level more than 20 cm below the surface, and in sandy soils less than 60-150 cm below the surface (depending on the grain size and loam content).
dry	sites with low moisture availability in the summer because of a low groundwater level and low moisture retention capacity of the soil; on sandy soils with a mean spring groundwater level of more than 60-150 cm below surface (depending on texture and loam content, which determine the capillary rise of groundwater).

Nutrient availability

Nutrient availability refers to the availability of macro-nutrients (N,P,K), or rather the availability of the limiting macro-nutrient in the rooting zone. Three classes are distinguished: low, moderate, and high nutrient availability. So far, the classes have not been described quantitatively, because a reliable and readily measurable parameter is lacking; concentrations of individual macro-nutrients obviously give little information about nutrient availability during the growing season (see Boeker, 1954; Knauer, 1972; Runhaar, 1989a; Wheeler et al., 1992). Only for grassland ecotopes can a general indication of the boundaries between classes be given, using parameters that are closely connected with nutrient availability: net biomass production (kg dry weight per hectare, per year) and N-mineralisation (kg N per hectare, per year):

low	net production on average less than 4000 kg; N-mineralisation on average less than 50 kg.
moderate	net production on average between 3000-8000 kg; N-mineralisation on average between 30 and 160 kg N.

| high | net production on average more than 7000 kg; N-mineralisation on average more than 160 kg N. |

Acidity

The acidity of the soil and groundwater influences vegetation in several ways. In acid soils the toxicity of Al^{3+}-ions seems to play an important role in determining species composition (see Clymo, 1962; Sparling, 1967), while on alkaline sites the availability of Fe can be limiting for many species (Grime and Hodgson, 1968). Furthermore, soil acidity influences bacterial activity in the soil and the solubility of, amongst others, phosphate.

acid	$pH-H_2O$ less than 4.5
weakly acid to neutral	$pH-H_2O$ between 4.5 and 6.5
alkaline	$pH-H_2O$ more than 6.5

Salinity

Three classes are distinguished: saline, brackish, and fresh. These classes have only been defined for sites in direct contact with surface water. They are based on the chloride concentration of the water:

saline	Cl^- more than 10,000 mg/l
brackish	Cl^- between 1,000 and 10,000 mg/l
fresh	Cl^- less than 1,000 mg/l

Dynamics

The characteristic 'dynamics of the ecosystem' is used to distinguish ecosystems in which the vegetation is exposed to physical stresses caused by sand drift, trampling, etc.

Vegetation structure

Vegetation structure differs from other factors, because it is not so much an operational site factor itself, but rather the result of the operational site factors 'vegetation management' and 'time'. Furthermore, vegetation structure determines the physiological factor 'light availability', which is especially important for understory plants.

| pioneer | open vegetations on initial soils or in places where the succession is hindered by physical or physiological stress; |
| grassland s.l. | low vegetations with herbs, bryophytes, and/or dwarf shrubs in situations where removal by mowing and grazing takes place; subdivided into bryophyte vegetations, grasslands s.s. (again subdivided into hayfields and meadows) and dwarf shrub vegetations; |

Appendix - continued	
tall herbaceous	high, dense herbaceous vegetations in places where little or no mowing or grazing takes place;
shrub	dominated by woody species that reach heights of 1-4 metres;
forest	dominated by woody species that reach heights of more than 4 metres; subdivided into pine forests and deciduous forests.

The application of quantitative methods of classification to strategic ecological survey in Britain   8

Robert G.H. Bunce

ABSTRACT - Objective methods of classification for vegetation were developed with the advent of computers in the 1960s. In the 1970s, these methods were applied by the Institute of Terrestrial Ecology to environmental data from maps, which are similar to vegetation data in that they are drawn from a continuously variable population. The present paper describes the application of these methods to produce a Land Classification for Britain and, also, for classifying data from representative vegetation samples. The procedure provides strata which enable quantitative estimates to be made of regional ecological resources and serves as a basis for modelling environmental issues.

Introduction

Strategic surveys of ecological resources are now the foundation of many impact assessments and a requirement for the development of rural policy by government. It is therefore essential that quantitative methods of data acquisition are used, so that agencies can be confident of the reliability of the results. The present paper describes the approach developed in Britain to answer such requirements. Quantitative procedures, mathematically based, are used at a variety of scales and differ according to the level at which questions are set. It is particularly important to select representative sites for detailed study by a strictly scientifically sound procedure, because the subsequent investment in such sites must be capable of being referred to the wider population for which it has been devised.

The mainly multivariate statistical techniques used have been developed primarily for vegetation analysis, but they are equally applicable to environmental data. Whilst there has been much discussion and comparison of such techniques (see Dale, 1988), it is more important to ensure a sound initial data base, than to concentrate on the comparison of techniques. Over the last few years the capture of data by digitising information from maps has changed dramatically, but it is essential to ensure that the objectives of any study determine the method of data collection rather than technical availability.

The present paper commences with a description of an environmental classification, in this case of Great Britain. The application of classification techniques to vegetation is then described, and examples are given.

The ITE Land Classification

The principle behind the ITE Land Classification is the concept that ecological characteristics, e.g. vegetation, are associated with physical factors, e.g. altitude. The former require detailed and expensive measurement, whereas the latter can be recorded from maps using automated procedures. One possibility is to use multiple regression to define relationships, but this becomes impractical.
Classification using standard units, usually a 1 square kilometre, from which environmental data are recorded can then be used to produce environmental strata, which are relatively homogeneous. These units can be sampled in the field for ecological parameters which are then related to the basic classification. Provided that correlations are sufficiently good, then estimates can be made with defined levels of statistical accuracy.

Initially, the system was tested in Cumbria (Bunce and Smith, 1978). Bunce and Heal (1984) describe the development of the classification of Great Britain. Subsequently, the system has been applied elsewhere, e.g. in Spain (Bunce et al., 1987).
The initial classification was created by applying TWINSPAN (Hill, 1979b) to environmental data (e.g. climate, topography) from 1212 1-km squares in a grid over Great Britain. Thirty-two Land Classes were identified showing distribution patterns within Britain as discussed below. The extended classification of Great Britain, from the 1212 squares to the approximately 240,000 km^2 in Great Britain, applied logistic regression and the discriminant function to a limited data

set of environmental variables, which was comparable to the initial analysis of all squares.

The classes represent the major environmental differences throughout Britain. Thus, classes 1-16 are predominantly English and lowland, whereas 17-32 are Scottish or upland. Within the main division, the former are divided primarily on climatic criteria whereas the latter are split by the analysis on altitudinal factors. Progressively more local attributes, e.g. slope and aspect, become important at the lower levels of the hierarchy. For instance, land class 6 occurs mainly in southwest England at low altitudes with broad, even slopes and many small valleys. By contrast, land class 18, whilst present in Exmoor and Dartmoor, is found mainly in Wales, northwest England and Scotland. Steep hillsides and rugged scenery predominate. Land class 23, which consists of extremely steep hillsides with ridges and scarps amongst the high mountains, is found mainly in Scotland. Land class 29 includes the indented coastlines of Scotland with uneven topography and complex scenery.

The first major survey based on the ITE Land Classification, which took place in 1977/78, was designed to define the main ecological characteristics of Britain. Eight squares were sampled at random from each of the 32 Land Classes giving a total of 256 squares distributed throughout Britain. Within each square, species data from higher plants and a restricted list of bryophytes were recorded from five random 200 m^2 quadrats and from two linear quadrats (1 m x 10 m) along streams, roadsides, and hedgerows. In addition, a map of the land cover and land use was made from 1: 10,000 Ordnance Survey maps. Further details are provided by Bunce and Heal (1984).

Classification of vegetation

The vegetation data mentioned above are in the form of species records from the quadrats. The procedure adopted uses a multivariate classification method to classify these quadrats into relatively homogeneous groups. These groups may then be used to obtain measures of their relative frequency in different types of landscapes.
The classification method used is TWINSPAN (Hill, 1979b), a polythetic hierarchical method widely used in numerical vegetation analysis and in many recent phytosociological analyses. The difference between the present approach and traditional phytosociology lies, therefore, in the method of data collection,

and not in the analytical method used. The units produced can be compared but random sampling will inevitably incorporate much vegetation which is regarded as heterogeneous in phytosociological terms. TWINSPAN also classifies the species into groups but, as Fernandez Prieto and Bueno Sanchez (1993) have pointed out, these groups are difficult to interpret. Agglomerative techniques are more appropriate for classifying species and, in the present project, maximal variance clustering was applied to the ordination scores derived from DECORANA (Hill, 1979a). It has been found to produce the best-defined species groups.

The species groups enable the vegetation types to be interpreted. They are also useful in defining the diversity within these vegetation types. In the classifications of both species and quadrats, the objective is to enable generalisations to be made within a structured framework. When subsequent change is to be examined, the comparisons are made within broad groups of vegetation types in order to remove background noise so that gradual changes can be determined.

The Land Classes were shown to be very different in their ecological characteristics. The main types of vegetation and landscapes were defined as described by Bunce et al. (1981), with the uplands having widespread semi-natural vegetation, whereas in the lowlands the majority of species resided in linear features.

The land cover data were converted to national estimates through the Land Class system by calculating the mean for the squares from each class and by then multiplying by the extent of the Land Class in Britain.
National figures for crops were compared with those available from official statistics, but those for semi-natural vegetation were not available elsewhere. When the former were compared with published figures, e.g. for wheat, the official statistics gave a figure of 1.17×10^6 whereas the figure from the Land Classes was 1.06×10^6. A variety of other comparisons was made, e.g. for forest and urban areas, which confirmed the validity of the sampling procedure.
Estimates of cover from the quadrat data also showed good agreement with the land cover measurements from mapped data, both giving estimates of moorland of 1.7×10^6 ha (Bunce and Barr, 1987).

The data from the 1978 survey have been used in several modelling exercises, such as the potential for wood energy plantations in Britain (Mitchell et al.,

1987), and in the development of a model to define the possible outcome of changes in the Common Agricultural Policy.

The 1984 land cover survey

The same 256 squares were resurveyed in 1984, together with a further 128 squares. The same basic categories of land cover and land use were recorded, but further detail was added on linear features and trees, as described in Barr et al. (1986).

The landscape changes were recorded between 1978 and 1984, with the main changes in crops being a shift away from barley and grass production to wheat and a large expansion of oilseed rape. Other changes in land cover were fragmentary but may be locally significant, although under 1% of the national cover. The overall trend was an increase in the intensity of use, whether within the grassland series, or in the conversion of upland vegetation to agriculture.

In comparison with total area, the losses of semi-natural vegetation through grassland improvement were small, but forestry continued to expand at about 20,000 ha/annum in the uplands.

The figures for linear features show that hedgerows declined by 28,000 km over that period; walls declined by 1400 km, but fences increased by 48,400 km.

The increasing concern about the changes in quality of the habitats making up the British countryside led to the setting up of a project to utilise the first two surveys as a basis for determining the ecological consequences of land use change. The project was mainly funded by the Department of the Environment (DOE). This project was a wide-ranging and exploratory research programme to identify recent and likely future trends in land use and to quantify the ecological changes that would result. The project involved communication and collaboration between various disciplines and research groups, and was designed to recommend techniques for improving the measurement and assessment of change, including both traditional ground survey methods and remote sensing.

As part of the programme, consideration of scenarios of land use change indicated that the probability of change between arable and grass was greatest in the Midlands and southwest England, with economic marginality being probably more important than the usual ecological definition of marginal land. However, it was concluded that ecologically significant change would occur *within* land cover types, e.g. permanent grasslands, which could not be covered by broad statistics. Ecological interpretation, therefore, required information on species composition. It showed that different landscapes have different potential for change related to the distribution of the species within them. In the lowlands, linear habitats, which occupy only some 5% of the land area, often contain over 80% of the species present and pressures on these features will cause changes, whereas the open fields have already lost most of their diversity. The implication of such distribution patterns for rural policy in the lowlands is that maximum gains in terms of future maintenance of diversity can be obtained by protecting, expanding, or diversifying such linear features.

The procedures of classification have enabled these resources to be quantified and defined.

Pilot studies were also carried out in a repeat survey of the quadrats from 1978, both in terms of analytical requirements and in the need for quality assurance of botanical data to ensure that reliable estimates of change could be made.

An independent analysis of the data base by the Unit of Comparative Plant Ecology (UCPE) at Sheffield University was carried out, based on theories concerning the likely response of species to alterations in management following from the growth strategy of species. Application of the theoretical framework of Grime et al. (1988) showed that the plants were prone to environmental stress. The study also demonstrated that the ITE data could be used within the framework of the UCPE strategies and would express in quantitative terms the shifts in management taking place at the landscape level. For example, in grasslands shifts in the increase in fertiliser were reflected by the increase in competitive ruderal species. Trends within the balance of species can, therefore, be used to indicate the processes involved.

The project also demonstrated that field survey and satellite imagery were complementary and that future surveys should utilise both approaches. Finally, exploratory work was carried out on a knowledge-based information system for geographically referenced rural data, which had the potential for presenting the results from such coordinated programmes.

The countryside survey 1990

The method for this survey was largely based on the previous project from the botanical point of view, with particular emphasis on linear features and the control of the quality of botanical survey data. The project was supported partly by basic funds from ITE and contract income from the Department of the

Environment (DOE), with some funding from the former Nature Conservancy Council.

The same 384 sites were covered as in 1984, with a further 128 included to increase the accuracy of estimates. These additional sample squares were allocated to the Land Classes according to the frequency of the classes in Great Britain. The resulting balance between classes, together with their improved definition, resulted in lower standard errors for national estimates.

Within each square the same quadrats were repeated as in 1978. In addition, three further quadrats were placed along roadsides and streamsides and directly beside boundaries adjacent to the five random quadrats. Finally, five 4 m^2 quadrats were placed in the centre of patches of semi-natural vegetation not covered by any of the above plots.

The land cover categories in the mapping procedure were largely the same as in 1984. The vegetation was recorded twice and checked for accuracy. Land cover maps were digitised using a Geographical Information System (ARC/INFO) to enable areal measurement of stock and change by overlay of boundaries recorded at different dates. Labels for the individual parcels form the basis for modelling, e.g. scenarios of climate change, and for pattern analysis.

The stock of land cover in 1990 was estimated, together with appropriate standard errors. The land cover categories were the same as those used in a baseline classification resulting from the DOE-funded project 'Comparison of land cover definitions'. The figures for these calculations will shortly be available and can be obtained from the author.

As to landscape features and habitats, it may be remarked that hedgerows are the only feature for which figures are currently available to date. The total hedge length in Britain was estimated to be 428,000 km in 1990 with a standard error of 28,300 km, and a net change of 121,000 km between 1984 and 1990 with a standard error of 11,600 km. Therefore, nearly 25% of hedgerows have undergone some sort of change. Complete removal of hedgerows amounted to 52,200 km or 9.5% of the total 1984 hedgerow length. Most change is associated with the management of hedgerows, with 111,500 km, or 20%, of hedgerows recorded in 1984 being recorded in 1990 as lines of trees or shrubs, or as relict hedgerows. Conversely, only some 25,000 km of new hedges in 1990 came from overgrown or derelict hedges. This suggests that hedgerows were subject to less active management in 1990 than in 1984.

As to the vegetation, the species composition as determined in the random samples enables the relative proportions of the main assemblages of species to be determined by multivariate analysis. At a high level within the hierarchy, the categories coincide with both the land cover categories, e.g. moorland, bog, or acid grassland, and also with phytosociological groups, although further analysis is required to quantify such comparisons. Work is currently in progress to characterise these major groups in terms of their species composition and on the ecological interpretation of the changes taking place.

An analysis of the data from hedgerows showed changes in the species composition since 1978 with a general trend towards a decline in diversity. Such losses are probably due to a decline in management in the centre of the hedgerow, causing plants to be shaded out, and intensification of agriculture along the edge of fields at the base of the hedge. The losses in hedgerows are therefore both in botanical quality and in quantity in terms of overall length.

Other applications and developments

The results of the analysis described above need to be coordinated so that an overall picture of the stock and change in the British countryside is obtained. Such analyses should involve other approaches to vegetation analysis, such as phytosociology and UCPE strategy theory, for which initial studies have already been carried out to demonstrate their usefulness and feasibility. Similarly, the vegetation data need to be related to the land cover categories derived from the satellite imagery and should be applicable in the models for change now being derived from climatic and agricultural policy studies.

Detailed analysis of biodiversity can also be used for input into probable future policy initiatives at the European level. The locational aspects of the data in the spatial distribution of vegetation patches in the landscape is an area of study which can explain the patterns observable in the landscape. The data also contain a wealth of information on the ecological character of the species individually and in assemblages, which enables interpretation and may be of major use to other botanists for reference purposes.

Discussion

The above analyses should be presented at the landscape level, so that policy advisors can develop appropriate policies for the countryside, e.g. for the maintenance of botanical diversity. Further analysis, therefore, must be carried out on botanical diversity in the British landscape at both patch and smaller scales.

Sensitive species and species that are indicative of change need to be identified. Current emphasis has been on change but not on the underlying processes and mechanisms. It is, therefore, necessary to investigate the latter so that models can be developed to predict changes in plant assemblages (compare Claessen et al., in chapter 10).

The relative importance of factors such as pollution management practices and socio-economic factors, needs to be determined. An understanding of the relationship between scale and plant diversity and their respective spatial distribution is fundamental to the management of the countryside. Some preliminary analyses have already been carried out, but the relationship, for example, of the species in linear features with their surrounding vegetation needs further study.

There is also a requirement for further tuning in satellite imagery and detailed ground cover information in order to develop more rapid procedures for monitoring change. Throughout all these procedures, classification is necessary to enable comparisons to be carried out and to remove background noise. The experience gained in the work described above demonstrates the effectiveness of objective methods, which may initially be more time-consuming but which will later provide major benefits in terms of consistency and reproducibility.

References

Barr, C.J., C.B. Benefield, R.G.H. Bunce, H.A. Ridsdale and M. Whittaker, 1986. *Landscape changes in Britain*. Institute of Terrestrial Ecology, Abbots Ripton.
Bunce, R.G.H. and C.J. Barr, 1987. The extent of land under different management regimes in the uplands and the potential for change. In: M. Usher and D.B.A. Thompson (eds.), *Ecological change in the uplands*. British Ecological Society special publication no. 7, Blackwell Scientific, Oxford, pp. 415-426.
Bunce, R.G.H. and O.W. Heal, 1984. Landscape evaluation and the impact of changing land-use on the rural environment: the problem and an approach. In: R.D. Roberts and T.M. Roberts (eds.), *Planning and ecology*. Chapman & Hall, London, pp. 164-188.
Bunce, R.G.H. and R.S. Smith, 1978. *An ecological survey of Cumbria*. Cumbria CC and Lake District Planning Board, Kendal.
Bunce, R.G.H., C.J. Barr and H.A. Whittaker, 1981. *The land classes of Great Britain: preliminary descriptions for users of the Merlewood method of land classification*.

Merlewood research and development paper no. 86, Institute of Terrestrial Ecology, Grange-over-Sands.

Bunce, R.G.H., R. Elena Rosello, and A. San Miguel, 1987. The potential for integration between various land uses in the Iberian peninsula. In: G. Grassi, B. Delmon, J.-F. Molle and H. Zibetta (eds.), *Biomass for energy and industry. 4th E.C Conference.* Elsevier Applied Science, London, pp. 1261-1266.

Dale, M.B., 1988. Knowing when to stop: cluster concept-concept cluster. *Coenoses* 3: 11-32.

Fernandez Prieto, J.A. and Bueno Sanchez, A., 1993. A new classification of the Muniellos Biological Reserve in north west Spain. *Vegetatio* 102: 47-68.

Grime, J.P., J.G. Hodgson and J.C. Hunt, 1988. *Comparative plant ecology: a functional approach to common British species.* Unwin Hyman, London.

Hill, M.O., 1979a. *DECORANA - a FORTRAN programme for detrended correspondence analysis and reciprocal averaging.* Section of Ecology and Systematics, Cornell University, Ithaca, New York.

Hill, M.O., 1979b. *TWINSPAN - a FORTRAN programme for arranging multivariate data in an ordered two way table by classification of the individuals and attributes.* Section of Ecology and Systematics, Cornell University, Ithaca, New York.

Mitchell, C.P., R.B. Tranter, P. Downing, O. Brandon, M.L. Pearce, R.G.H. Bunce and C.J. Varr, 1987. *Growing wood for energy in Great Britain.* Report to the Energy Technology Support Unit, Harwell.

A flexible multiple stress model: who needs a priori classification?

9

Joris B. Latour, Rudo Reiling and Jaap Wiertz

ABSTRACT - In this contribution we shall review a multiple stress model for vegetation (MOVE), which predicts the effects on ecosystems of stresses imposed by different environmental problems. The model consists of two modules: one for soil and one for vegetation. In the soil module changes in abiotic site factors (e.g. soil moisture, soil nutrient availability, and soil acidity) are predicted for various environmental scenarios. In the vegetation module the ecological response functions of species (plotted in a multi-dimensional hypervolume according to Hutchinson) are used to predict the occurrence probability of species for each combination of the abiotic site factors. By linking the two modules, the occurrence probability of species can be predicted in each geographical unit. The model can be used for scenario studies, standard-setting, and the identification of key problems and problem areas. Examples given are based on specific pilot studies.

Introduction

Generally, changes in ecosystems, populations, and species are attributed to combinations of environmental stresses (e.g. Baldock, 1990). Studying the combined effect of stresses is complex since the effects of different stresses are often not simply additive but synergistic or antagonistic (e.g. European Inland Fisheries Advisory Commission, 1987), and since combined effects often occur on non-linear time scales (e.g. 'Chemical time bombs'; Stigliani, 1991).

Various approaches can be used for modelling multiple stress: functional mechanistic models, models that use ecosystem classifications, and models based on characteristics of species. In most functional mechanistic models that have been used for this purpose a trade-off has been made, between the geographical scale of the model, the number of ecosystems it refers to, and the complexity of processes described. It will be difficult to cover the many different ecosystems and various environmental problems in the Netherlands with functional mechanistic models.

Ecosystem classification systems have been used in several regional and national studies, particularly on the topic of desiccation (e.g. Witte et al., 1989; Witte et al., 1992; Runhaar et al., 1985). Optionally, such systems may also be used for multiple stress modelling. Classified ecosystems, however, generally do not respond as coherent entities to changing conditions. The response entity is, in principle, species, which all respond differently to large environmental changes. Non-analogue ecosystems could emerge in the dynamic response processes (Peters and Darling, 1985). In modelling ecosystem response to climatic changes rigid ecosystem classifications have also been criticized (Leemans, in press). Leemans (1992), therefore, uses models that are based either on species defined by their physiological characteristics or, if not possible, on functional types.

Latour and Reiling (1991; in press (b)) have designed a species-centred, probabilistic multiple stress model for plant species (MOVE). In this model, responses to environmental change are described for each species individually. An advantage of species-centred models is that they can be used directly to predict changes in ecological objectives. In the Netherlands ecological objectives are currently being worked out using target species. Target species have been selected for various aquatic (V&W, 1989) and terrestrial ecosystems (LNV, 1990). Both functional mechanistic models and models based on ecosystem classification are less suitable for the prediction of changes in populations of single species. Another advantage of species-centred models is that results can be validated easily in the field using monitoring systems. Monitoring ecosystem functioning or classified ecosystems is more laborious than monitoring single species.

In this chapter we shall review the model, MOVE, discussing its use for multiple stress modelling on a national scale and highlighting such applications as prediction, determination of ecological standards, and the ranking of environmental threats. The model has not yet been completed. The examples given are based on specific pilot studies.

The multiple stress model, MOVE

The Multiple stress mOdel for VEgetation (MOVE) (Latour and Reiling, 1991; Latour and Reiling in press (b)) predicts probability of occurrence of individual plant species as a function of acidification, desiccation, and eutrophication scenarios on a national scale. The model consists of a soil module predicting the abiotic effects and a vegetation module predicting the corresponding biotic effects (Figure 9.1).

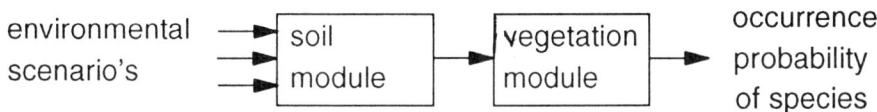

Figure 9.1 Systematic representation of the multiple-stress model for vegetation, MOVE

In the soil module, changes in abiotic soil factors (soil moisture, soil acidity, and nutrient availability) relevant to the occurrence of plant species are predicted. The input consists of acid and nitrogen deposition, and changes in groundwater level. A dynamic soil model (e.g. Simulation Model for Acidification Regional Trends — SMART — De Vries et al., 1989) will be revised for prediction of unfertilized, Dutch ecosystems. The following processes will be included in this model: net uptake and net immobilization of nitrogen; weathering of carbonates, silicates and Al oxides and hydroxides; cation exchange, CO_2 equilibria, and nutrient dynamics. In fertilized systems, the nitrogen load is linked to the vegetation module without dynamic modelling of the soil processes. The nitrogen load in the Netherlands is highly correlated with plant-species occurrence in fertilized systems (Van Strien, 1991).

The vegetation module predicts the probability of species occurrence as a function of three abiotic soil factors: soil acidity, nutrient availability and soil moisture (Figure 9.2). With regression statistics the occurrence probability of

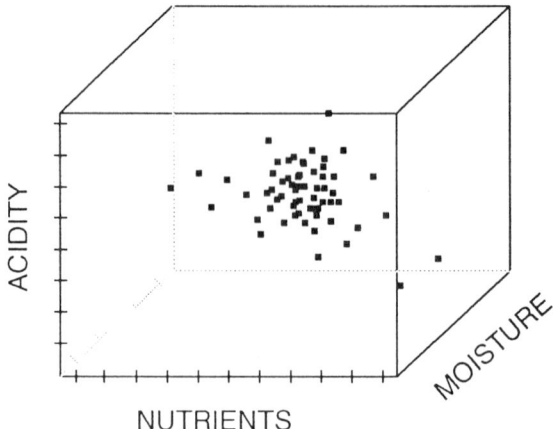

Figure 9.2 Probability of species occurrence shown in a matrix with dimensions defined by variables related to acidification, eutrophication and desiccation. The occurrence of a certain species in the matrix is indicated with small squares

a species can be calculated for each combination of soil factors or for each environmental variable separately, resulting in species-response curves.

Species-response curves of 700 plant species have been determined for soil moisture, nutrient availability and soil acidity (Wiertz et al., 1992) using Gaussian logistic regression models. Regression was based on an extensive database developed for a revision of the Dutch classification of plant communities (Schaminee et al., 1989). This database consists of 17,000 vegetation relevés. No information on abiotic site factors of the vegetation relevés was available. These were assessed in retrospect with Ellenberg indication values (Ellenberg, 1979). Ellenberg indication values indicate the relationship between the occurrence of plant species and nutrient availability, acidity, soil moisture, salinity, and temperature. These values have been assigned to most plant species of western and central Europe, and have been completed for the Netherlands (Wiertz, 1992). The abiotic site factors of each vegetation relevé are assessed by averaging the indication values of all the species recorded. Differences in abundance of species have not been considered. Calculated averages, in Ellenberg indication values, are used as a semi-quantitative assessment of the abiotic site factors. Next, the occurrence frequency of each species is established as a function of the averages of vegetation relevés. The occurrence frequency is described with Gaussian logistic regression models (Jongman et al., 1987). Since this analysis used floristic information to assess the abiotic site factors, any (historical) vegetation sample can be included in the analysis, extending the database. Moreover, such an analysis excludes potential bias caused by high temporal and spatial variation in the actual measurements of abiotic site factors. Deduction of values for the abiotic site factors from the vegetation sample guarantees ecological relevance.

Species occurrence has been described as being significant for 95% of the species using unimodal and linear regression models. Most of the significant models were unimodal. Linear models were found for nutrient availability (4%) and salinity (20%).

Ellenberg indication values can be calibrated with quantitative values for the abiotic site factors using combined samples of vegetation and environmental variables. This calibration connects the soil module with the vegetation module. Calibration for soil acidity, nutrient availability and soil moisture is now in progress for the Netherlands.

Setting ecological standards

Risk assessment (EPA, 1990; VROM, 1989) is used for various environmental problems to quantify ecological standards that correspond to similar protection levels for ecosystems. Concentrations of various compounds protecting 95% of the species of an ecosystem (VROM, 1991) — the maximum tolerable concentrations (MTC) — have been calculated by extrapolation of single-species toxicity data (e.g. No Observed Effect Concentrations) to an ecosystem. Various extrapolation techniques have been proposed (see Slooff, 1992, for an overview). Similarly, Van der Eerden et al. (1992) calculated critical levels for NH_3 and SO_2 based on NOECs determined in laboratory tests.

Using the multiple stress model for vegetation (MOVE), we have recently adopted the risk assessment for eutrophication and desiccation (Latour and Reiling, 1992). The method uses percentiles of the species-response curves as a measure of risk at the species level, analogue to NOECs. Next, the percentage of species protected according to these percentiles is calculated for each value of the environmental variable. For instance, the 5 and 95 percentiles of the species-response curves are used as NOEC-like measures of the risk at the species level (Figure 9.3). The 5 percentile corresponds with a reduced occurrence probability due to 'limitation', the 95 percentile due to 'intoxication'.

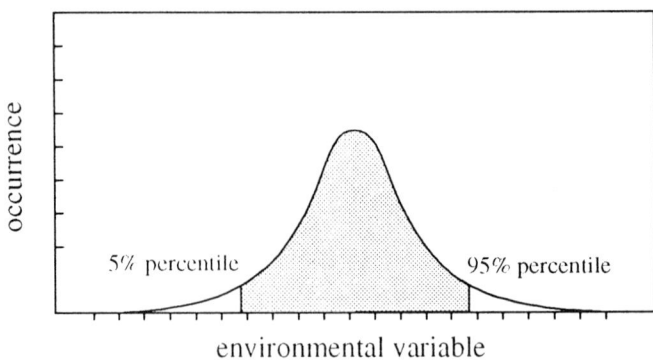

Figure 9.3 A species-response curve showing the occurrence probability of a species as a function of an environmental variable using the 5 and 95 percentiles

Next, the percentage of species protected is plotted for each value of the environmental variable. Species are considered protected if the 5 percentile of the species is lower and the 95 percentile higher than the environmental variable. The 95% protection level of all species can be established from this relationship.

We have calculated the potential number of species of grassland in the province of South Holland as a function of nitrogen load based on species-response curves of 275 plant species from Clausman et al. (1987). Nitrogen load ('N'), in kg N ha^{-1} yr^{-1}, is calculated from Ellenberg indication values ('E') with: N = (E-2.6)/0.016 based on linear calibration (Snedecor and Cochran, 1989) of optima of 42 plant species expressed in Ellenberg values (Clausman et al., 1987) with optima expressed in nitrogen load (Van Strien, 1991). The maximum tolerable nitrogen load, corresponding with a protection of 95% of the species, is 60 kg N ha^{-1} yr^{-1} (Latour and Reiling, 1992). This is one order of magnitude less than current loads of 450 kg N ha^{-1} yr^{-1} (CBS unpublished data), resulting in a protection of about 25% of the plant species.

In the Netherlands, ecological targets are currently being specified using a system of 100 so-called nature conservation target types (LNV, 1992). These target types are specified for areas in the ecological network of the Netherlands (L&V, 1989). The target types are defined as assemblages of plant communities characterized by target species, abiotic conditions and minimum surface area. We are currently assessing ecological standards of each of these nature conservation target types for soil pH, soil moisture and nutrients, using the risk assessment as described earlier. In addition, various modifications of the extrapolation methods are being explored.

Critical loads for nitrogen, based on changes in species composition

The critical load concept is used as a scientific tool for planning far-reaching and cost-effective emission controls in the field of acidification (Kämäri et al., 1992). Critical loads are calculated with mass balance models (e.g. Svedrup et al., 1990; De Vries and Kros, 1991) that describe changes in critical chemical values for ion concentrations or ion concentration ratios in forest soils, groundwater or tree needles. These critical values are indicative of the ecosystem status (De Vries and Heij, 1991).

The species composition of ecosystems may alter, however, even if the critical values for ion concentrations and ion ratios are not exceeded. Bobbink et al. (1992) assessed critical loads for ecosystems based on changes in species composition of flora and fauna. Such changes are derived from various sources of information, such as: field and laboratory experiments or comparison of flora and fauna composition in time and space. It is not possible, however, to ascertain if empirically assessed critical nitrogen loads protect various ecosystems equally. Moreover, the current empirically assessed critical nitrogen loads do not incorporate regional differences, for example, in soil buffering capacity and sensitivity of species.

As stated earlier, we used the concept of risk assessment to quantify ecological standards corresponding to similar protection levels for ecosystems. Recently, we have proposed calculating critical loads based on changes in species composition systematically and quantitatively using MOVE. As an example we calculated critical loads for lowland wet heathland (Latour and Reiling, 1993). Depending on the choice of the desired occurrence probability, critical loads were between 13 and 37 Kg N ha^{-1} yr^{-1}. Currently, critical loads are being calculated for several other ecosystems.

Prediction

The occurrence probability of a species, groups of species or all species can be predicted for scenarios for acidification, eutrophication and desiccation. As an example, the probability of occurrence for grassland plant species in fertilized grasslands in the Netherlands in 1989 and in 2010 has been assessed as a function of nitrogen load. Ecological response curves of 275 grassland plant species found in South Holland (Clausman et al., 1987) were included in this analysis. Species were considered protected if the 95 percentiles of the ecological response curves were not exceeded. It is assumed that response curves in South Holland do not differ systematically from ecological response curves for other areas in the Netherlands.

Data on nitrogen load were based on agricultural statistics (LEI-DLO, 1991). The nitrogen load for 2010 was based on a scenario in Maas (1991). These statistics include average nitrogen loads from manure and fertilizer for each soil type in each region. The geographic distribution of grasslands was derived from satillite imagery (Landsat-TM). The percentage of species for which the 95 percentiles are not exceeded is calculated for each region. Results are expressed as potential species diversity. In most regions situated on sandy soils of the central, eastern and southern parts of the Netherlands, potential diversity at present is less than 10% of the possible 275 species (Plate 9.1 and 9.2). In most regions situated on peat and clay soils in the western and northern parts of the Netherlands the potential diversity varies between 10 and 30%. In some regions potential diversity is higher, up to 60%. If policy objectives for reduction of nitrogen loads are implemented (Maas, 1991) the potential diversity will increase by on the average 10% in 2010 (Plate 9.2).

Locally, potential diversity may deviate from regional averages since the nitrogen load will obviously vary within regions. Data on the variance of nitrogen load within regions and on individual farms is not currently available. Consequently, the actual diversity will deviate from potential diversity. Actual diversity can be mapped using extensive floristic databases (Groen et al., 1992).

Currently, a feasibility study on using the MOVE model for multiple stress modelling on a European scale is being conducted. For a limited number of species the probability of occurrence in Europe is determined by the combined effect of global (climatic change), continental (acidification) and regional (desiccation and eutrophication) environmental problems.

Ranking environmental threats to set abatement priorities

In comparative threat analysis, threats imposed by different environmental problems are measured and compared. This type of analysis is used to assess the overall threat to the environment and to identify the most serious threats ('key problems') and areas which are particularly threatened ('problem areas'). Results can be used to set cost-effective abatement strategies. Elementary to comparative threat analysis is a common yardstick with which stresses imposed by different environmental problems can be compared. The amount by which the carrying capacity is exceeded by environmental loads may be used as such a yardstick. Latour and Reiling (in press (a)) conclude, on the basis of one national and two European case studies, that comparative threat analysis is hampered by a diversity of methods describing the carrying capacity for each environmental problem. A solution can be found in a risk-based comparative threat analysis using MOVE.

We recently compared threats imposed by acidification and eutrophication in a case study on undergrowth in nutrient-poor deciduous forest (Latour et al., 1993). Characteristic species for such nutrient-poor acidic deciduous forest are implied in the *Quercion robori-petraeae* and *Fago-Quercetum* (Loopstra and Van der Maarel, 1984). Ecological response curves of 13 plant species were derived from Wiertz et al. (1992). Calibration of Ellenberg indication values for pH was based on Gremmen (1987) and Ellenberg et al. (1991). The soil pH was calculated from the Ellenberg indication value for acidity (EpH) as: soil pH=0.45 EpH + 2.7. The nitrogen load, N (kg N ha^{-1} a^{-1}) is calculated from Ellenberg indication values for nutrients (E) with: N = E∗30 - 60 (Gremmen, 1987).

Data concerning actual pH of soils were derived from De Vries and Leeters (1993), Bongers et al. (1989), and Van den Berg et al. (1990). Sixty-eight locations in these databases relating to nutrient-poor, dry deciduous forest were included in the analysis. Nitrogen availability of plant species in nutrient-poor deciduous oak and beech forests was assessed as the sum of nitrogen deposition and gross nitrogen mineralization. Deposition levels of nitrogen were derived from Erisman (1991). Gross nitrogen mineralization is the product of leaf mass and leaf N content. Data on leaf mass and leaf nitrogen content under various N deposition levels were derived from De Vries et al., (1990).

Table 9.1 gives three possible ecosystem protection levels for soil pH and nitrogen availability calculated as average 5, 10 and 20 percentiles of the ecological response curves of 13 herbaceous species. The protection level based on the 20 percentiles represents the 'most strict' protection, since the 20 percentiles are closer to the optimum of the species than the 5 and 10 percentiles. The protection level based on the 20 percentile values therefore refers to higher pH values and lower nitrogen availability than the 5 and 10 percentiles. In comparative threat analysis we assume that the ecosystem is protected if the nitrogen availability load is lower and the soil pH higher than the protection levels.

Table 9.1 Ecosystem protection levels for soil pH and nitrogen availability calculated for deciduous forests on nutrient-poor, dry, acidic sandy soils as the averaged 5, 10 and 20 percentiles of the ecological response curves of 13 herbaceous species (Latour et al., 1993)

Protection levels	Soil pH	Nitrogen availability (kg N ha^{-1}a^{-1})
Optimum	4.26	46
20 percentile	3.84	76
10 percentile	3.61	91
5 percentile	3.43	104

Table 9.2 Comparison of acidification risks (soil pH) and eutrophication (N availability) for deciduous forests on nutrient-poor, dry, acidic soils (Latour et al., 1993). The percentages of forest stands ($n=68$) are given in which the optimum, 20, 10, and 5 percentile protection levels are exceeded because of soil pH and nitrogen availability, respectively. Confidence intervals are given in parentheses

	Measured soil pH (%)	Modelled nitrogen availability (%)
Optimum	91 (82-97)	100 (95-100)
20 percentile	75 (63-85)	82 (71-90)
10 percentile	38 (27-50)	71 (58-81)
5 percentile	15 (7-25)	51 (39-64)

Table 9.2 shows the percentage of locations in which the protection levels are exceeded because of nitrogen availability and soil pH. There are more locations in which species are at risk as a result of nitrogen availability than as a result of soil pH. Differences are significant for the 5 and 10 percentile levels, as indicated by the confidence intervals (Table 9.2). Excessive nitrogen availability (eutrophication) thus represents a more serious threat to plant species of deciduous forests growing on nutrient-poor, dry, acidic sandy soils than soil pH (acidification).

Discussion

We have reviewed a species-centred modelling approach that can be used for prediction, setting ecological standards and ranking environmental problems.

The functional soil model, SMART, to be used in the MOVE soil module, will be linked with the probabilistic species-centred vegetation module. The latter is based on an extensive database, which may be extended and optimized in the future. Both abiotic and biotic modules may be used to indicate ecosystem stress. Changes in nutrient cycling and increased leaching of nitrogen, as modelled in the soil module, are often used as ecosystem health indicators (e.g. Rapport, 1989; Ulrich and Bredemeijer, 1990). Changes in species composition as modelled in the vegetation module may provide an early warning of ecosystem collapse (Rapport et al., 1985). Changes in species composition are also relevant if species are target species (V&W, 1989; LNV, 1990).

The quality of the regression analysis depends on the database used for the analysis. Wiertz et al., (1992) point out that the database of Schaminee contains a reasonable number of relevés of almost every plant community (>20 per alliance) and shows an equal distribution of the site factor values over nearly all the classes. However, as botanists have a preference for sampling in ecologically interesting sites, the database will contain relatively few relevés of less interesting sites. In the future we shall test to what extent this influences the results and to what extent this can be corrected.

Several constraining factors are responsible for the shape of the ecological response curve. In the case of a response curve for nutrients, the 'nutrient-poor' side of the ecological amplitude may reflect physiological constraints of a species' nutrient metabolism in combination with reduced abilities of species to compete. The 'nutrient-rich' side may be dictated by nutrient intoxication, in combination with competition. We consider such competition as an integral part of the nutrient stress in real-world circumstances.

MOVE predicts whether or not the conditions conducive to a certain species occurrence are present as a function of (general) environmental policy. Whether or not a species actually occurs at a given location evidently depends on many environmental factors that are not incorporated in the model. The model, for instance, does not take into consideration local differences in management practices. In view of this, field validation of the vegetation module needs more elaboration.

Using MOVE, prediction at the ecotope or plant community level is possible after aggregating the response curves of individual species to ecological response curves for groups of species. There are several ways this can be achieved, for instance, by averaging the separate response curves or by selecting the most sensitive dose-response curve species.

Other models that use dose-reponse curves of ecotopes, rather than of single species, may, in a separate analysis, have used characteristics of individual species to identify the dose-response curve of the ecotopes (see, for example, Claessen et al., chapter 10). The way the species characteristics have been used in such an analysis is, however, often not indicated clearly. Moreover, such dose-response curves cannot be used if the ecosystem classification is changed.

The vegetation module of MOVE strongly resembles the regional species-centred model ICHORS (Barendregt et al., 1986). This model predicts the occurrence probability of each individual species as a function of 22 chemical and physical variables of soil, groundwater, and surface water, using samples in which abiotic site factors and vegetation have been sampled simultaneously. It is, however, unlikely that these models can be developed for the Netherlands in the near future because extensive field inventories are needed as input for their stepwise logistic regression analysis. Moreover, these regional models do not model changes in abiotic site factors, as is done in the abiotic site module of MOVE. Instead, they use predefined values for the abiotic site factors as input data. These regional models can, of course, be used for specific regions for which they are valid to improve the relatively more general vegetation module of MOVE.

Currently, the calibration of Ellenberg indication values in abiotic site factors is being quantified, and the SMART model adjusted. Moreover, the procedure of standard setting is currently under study. The selection of percentiles and the species protection level will be underpinned and the results compared with known ecological standards of a limited number of well-researched plant communities. In this chapter, the primary intention of using the 95% protection levels of all species and the 5 and 95 percentiles of species-response curves is to illustrate the method. The results of these cases are therefore only preliminary.

We believe that MOVE can be extended to other European countries. The soil model, SMART, was developed on a European scale in the context of critical load studies. The Ellenberg indication values, needed in the vegetation module, have been assigned for most European species. A prerequisite for an analysis with MOVE in other countries, however, is a digital database of vegetation relevés with which ecological response curves can be calculated, and a set of relevés with abiotic measurements for the calibration of the Ellenberg values. Auxiliary data required include soil maps, deposition data and flora atlases.

References

Baldock, D., 1990. *Agriculture and habitat loss*, WWF International CAP Discussion Paper, report no. 3, 67 pp.
Barendregt, A., M.J. Wassen, J.T. de Smidt and E. Lippe, 1986. Ingreep-effect voorspelling voor waterbeheer. *Landschap* 1: 41-55.
Bobbink, R., D. Boxman, E. Fremstad, G. Heil, A. Houdijk and J. Roelofs, 1992. Critical loads for nitrogen eutrophication of terrestrial and wetland ecosystems based upon changes in vegetation and fauna. In: P. Grennfelt and E. Thornelof, (eds.), *Critical loads for nitrogen. Report from a workshop held at Lokeberg, Sweden 6-10 april 1992*, Nordic Council of Ministers, Copenhagen.
Bongers, A.M.T., R.G.M. De Goede, F.I. Kappers and M. Manger, 1989. *Ecologische typologie van de Nederlandse bodem op basis van de vrijlevende nematodenfauna. Verslag van het project 'Het functioneren van nematoden in bodemecosystemen'*, RIVM, Bilthoven, report no. 718602002.
CBS, unpublished data, *Annual agricultural statistics*, Centraal Bureau voor de Statistiek, The Hague.

Clausman, P.H.M.A., A.J. Den Held, L.M. Jalink and J. Runhaar, 1987. *Milieuindicaties van vegetaties (TOEWIJS)*. Dienst Ruimte en Groen van de Provincie Zuid-Holland, The Hague, The Netherlands, deelrapport 2, 83 pp.

De Vries, W., and G.J. Heij, 1991. Critical loads and critical levels for the environmental effects of air pollutants. In: G.J. Heij and T. Schnieder (eds.), *Acidification research in the Netherlands*. Elsevier, Amsterdam, pp. 205-214.

De Vries, W., and J. Kros, 1991. *Assessment of critical loads and the impact of deposition scenarios by steady state and dynamic soil acidification models*, Winand Staring Centre, Wageningen, report no. 38, 61 pp.

De Vries, W. and E.E.J.M. Leeters, 1993. *Effects of acid deposition on 150 forest stands in The Netherlands. Relationship between forest vitality and the chemical composition of foliage, humus layer and soil solution*. DLO Winand Staring Centre for Integrated Land, Soil and Water Research, Wageningen, report no. 69.1.

De Vries, W., A. Hol, S. Tjalma and J.C.H. Voogd, 1990. *Literatuurstudie naar voorraden en verblijftijden van elementen in bosecosystemen*. Winand Staring Centre, Wageningen.

De Vries, W., M. Posch and J. Kämäri, 1989. Simulation of the longterm soil response to acid deposition in various buffer ranges, *WASP* 48: 349-390.

Ellenberg, H., 1979. Zeigerwerte der Gefäßpflanzen Mitteleuropas. *Scripta Geobotanica* 9: 1-122.

Ellenberg, H., H.E. Weber, R. Dull, V. Wirth, W. Werner and D. Paulissen (ed.), 1991. *Indicator values of plants in Central Europe*. Erich Goltze, Gottingen.

EPA, 1990. *Reducing risk, setting priorities and strategies for environmental protection*. Report of the Science Advisory Board, Relative Risk Reduction Strategies Committee. US-EPA, Washington D.C..

Erisman, J.W., 1991. *Acid deposition in the Netherlands*. RIVM, Bilthoven, report no. 723001002.

European Inland Fisheries Advisery Commission, 1987. *Working party on water quality criteria for European freshwater fish, Revised report on combined effects on freshwater fish and other aquatic life of mixtures of toxicants in water*, EIFAC Techn Pap report no. 37 Rev. 1, 75 pp.

Gremmen, N.J.M 1987. *Natuurtechnisch model voor de beschrijving en voorspelling van effecten van veranderingen in waterregime op de waarde van een gebied vanuit natuurbehoudsstandpunt, uitgangspunten en modelconcept*. Studiecommissie Waterbeheer Natuur, Bos en Landschap, report no. 1e, 68 pp.

Groen, C.L.G., R. Van der Meijden, R. Huele and M. Van 't Zelfde, 1992. *FLORBASE, een bestand van de Nederlandse flora, periode 1975-1990*. CML, Leiden, report no. 91.

Jongman, R.H.G., C.J.F. Ter Braak and O.F.R. Van Tongeren, 1987. *Data analysis in community and landscape ecology*. Pudoc, Wageningen, The Netherlands, p. 299.

Kämäri, J., M. Amann, Y.W. Brodin, M.J. Chadwick, A. Henriksen, J.P. Hettelingh, J. Kuylenstierna, M. Posch and H. Sverdrup, 1992. The use of critical loads for the assessment of future alternatives to acidification, *Ambio* 21: 377-386.

L&V, 1989. *Natuurbeleidsplan, beleidsvoornemen*. Ministerie van Landbouw en Visserij, The Hague, The Netherlands, 179 pp.

Latour, J.B., and R. Reiling, 1991. *On the Move, concept voor een nationaal effecten model voor de vegetatie (MOVE)*. RIVM, Bilthoven, The Netherlands, report no. 711901003, 23 pp.
Latour, J.B., and R. Reiling, 1992. *Ecologische normen voor vermesting, verzuring en verdroging. Aanzet tot een risicobenadering*, RIVM, Bilthoven, report no. 7119001003.
Latour, J.B., and R. Reiling, 1993. Critical loads for nitrogen based on changes in species composition, Perspectives for a risk assessment. Contribution to the CCE Mapping Workshop, Madrid, Spain, 15-19 march 1993.
Latour, J.B., and R. Reiling, in press (a). Comparative threat analysis, three case studies. *Environment Monitoring and Assessment*.
Latour, J.B., and R. Reiling, in press (b). MOVE, a multiple-stress model for vegetation. *Sci. Tot. Environ*.
Latour, J.B., R. Reiling and F. Bekhuis, 1993. Comparative threat analysis using a risk assessment. In: T. Schneider (ed.), *Comparative Risk Analysis and Priority Setting of Air Pollution Issues*, June 7-11, Keystone, Colorado.
Leemans, R., 1992. Modelling ecological and agricultural impacts of global change on a global scale. In: *Journal of Scientific & Industrial Research* 51, aug./sept., 709-724.
Leemans, R., in press. The use of plant functional type classifications to model the global land cover and simulate the interactions between the terrestrial biosphere and the atmosphere. In: T.M. Smith, H.H. Shugart and F.I. Woodward (eds.), *Plant Functional Types Classifications*. Cambridge University Press, Cambridge.
LEI-DLO, 1991. *Berekeningen ten behoeve van de Nationale Milieuverkenningen 2 1990-2010 in opdracht van het RIVM*. Landbouw Economisch Instituut, The Hague.
LNV, 1990. *Natuurbeleidsplan, regeringsbeslissingen*. Ministerie voor Landbouw, Natuurbeheer en Visserij, The Hague, Tweede Kamer vergaderjaar 1989-1990, 21149, nrs. 2-3, 272 pp.
LNV, 1992. *Meerjarenprogramma Natuur en Landschap*. Ministerie voor Landbouw, Natuurbeheer en Visserij, Tweede Kamer vergaderjaar 1991-1992, 22303, nrs 1-2, 117 pp.
Loopstra, I.L., and E. Van der Maarel, 1984. *Toetsing van de ecologische soortengroepen in de Nederlandse flora aan het systeem van indicatiewaarden volgens Ellenberg*. De Dorschkamp, Wageningen, report no. 381, 143 pp.
Maas, R. (ed), 1991. *National Environmental Outlook 2 1990-2010*. RIVM, Bilthoven, The Netherlands, 550 pp.
Peters, R.L., and J.D.S. Darling, 1985. The greenhouse effect and nature reserves. In: *Bioscience* 35/11, 707-717.
Rapport, D.J., 1989. What constitutes ecosystem health? *Perspect. Med. Biol*. 30: 120-133.
Rapport, D.J., H.A. Regier and T.C. Hutchinson, 1985. Ecosystem behaviour under stress. *Am. Nat*. 125: 617-640.
Runhaar, J., R.A.M. Stevers and G. Baarse, 1985. Beleidsanalyse kustverdediging texel. *Landschap* 2: 88-98.
Schaminee, J.H.J., V. Westhoff and G. Van Wirdum, 1989. Naar een nieuw overzicht van de Plantengemeenschappen van Nederland. *De Levende Natuur* 90: 204-209.

Slooff, W., 1992. *RIVM Guidance Document, Ecotoxicological effect assessment. Deriving Maximum tolerable concentrations from single-species toxicity data.* RIVM, Bilthoven, The Netherlands, report no. 719102018, 49 pp.

Snedecor, G.W., and W.G. Cochran, 1989. *Statistical methods.* Iowa State University Press, Ames, Iowa, p. 503.

Stigliani, W.M., 1991. *Chemical Time Bombs, definition, concepts and examples.* IIASA, Laxenburg, Austria, 23 pp.

Svedrup, H.U., W. De Vries and A. Henriksen, 1990. *Mapping critical loads, A guidance manual to criteria, calculation, data collection and mapping.* Nordic Council of Ministers, Copenhagen, 14 pp.

Ulrich, B.B.C., and M. Bredemeijer, 1990. Ecological indicators of temperate forest ecosystem condition. In: D.H. McKenzie, D.E. Hyatt and V.J. McDonald (eds.), *Ecological Indicators.* Fort Lauderdale, Miami, U.S.A.

V&W, 1989. *Derde Nota Waterhuishouding, Water voor nu en later.* Ministerie van Verkeer en Waterstaat, The Hague, Tweede Kamer vergaderjaar 1988-1989, 21250, nrs. 1-2, 297 pp.

Van den Berg, S., R.G.M. De Goede and F.I. Kappers, 1990. *Fysisch-chemische analyses van bodemmonsters en topografische beschrijving van bemonsterde locaties t.b.v. het project bodemclassificatie.* RIVM, Bilthoven, report no. 718819001.

Van der Eerden, L.J., T.A. Dueck, A.C. Posthumus and A.E.G. Tonneijck, 1992. Assessment of critical levels for air pollutant effects on vegetation, some considerations and a case study on NH3. In: UNECE Workshop on Critical Levels, U.K. Egham, 23-26 March 1992.

Van Strien, A., 1991. *Maintenance of plant species diversity on dairy farms.* Ph.D. thesis, Leiden University.

VROM, 1989. *National Environmental Policy Plan - to choose or to loose.* Ministry of Housing, Physical Planning and Environment, The Hague, Tweede Kamer vergaderjaar 1988-1989, 21137, nos. 1-2, 257 pp.

VROM, 1991. *Notitie Milieukwaliteitsdoelstellingen bodem en water.* Tweede Kamer vergaderjaar 1990-1991, nr. 21990, The Hague, 38 pp.

Wiertz, J., 1992. *Schatting van ontbrekende vocht- en stikstofgetallen van Ellenberg (1979).* IBN, Wageningen, report no. 92/7, 32 pp.

Wiertz, J., J. Van Dijk and J.B. Latour, 1992. *De MOVE-vegetatie module. De kans op voorkomen van 700 plantesoorten als functie van vocht, pH, nutrienten en zout.* IBN/RIVM, Wageningen/Bilthoven, report no. IBN 92/24; RIVM 711901006, 138 pp.

Witte, J.P.M., F. Klijn, F.A.M. Claessen, C.L.G. Groen and R. Van der Meijden, 1992. A model to predict and assess the impacts of hydrologic changes on terrestrial ecosystems in the Netherlands, and its use in a climate scenario. *Wetlands Ecology and Management* 2: 69-83.

Witte, J.P.M., F.H.M. Van der Ven, F.A.M. Claessen, C.R.A. Overmars, W. Stortelder, H.W.J. Van der Valk, R. Van der Meijden and G. Veenbaas, 1989. Predicting the effects of national water management on terrestrial ecosystems in the Netherlands. In: Closing the gap between theory and practise, Proceedings of Baltimore Symposium, May 1, 1989, IAHS publication no. 180.

Ecosystem classification and hydro-ecological modelling for national water management   10

Frans A.M. Claessen, Frans Klijn, J. Flip (P.) M. Witte and J. Gerard Nienhuis

ABSTRACT - The Netherlands is confronted with a deterioration of vegetations of wet and moist sites due to a lowering of the groundwater level, a diminished intensity of seepage, and the inlet of surface water of poor quality. This deterioration is called 'drought damage'. Its recognition is now leading to a revision of the ideas on water use and water management.
In support of the Netherlands' policy on surface water and groundwater management, a policy analysis was carried out. For this analysis, the hydro-ecological model DEMNAT (Dose-Effect Model for terrestrial NATure) was developed for predicting the ecological effects of changes in water management.
The model consists of three parts: a geographical database, dose-effect functions, and a nature valuation system. All three refer to ecosystems, which requires a classification.
Instead of developing one ecosystem classification specific for this case, we combined an existing classification of ecotopes with a specially developed classification of ecoseries. The ecotopes classification relates the species composition of the vegetation to operative site factors, such as acidity or moisture availability. The ecoseries classification relates these site factors to conditioning soil and groundwater characteristics, such as texture, organic matter content, and groundwater level. The ecoseries also determine the response of the operative site factors to changes in the local hydrology. The combination of ecotopes and ecoseries provided the ecosystem types for which the calculations were carried out. In mapping terms, an overlay of ecotopes on ecoseries was produced. For all relevant combinations of ecotopes on ecoseries, dose-effect functions were defined for a number of hydrological doses.
We present some results of the environmental impact assessment, carried out with DEMNAT for a number of groundwater extraction scenarios. We predicted the changes in nature value of 15 selected ecotope types for gridcells of 1 km^2.

Introduction

Since about 1950, much intervention in the Netherlands' water management has resulted in severe damage to terrestrial ecosystems at wet and moist sites. These interventions caused lowered groundwater levels, diminished intensity of upward seepage, and the inlet of surface water of poor quality to compensate for water shortages. The deterioration of the ecosystems is called drought damage or 'desiccation'.[1]
In the eighties 75% of wet and moist sites were found to suffer from drought damage (Braat et al., 1989). The main interventions which caused this deterioration of nature values were an increased drainage of agricultural and urban areas, an intensified regulation of surface water levels, a doubling of the groundwater extraction for public water supply since the fifties, and an increased sprinkling irrigation in agriculture from both groundwater and surface water.

These interventions changed the hydrological conditions within and around natural sites. In general, they resulted in a lowering of the groundwater levels between 0.2 and 1.0 metre, as well as in a lowering of the surface water level between 0.1 and 0.5 metre (Rolf, 1989). Also, changes occurred in the quality of both groundwater and surface water, involving a lowering of the pH, or an increase in the availability of nutrients in the rooting zone. Consequently, only small areas remain with wet or moist, nutrient-poor, pH-buffered conditions.

The recognition of drought damage as an important environmental problem is now leading to a revision of the ideas on water use and water management. The revised ideas are made explicit in current policy plans concerning the national policy on water management in general and public and industrial water supply in particular.
In scientific support of the national policy on surface water and groundwater management, a policy analysis was carried out (Claessen and Witte, 1991). For the prediction of the ecological effects of water management alternatives, a Dose-Effect Model for terrestrial NATure (DEMNAT) was developed. This model is suited for determining changes in the nature value of terrestrial ecosystems resulting from changes in abiotic conditions following water

---

[1] Drought damage and 'desiccation' are both inadequate translations of the Dutch 'verdroging'. 'Verdroging' is one of the themes for the Netherlands' environmental policy used to indicate a group of related environmental problems resulting from water shortages due to groundwater extraction and enhanced drainage.

management interventions. For practical reasons, DEMNAT focuses on the species composition of the vegetation, because this is the most affected component of terrestrial ecosystems.

Recently, DEMNAT was further developed to enable application in an environmental impact assessment for the National Policy Plan on Public Drinking Water Supply and Industrial Water Supply (Beugelink et al., 1992; 1993).

After this general introduction on the policy context of DEMNAT's development, we shall give a short overview of the concepts and tools. Then we shall go into the requirements and use of ecosystem classifications in hydro-ecological modelling in more detail. In the next section, the DEMNAT model is described, followed by an illustration of its application in the environmental impact assessment for the National Policy Plan on Public Drinking Water Supply and Industrial Water Supply. We shall end with a discussion.

Concepts and tools

Predictive modelling was the predominant approach to all policy questions concerning the effects of various water management scenarios, in an attempt to quantify these effects. With regard to the *ecological* effects of water management scenarios, the study was limited to vegetation as the most affected component of wet and moist ecosystems. The results of the predictive modelling were expressed in both objective terms, such as loss of surface area and species richness, and in normative terms, i.e. nature value.

The *hydro-ecological* modelling was based on the following chain of processes (see also Figure 10.1):

1. Water management interventions result in changes in local hydrological conditions, namely the spring groundwater level (SGL), the intensity of upward seepage, the surface water level, and the share of allochtonous surface water of poor quality.
2. The local hydrological changes result in changing moisture content, the mineralization of organic matter and soil acidification. These processes are controlled by *conditioning* factors, such as soil texture, organic matter content, $CaCO_3$ content, groundwater level, and the chemical composition of seepage water.

3. These processes result, in turn, in changes in *operative* site factors, namely the moisture availability, the nutrient availability and the acidity in the rooting zone.
4. Finally, the changes in operative site factors result in changes in the species composition of the vegetation. These changes are expressed in terms of relative species richness or 'completeness', and in terms of relative nature value.

For the various steps in this chain of effects, different models were used or newly developed.

First, a national groundwater model ('LGM'; Pastoors, 1992; 1992a) was developed to calculate the changes in groundwater levels, the changes in the surface water levels of small bodies of water, and the changes in the upward seepage of groundwater, all due to changes in groundwater withdrawal. These calculations are carried out per gridcell of 1 km^2.

Second, a national surface water distribution model (DM-DEMGEN) was used to estimate the changes in water quality inside 80 water management districts, due to the inlet of polluted water from the rivers Rhine and Meuse (Pakes et al., 1992). It was used to calculate the ratio between the volumes of inlet water and the resident volume of surface water within a district during the growing season. The higher the share of inlet water, the larger the adverse effect on surface water quality.

The forementioned two models are hydrological models, primarily used for calculating local changes in hydrological conditions. The third model, then, is DEMNAT (Witte, 1990; Witte et al., 1992a and b; Nienhuis et al., 1992), intended for predicting the biotic response to the local hydrological changes. For DEMNAT, the hydrological changes at site level function as dose, whereas the effects are expressed in terms of changed species composition, as well as in terms of changed nature value.

Before elaborating further on DEMNAT, we shall elucidate the role of the ecosystem classifications we used for the hydro-ecological modelling.

Ecosystem classification for DEMNAT

Hydrological changes cannot be translated into changes in the species composition of the vegetation directly, because a number of processes in abiotic components are also involved. Some of these, such as mineralization or acidi-

ECOSYSTEM CLASSIFICATION FOR WATER MANAGEMENT 203

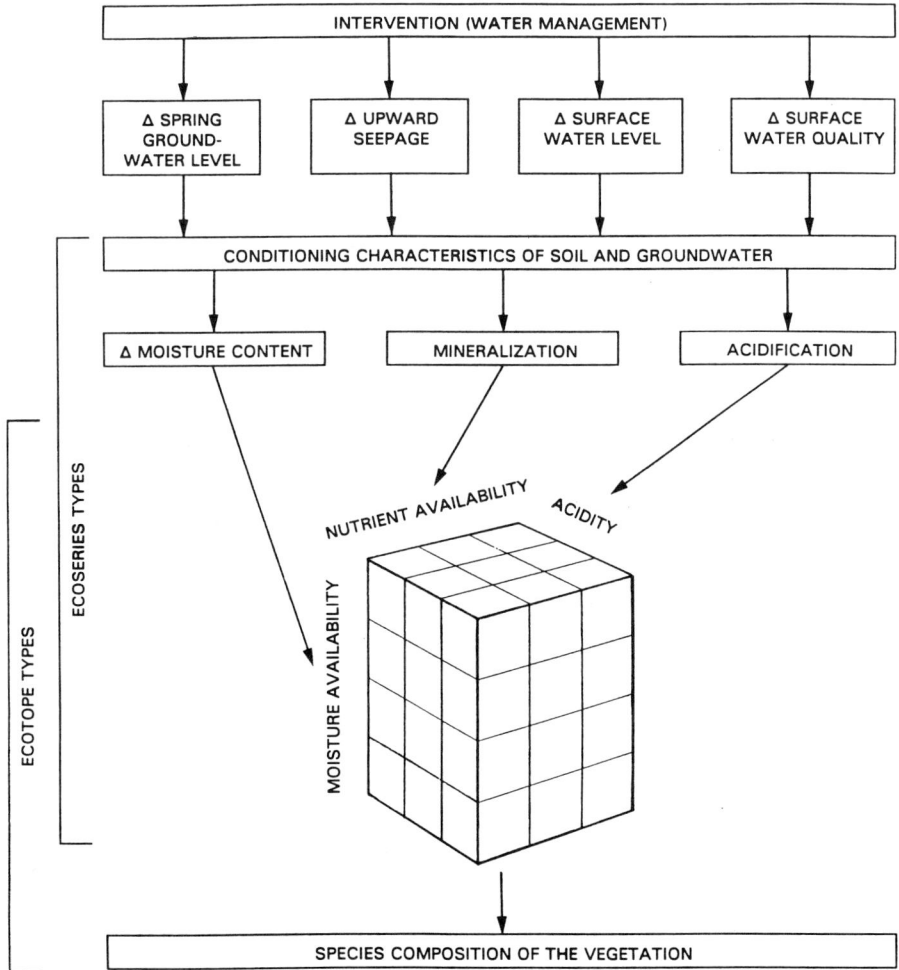

Figure 10.1 Process chain resulting from interventions in water management, distinguished for predicting responses of operative site factors and species composition of the vegetation (adapted from Van Beusekom et al., 1990). The place of ecoseries and ecotopes has been indicated on the left (Klijn et al., 1992)

dification, may aggravate the effects of groundwater lowering, whereas others, such as water retention or capillary rise, diminish the effects.

Because of these intermediate abiotic processes, *two related classifications* were used in the model as an *interface* between the hydrological doses on one side and the biotic response on the other. We combined an existing classification of *ecotopes* (Stevers at al., 1987; Runhaar et al., 1987) with a specially developed classification of *ecoseries* (Klijn et al., 1992). The ecotopes classification relates the species composition of the vegetation to operative site factors, such as acidity or moisture availability. The ecoseries classification relates these operative site factors to conditioning soil and groundwater characteristics, such as texture, organic matter content, or groundwater level. Since the ecoseries are classified on such conditioning characteristics, they also determine the response of the operative site factors to changes in the local hydrology (see Figure 10.1, left margin). The ecotopes determine the response of the species composition to the changes in the operative site factors (see Figure 10.1, left margin).

As to the question of ecosystem classification, we should notice that ecotopes are generally considered as being defined by both conditioning *and* operative characteristics. However, for classification purposes, one should select a limited number of classification characteristics. Moreover, for practical nationwide use, a classification should also be simple. To this end, the existing ecotope classification (see Runhaar and Udo de Haes in chapter 7) took only operative factors into account as classification characteristics. As this classification proved practicable for impact assessment, we chose to use it in combination with another relatively simple classification, which emphasizes the relation between conditioning and operative characteristics. Therefore, instead of developing one new classification, for practical reasons we chose to combine two related classifications. Only the combination of the two classifications yields the ecosystem types for which all further calculations were carried out. In mapping terms, this combination can be regarded as an overlay of an 'ecological' vegetation map and an 'ecological' soil/groundwater map: the 'true' ecotopes.

After this explanation, we shall have a further look at the classifications of ecotopes and ecoseries we used.

Ecotopes classification

Ecotopes were defined as spatial units that are homogeneous to vegetation structure, succession stage, and operative site factors which determine the species composition of the community, in this case, the vegetation.
The abiotic operative site factors are used for defining abiotic sites. The most important site factors were found to be moisture availability, nutrient availability, acidity, and salinity (Stevers et al., 1987; Runhaar et al., 1987). These are used as classification characteristics. All the relevant combinations of classes that can be encountered in reality were distinguished as *abiotic site types*.
These abiotic site types determine the 'potential vegetation', in terms of series of successional stages. Within these site types, a further distinction in *ecotope types* is based on vegetation structure/succession stage. Thus, for example, pioneer vegetation, grassland, tall herbaceous vegetation, shrubland, and woodland are distinguished on sites with similar abiotic site conditions.
For more details about this ecotope classification we refer to chapter 7, where Runhaar and Udo de Haes treat it elaborately.

For the predictive modelling with DEMNAT we clustered the ecotope types into a smaller number of ecotope groups according to similarity of vegetation structure. Thus, only aquatic, herbaceous, and woody vegetations were distinguished. The main reason for this clustering was a practical one. It was found to be impossible to acquire sufficient data to draw up a nationwide picture of all the distinctions regarding vegetation structure. For reasons of simplicity, we shall only use the terms ecotope and ecotope type.

Ecoseries classification

An ecoseries classification was developed especially for DEMNAT (Klijn et al., 1992) in order to relate operative site factors to conditioning site factors because the conditioning site factors control the response of ecosystems to changes in local hydrological conditions. In fact, the operative site factors used to classify ecotope types can be regarded as in any case depending on the interaction of conditioning site characteristics, such as soil texture, organic matter content, $CaCO_3$ content, groundwater level and the chemical composition of upward seepage.
Against this background, ecoseries were defined as spatial units that are homogeneous to those abiotic ecosystem characteristics that control the

operative site factors determining the species composition of the vegetation, and/or that control the effects of abiotic processes resulting from interventions and depositions in the environment.

In order to select the relevant abiotic ecosystem characteristics referred to in this definition, the relations between operative and more conditioning site characteristics were investigated. Also, the processes related to lowering the groundwater and associated interventions were scrutinized.

From these analyses a large number of relevant ecosystem characteristics were derived. In theory, all would do as classification characteristics. As this would be very impractical, the long list was arranged in groups on the basis of correlative and causal relations between the individual characteristics, resulting in the selection of seven classification characteristics which were considered sufficient for a comprehensive classification:
- parent material/texture
- profile differentiation
- organic matter content
- $CaCO_3$ content
- enrichment with iron-oxides
- groundwater level
- salinity/cation composition of upward seepage

These classification characteristics were selected with two considerations in mind:
- Primarily, there should be no doubt about the ecological relevance, which is to be considered as an unambiguous relation with the operative site factors of the ecotope classification;
- Secondly, it should be possible to relate the characteristics unambiguously to available geographical data. This consideration is mainly of practical importance.

For each classification characteristic a limited number of classes was defined. The combination of the classes yielded the ecoseries classification, which can be interpreted as a combination of an ecological soil classification, an ecologically relevant classification of groundwater levels, and an eco-hydrological classification of groundwater balances. The classification of groundwater level is mainly based on the Mean Spring Groundwater Level (MSGL).

DEMNAT

The prediction model in its present state consists of three different parts, namely a geographical database of the Netherlands concerning the spatial distribution of the ecosystem types, dose-effect functions, and a nature valuation system. These parts will be described successively. We shall refer to the most recent version of DEMNAT, as it was used for the environmental impact assessment for the national public water supply policy. This means that a number of points made below are specific for this nationwide study, but are not inherently connected with DEMNAT's structure as such. DEMNAT may be used with other data for regional applications as well.

Geographical database

One of the main problems we encountered was in assessing the impact of changes in the water management at a national scale on terrestrial ecosystems at a local scale, i.e., ecotope scale level. Additional complicating factors originated from the need to make predictions for the entire country and the availability of ecological data for such a large area.

The hydrological doses could be calculated with a spatial accuracy of about one 1 km^2. For the distribution of inlet water, this required a scaling down of the calculations for the 80 water management districts. A higher accuracy than 1 km^2 would hence not be achievable.
A 1 km^2 grid would also allow representing the whole country by only about 36,000 gridcells. This was considered manageable.
Finally, data on the relevant ecological characteristics were also available or could be gathered for 1 km^2 gridcells. It concerned data on soil, groundwater, and species composition.
It was therefore decided that the geographical *database* was to contain data for a 1 km^2 grid over the Netherlands. However, all the *calculations* were carried out for all combinations of hydrological dose, ecoseries, and ecotope, in other words, for all combinations of hydrological conditions, soil conditions and vegetation *within* the gridcells. Hence, a number of calculations were carried out for each gridcell. We shall go into this further in the next section on dose-effect functions.

As to the geographical distribution of ecoseries, this was established from an existing soil data base of the Netherlands which was developed for the Landscape Ecological Mapping of the Netherlands (Canters et al., 1991; De Waal, 1992). The 230 soil legend units were generalized by aggregation into 49 'ecological soil types'. The combination of these ecological soil types with six groundwater level classes results in ecoseries types. For each square kilometre, the surface of all ecoseries is given in hectares.

The ecoseries are of course defined in terms of conditioning soil and groundwater characteristics, because such characteristics were used for the classification. However, a *translation* in terms of operative abiotic site factors is possible, due to the relation between conditioning and operative characteristics. The translation should then be based on so-called *site-diagrams*, which were drawn up for all ecoseries types based on expert judgement. These site-diagrams express the estimated distribution of abiotic site types for each ecoseries type in percentages.

Such a translation may yield maps which show the potential distribution of abiotic site types, or the 'potential vegetation', under undisturbed circumstances. Plate 10.1 is an example of such a map, in this case of wet, moderately rich sites.

The geographical distribution of ecotopes was determined by using the presence of species in the 1 km^2 gridcells as a diagnostic characteristic. The *ecological species groups* that are specified for each ecotope type (see Runhaar and Udo de Haes in chapter 7) enable such a use of floristic data for estimating the occurrence of ecotopes (see also Groen et al. in chapter 13). The theoretically better solution of using vegetation maps and inventories or direct mapping of ecotope types was found impossible, because of insufficient data.

A floristic database was set up for a first application in DEMNAT. This so-called FLORBASE is still being extended (see also Groen et al. in chapter 13). From FLORBASE, maps of 15 selected ecotopes of wet and moist sites were derived, using 1187 of the species (Witte and Van der Meijden, 1992).

The present (1975-1990) and past (before 1950) distribution and relative species richness of ecotopes were established for the 15 selected types. Plate 10.2 shows the present distribution and relative species richness, in three classes, of ecotope K27, i.e., herbaceous vegetations on wet, moderately rich sites.

A comparison of Plates 10.1 and 10.2 reveals that the present distribution of ecotopes of this type corresponds to the estimated potential distribution of the abiotic site it belongs to. This is remarkable, since marked declines in species richness have occurred during the last decades, as was found by comparing the present situation with the pre-1950 situation. Similar correspondences were found for the other abiotic site types.

In the geographical database, a connection was made between the *presence* of ecotopes of a certain type and the *surface area* of ecoseries. Such a connection is needed to determine the share of the various ecosystem types to be used for the calculations in each gridcell.

Dose-effect functions

Dose-effect functions were defined by Van der Linden et al. (1992) for all existing combinations of ecoseries types and ecotope types for a number of relevant doses, i.e., changes in local hydrology. These functions describe the decrease in the relative richness of the species belonging to the same ecological species group in relation to a range of hydrological doses.

The dose-effect functions were built up stepwise. First, for each ecoseries type the hydrological doses were transformed into changes in separate operative site factors. Then, for each ecotope type the changes in relative species richness were estimated as a consequence of these changes in operative site factors. Finally, the thus-found dose-effect functions were compared with available empirical data (see also Runhaar and Udo de Haes in chapter 7).

Figure 10.2 shows a number of dose-effect functions for a lowering of the spring ground water level (SGL) for an ecotope type on five different ecoseries. In this case, the dose-effect functions express the relation between relative species richness and SGL.

Figure 10.3 shows, more schematically, the difference in response between, respectively, a lowering and a raising of the groundwater level. In the latter case, the recovery of the relative species richness is not 100% within a period of 10 to 20 years, because either the recovery of the abiotic site conditions is not quite reversible, or species may not be able to establish themselves again after complete disappearance.

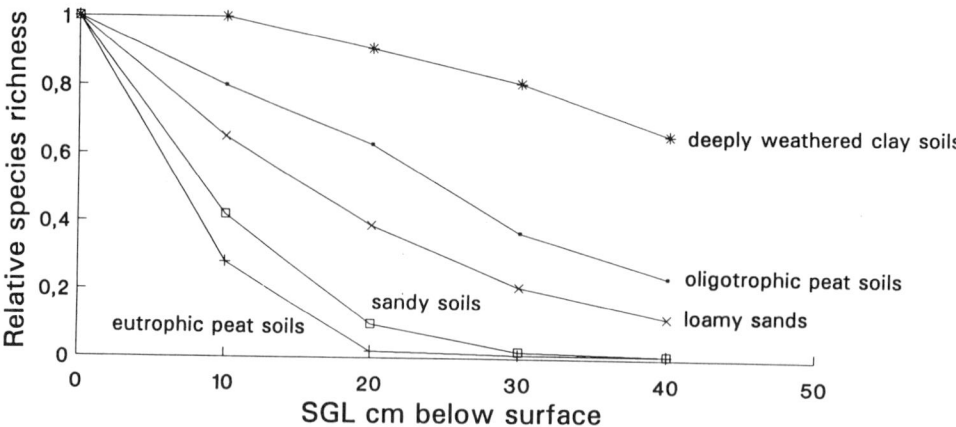

Figure 10.2 Dose-effect functions for lowering of the spring groundwater level (SGL) and the resulting decrease in relative species richness for the ecotope type K21 (herbaceous vegetations of wet, nutrient-poor, acid sites) on 5 different ecoseries named after the main soil type (Van der Linden et al., 1992)

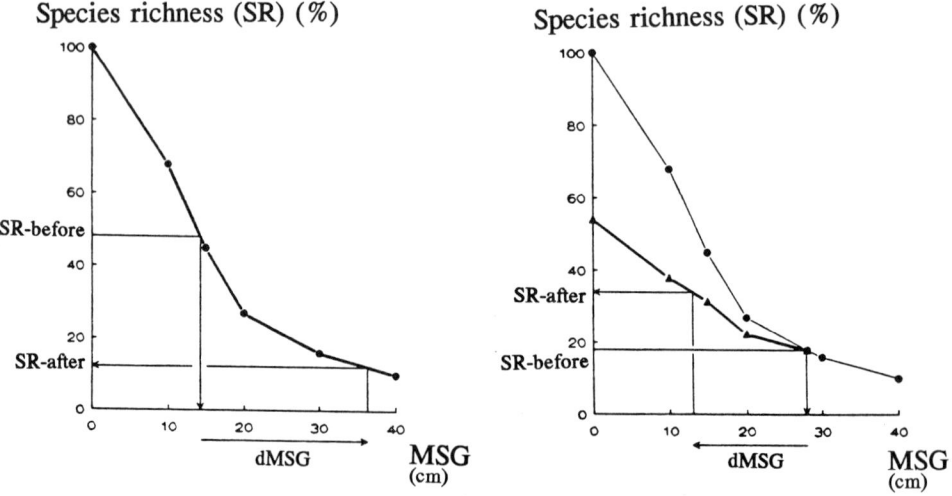

Figure 10.3 Dose-effect functions for lowering (left) and raising (right) the spring groundwater level (SGL), illustrating a hysteresis due to retarded recovery

Despite the fact that dose-effect functions were specified for all combinations of ecotopes on ecoseries, some trials proved that biased outcomes of the predictive modelling occurred because some characteristics of local situations were not covered by the classifications. Such biased outcomes might result, for example, from the presence of perched water tables, i.e., shallow groundwater levels that occur above impermeable layers in the soil. In such cases, groundwater lowering in the region will not affect the local groundwater level, nor the vegetation. Subdued response could also occur if, for example, ditches are so frequent that plant species can easily shift down the ditch-banks when the water level is lowered. Process-response conditioning of this kind was handled either separately in DEMNAT, or by improving the dose-effect functions. In this context, the following additional characteristics were specified for all ecoseries:
- probability of occurrence of upward seepage;
- probability of sufficient available soil water in the profile to bridge dry periods;
- probability of occurrence of a perched water table; and
- abundance of ditches.

The specification of these characteristics enabled corrections to be carried out.

Nature valuation system

A specific procedure was developed for nature valuation. This should primarily enable the expression of the predicted effects in normative terms but, secondly and even more importantly, enable a summation of all the results for each gridcell in one unifying term (Witte and Van der Meijden, 1992).

The valuation was based on the following commonly used criteria:
- national rarity
- international rarity
- completeness
- surface area per gridcell

National and international rarity are used to assess the value of the different *ecotope types*. This needs to be determined only once for each type.

In contrast, completeness is used to determine the 'floristic' quality of an actually found *ecotope* within a gridcell. It is calculated on the basis of the actual number of indicator species relative to an upper and lower threshold value for species number. These threshold values have been set specifically for each ecotope type. The lower threshold is set to achieve as much certainty as possible about the real occurrence of the ecotope as a spatial unit, in other words, to avoid noise. In contrast, the upper threshold is set in relation to what is regarded as a 'completely developed' ecotope of the given type. Hence, the completeness can be regarded as a measure of relative species richness, in this case, relative to the thresholds. Completeness was determined from FLORBASE (Witte and Van der Meijden, 1992).

Also, the surface area on which an ecotope can be found in reality is specific for each gridcell. Hence, completeness and surface area of a suitable abiotic site type are used to calculate values for individual gridcells.

The sum of the values for all ecotopes established in a gridcell, weighed according to the value of the respective types to which they belong, is a measure of the total value.

Naturally, a comparison of the present nature value with the predicted nature value for a number of scenarios yields changes in nature value for each gridcell. Next, the values for all gridcells of the entire country, or any region, can be translated into overall scores.

The application of DEMNAT for the Netherlands' policy on public and industrial water supply

From 1991 to 1993 an environmental impact assessment was carried out for the National Policy Plan on Drinking and Industrial Water Supply on behalf of the Ministry of Housing, Physical Planning, and the Environment (Beugelink et al., 1992; 1993).

The ecological impacts of various scenarios for groundwater extraction in the next decades were analysed. The scenarios are described in Table 10.1, and Figure 10.4 shows the extraction sites for public water supply. The majority of the scenarios involve a decrease in groundwater extraction, which reflects the political desire to survey the possibilities to counteract drought damage.

Plate 10.3 shows the model area, which covers about 75% of the country. Approximately 90% of the extracted quantity of groundwater comes from the modelled area. Furthermore, this figure shows the rise of the groundwater level as calculated for scenario 3, i.e. a 50% decrease of the extraction for drinking water. In this scenario, in some regions the level can rise from several decimetres up to a metre.

ECOSYSTEM CLASSIFICATION FOR WATER MANAGEMENT 213

□ g.w. extraction from phreatic aquifers
• g.w. extraction from semi-confined aquifers
▼ g.w. extraction at river bank infiltration sites

Figure 10.4 Extraction sites for public water supply in the Netherlands

Table 10.1 Scenarios for groundwater extraction in the Netherlands

scenario	description
1	increase of groundwater extraction at all sites for public water supply by 25% compared to the year 1988
2	decrease of groundwater extraction at all sites for public water supply by 25% compared to the year 1988
3	decrease of groundwater extraction at all sites for public water supply by 50% compared to the year 1988
4	decrease of groundwater extraction at all sites for industrial water supply by 50% compared to the year 1988
5	termination of groundwater extraction at all sites for industrial water supply
6	termination of groundwater extraction from phreatic aquifers at all sites for drinking water supply
7	termination of groundwater extraction from semi-confined aquifers at all sites for drinking water supply
8	termination of all groundwater extraction at river bank infiltration sites for drinking water supply

Plate 10.4 shows the present nature value, i.e., for the period 1975-1990, totalled for all 15 ecotope types taken into account. Plate 10.5 illustrates the change in total nature value for scenario 3, relative to the actual situation. The present nature value of all ecotopes within the model area equals 162,000 nature value units (nvu). In the past, i.e. around 1950, this was 244,000 nvu.

The consequences of the eight scenarios on total nature value within the model area are summarized in Table 10.2. It demonstrates that a 25% increase of the groundwater extraction for drinking water supply, corresponding to scenario 1, would lead to a further reduction of the present nature values by 3.5% nationwide.

Table 10.2 The predicted consequences of the eight scenarios (first two columns) in terms of absolute changes of nature value units (dN [nvu]), as percentage of the present nature value (dN [%]), and relative to the extracted amount of groundwater

scenario	dQ [10^6 m^3/yr]	dN [nvu]	dN [%] i.r.t. present situation	dN/dQ [nvu/10^6m^3/yr]
1	+ 180	- 5,600	- 3.5	- 31
2	- 180	4,700	2.9	26
3	- 360	8,900	5.5	25
4	- 130	1,500	1.0	11.5
5	- 260	3,300	2.0	13
6	- 220	8,600	5.3	39
7	- 480	7,600	4.7	16
8	- 25	200	0.1	8

In the very hypothetical case that all groundwater extractions would be terminated, i.e. scenarios 5, 6, 7, and 8 combined, an increase in nature value of about 12% compared to the present situation could be expected.

Figure 10.5 shows the relationship between the change in extraction of groundwater per year and the total change in nature value (Beugelink et al., 1992). The figure shows a linear relationship per type of extraction as represented by different symbols. Reducing the extraction of groundwater for drinking water supply seems more profitable than reducing groundwater extraction for industrial purposes. This phenomenon can be explained by the fact that both drinking water extraction sites and nature conservation areas are often located at far distances from highly urbanized or industrialized areas. Consequently, they may be close together or overlapping. In contrast, industrial activities and industrial extraction sites are usually situated in or close to urban areas with few nature values.

Finally, Figures 10.6 and 10.7 show the contribution to the total present nature value of each ecotope type, as well as the change in nature value for scenario 3 for each ecotope type.

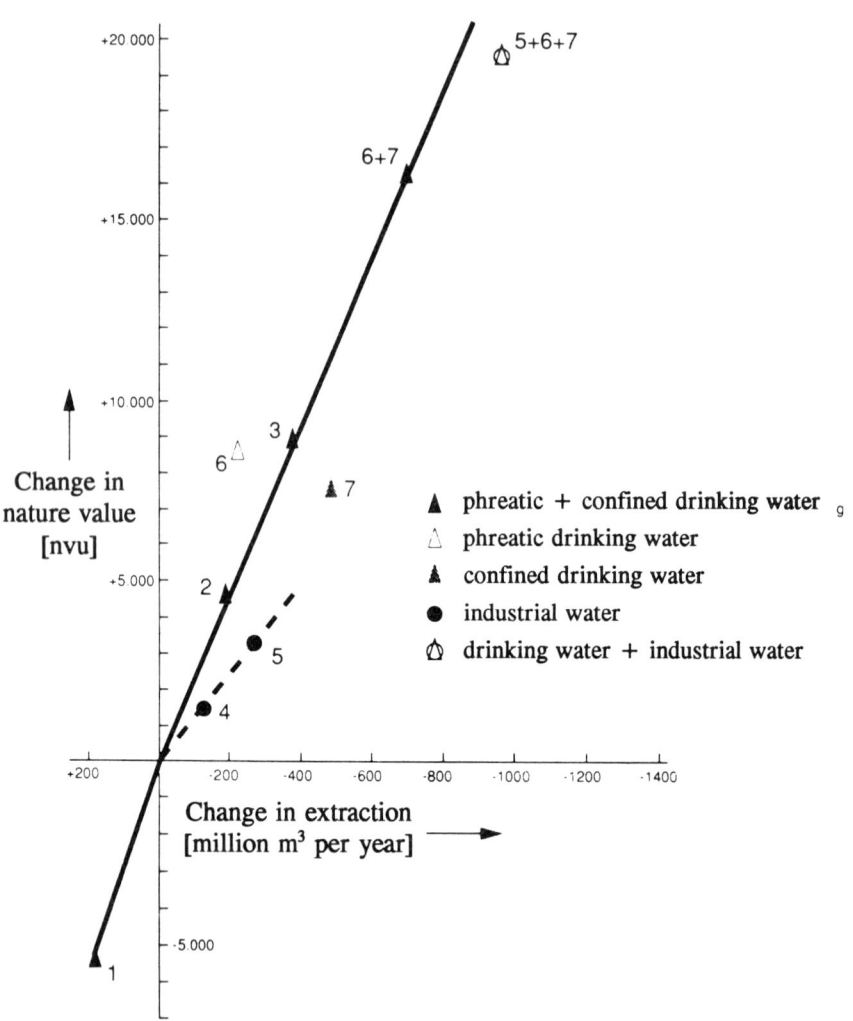

Figure 10.5 The relationship between the change in the nationwide extraction of groundwater per year and the total change in nature value for the 15 ecotope types in the entire modelled area (Beugelink et al., 1992)

It appears that at present the total nature value of all 15 ecotope types is determined mainly by those of more common type, i.e., ecotopes of nutrient-rich sites. These form the largest share in Figure 10.6, as demonstrated by the ecotope types A18, K28 and H28, respectively aquatic, herbaceous and woody

ECOSYSTEM CLASSIFICATION FOR WATER MANAGEMENT 217

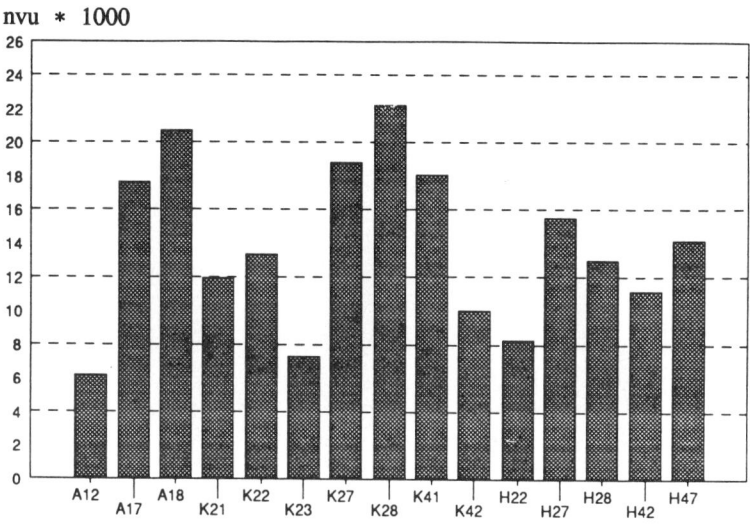

Figure 10.6 Present nature values of the 15 ecotope types. The codes used to indicate the ecotope types are explained in the legend below

	Ecotope type	
Code	Abiotic site type	Vegetation
A12	Nutrient-poor, acid and neutral waters	semi-aquatic and helophytic veget.
A17	Moderately nutrient-rich waters	as A12
A18	Very nutrient-rich waters	as A12
K21	Wet, nutrient-poor, acid soils	pioneer vegetations and grasslands
K22	Wet, nutrient-poor, weakly acid soils	as K21
K23	Wet, nutrient-poor, alkaline soils	as K21
K27	Wet, moderately nutrient-rich soils	as K21 and tall herb vegetations
K28	Wet, very nutrient-rich soils	as K21 and tall herb vegetations
K41	Moist, nutrient-poor, acid soils	as K21
K42	Wet, nutrient-poor, weakly acid soils	as K21
H22	Wet, nutrient-poor, weakly acid soils	woodlands and shrublands
H27	Wet, moderately nutrient-rich soils	as H22
H28	Wet, very nutrient-rich soils	as H22
H42	Moist, nutrient-poor, weakly acid soils	as H22
H47	Moist, moderately nutrient-rich soils	as H22

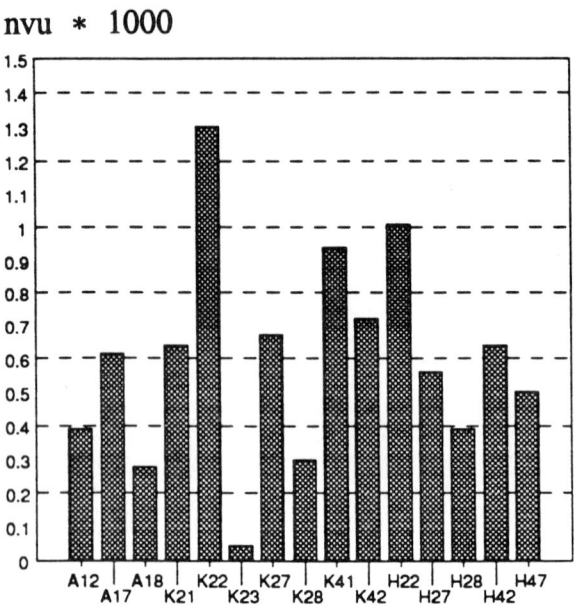

Figure 10.7 Change in the nature values of the 15 ecotope types, calculated for scenario 3. The same legend applies as for Figure 10.6

vegetations of wet, nutrient-rich sites. This need not be a surprise for the heavily eutrophicated Netherlands.

In contrast, Figure 10.7 illustrates that the *change* in nature value is determined by the rarer, highly valued ecotope types. Examples of such ecotope types are K22 and H22, respectively herbaceous and woody vegetations of wet, nutrient-poor, weakly acid sites. These are among the rarest site types at present, which implies a high value for the ecotope type as such. This partly explains their influence on the total change. However, they are also among the most threatened abiotic site types, as it is obvious that they are the most likely to suffer not only from drought damage, but also from eutrophication and acidification. One would expect K23, the herbaceous vegetations of wet, nutrient-poor, alkaline sites, to have a similarly high share in the change of nature value, but unfortunately the coastal dune area, in which ecotopes of this type are almost exclusively found, was outside the modelled area.

Discussion

The example given above concerning the application of DEMNAT for the national public water supply demonstrates that it is a practical, hydro-ecological prediction model for environmental impact assessment. It yielded useful results showing clearly distinguished differences between the ecological impact of different groundwater extraction scenarios for the next decades.

We should not forget, however, that in the present study only very simple scenarios have been simulated. These give some insight into the general relation between groundwater extraction and nature values at a national scale. Due to these simple scenarios, as Figure 10.5 demonstrated, the change in nature value shows a linear relationship with the increase or reduction of each type of groundwater extraction. i.e., at a national scale.

However, in its present form, DEMNAT already allows for a much more sophisticated approach to the optimization of groundwater withdrawal, including a reallocation of wells from one region to another. Both the geographical database with its 1 km^2 grid and the differentiated dose-effect functions allow for a much more detailed impact assessment.

On the other hand, we would certainly not claim that DEMNAT is perfect yet. We are well aware of a number of shortcomings, partly fundamental and partly practical.

Fundamental shortcomings are bound to the fact that neither the ecoseries classification, nor the dose-effect functions have been thoroughly tested yet, especially with respect to recovery. The ecoseries classification should be tested to its ecological relevance, by precisely determining the correspondence with site factors in the field. The dose-effect functions should be tested as to how far they correspond with real changes following water management measures (Van der Linden et al., 1992). Unfortunately, this would require at least years, but perhaps even decades, of measurements for which we cannot wait.

Practical shortcomings are mainly the result of lack of data. For a national study, especially the lack of detailed geographical information on vegetation, which also adequately covers species composition, is a major drawback. This is especially complicated by the fact that the main nature values are often constituted by very small spatial units, such as ditch banks, etc. (see also Brandt et al., chapter 12), which are difficult to map. It seems that the approach followed so far, i.e., estimating ecotope occurrence from floristic data, is practically the only possibility (see also De Blust et al. in chapter 11 and Groen et al. in chapter 13). This, however, implies a lack of knowledge on the abundance in which species occur, and no knowledge of the fact whether species of the same ecological species group really occur in one spatial unit (Witte et al., 1992a).

As to the ecosystem classifications used, we should first stress that an adequate ecosystem classification is indispensable for predictive modelling of this kind. It should, as Udo de Haes and Klijn pointed out in chapter 1, allow for quantifying the response of the various ecosystem types to interventions, for example, in water management. Also, it should allow for a quantification of the nature value.

In the case of DEMNAT, the ecosystem's response could be calculated due to the fact that the ecoseries classification enabled quantifying the change in operative site factors, whereas the ecotope classification enabled quantifying the subsequent response of species richness.

For assessing the nature value and changes in it, the availability of floristic data seems almost indispensable. Thanks to the ecological species groups that are specified for all ecotope types, the floristic database FLORBASE could be used to — at least — *estimate* floristic nature values in terms of completeness of ecosystems. A further quantification of the nature value of a grid-cell by combining the completeness with surface areas proved difficult, because we had no access to vegetation maps. In this context, the abiotic site maps derived from the ecoseries maps can be regarded as no more than a very rough estimate. An estimate, which owes its confidence only to the fairly good correspondence between actual distribution of ecotopes as derived from the floristic database on the one hand, and potential distribution as derived from the ecoseries maps by means of expert judgement on the other hand.

Therefore, despite the fact that the ecotopes and ecoseries classifications each cover only part of the 'ecotope' concept generally recognized, their combination enables defining dose-effect functions, and their spatial overlay seems to yield a fairly realistic picture of ecological land units at the scale level of ecotopes.

References

Beugelink, G.P., F.A.M. Claessen and J.H.C. Mülschegel, 1992. *Effecten op natuur van grondwaterwinning ten behoeve van Beleidsplan Drink- en Industriewatervoorziening en MER.* (with english summary: Effects of groundwater extraction on nature for the National Policy Plan on Drinking and Industrial Water Supply and EIA). Special series 'Onderzoek Effecten Grondwaterwinning' report no. 16, Bilthoven.

Beugelink, G.P., F.A.M. Claessen and J.H.C. Mülschegel, 1993. Calculating the effects of groundwater abstraction on natural vegetation. In: *Annual report 1992*. National Institute of Public Health and Environmental Protection, Bilthoven.

Braat L.C., A.R. Van Amstel, A.C. Garritsen, C.R. Van Gool, N. Gremmen, C.L.G. Groen, H.L.M. Rolf, J. Runhaar and J. Wiertz, 1989. *Verdroging van Natuur en Landschap in Nederland.* Ministerie V&W, Lelystad.

Canters, K.J., C.P. den Herder, A.A. de Veer, P.W.M. Veelenturf and R.W. de Waal, 1991. Landscape-ecological mapping of the Netherlands. *Landscape Ecology* 5/3: 145-162.

Claessen, F.A.M. and J.P.M. Witte, 1991. National water management strategies for conservation and recovery of terrestrial ecosystems. In: O. Ravera, *Terrestrial and Aquatic ecosystems: Pertubation and Recovery.* Ellis Horwood Limited, New York etc. pp. 526-534.

De Waal, R.W., 1992. *Landschapsecologische kartering van Nederland: Bodem en grondwatertrappen.* LKN rapport nr.2, SC-DLO-rapport 132, Wageningen.

Klijn F., A. ter Harmsel and C.L.G. Groen, 1992. *Ecoseries 2.0 Naar een ecoserieclassificatie ten behoeve van het ecohydrologisch voorspellingsmodel DEMNAT-2.* (with english summary: Ecoseries classification for the hydro-ecological prediction model DEMNAT-2). CML report no. 85, Leiden/ Special series 'Onderzoek Effecten Grondwaterwinning' report no. 5, Bilthoven.

Nienhuis J.G., J.B.S. Gan and R. Lieste, 1992. *Het ecohydrologisch voorspellingsmodel DEMNAT-2, technische modelbeschrijving.* Special series 'Onderzoek Effecten Grondwaterwinning' report no. 2, Bilthoven.

Pakes, U., R.H. Van Waveren and F.A.M. Claessen, 1992. *Berekening invloed systeemvreemd water met DEMGEN.* (with english summary: Calculation of the influence of allochtonous water with DEMGEN). RIZA werkdocument 92-117X, Lelystad/ special series 'Onderzoek Effecten Grondwaterwinning' report no. 8, Bilthoven.

Pastoors, M.J.H., 1992. *Landelijk Grondwater Model (LGM); conceptuele modelbeschrijving.* (with english summary: National Groundwater Model; conceptual model description). RIVM report no. 714305004/ special series 'Onderzoek Effecten Grondwaterwinning' report no. 10, Bilthoven.

Pastoors, M.J.H., 1992a. *Landelijk Grondwater Model (LGM); berekeningsresultaten.* (with english summary: National Groundwater Model; results of calculations). RIVM report no. 714305005/ special series 'Onderzoek Effecten Grondwaterwinning' report no. 12, Bilthoven.

Rolf, H.L.M., 1989. *Verdroging van Natuur en Landschap in Nederland. Technisch rapport: Verlaging van de grondwaterstanden in Nederland, analyse periode 1950-1986.* Ministerie van Verkeer en Waterstaat.

Runhaar, J., C.L.G. Groen, R. Van der Meijden and R.A.M. Stevers, 1987. Een nieuwe indeling in ecologische groepen binnen de Nederlandse flora. *Gorteria* 13: 277-359.

Stevers, R.A.M., J. Runhaar, H.A. Udo de Haes and C.L.G. Groen, 1987. Het CML-ecotopensysteem, een landelijke ecosysteemtypologie toegespitst op de vegetatie. *Landschap* 4/2: 135-150.

Van Beusekom C.F., J.M.J. Farjon, F. Boekema, B. Lammers, J.G. de Molenaar, W.P.C. Zeeman, 1990. *Handboek grondwater beheer voor Natuur, Bos en Landschap.* SDU, The Hague.

Van der Linden, M., J. Runhaar and M. Van 't Zelfde, 1992. *Effecten van ingrepen in de waterhuishouding op vegetaties van natte en vochtige standplaatsen.* (with english summary: Effects of interventions in water management on vegetations of moist and wet sites). CML report no. 86, Leiden/ Special series 'Onderzoek Effecten Grondwaterwinning' report no. 7, Bilthoven.

Witte, J.P.M., 1990. *DEMNAT: aanzet tot een landelijk ecohydrologisch voorspellingsmodel.* DBW/RIZA nota 90.057, Lelystad.

Witte, J.P.M., F. Klijn, F.A.M. Claessen, C.L.G. Groen and R. Van der Meijden, 1992a. A Model to Predict and Assess the Impacts of Hydrological Changes on Terrestrial Ecosystems, and its use on a Climate Scenario. *Wetlands Ecology and Management* 2, no. 1/2: 69 - 83.

Witte, J.P.M., C.L.G. Groen and J.G. Nienhuis, 1992b. *Het ecohydrologisch voorspellingsmodel DEMNAT-2, conceptuele modelbeschrijving.* CML report 89, Leiden/ RIVM report 714305007, Bilthoven.

Witte, J.P.M. and R. Van der Meijden, 1992. *Verspreiding en natuurwaarden van ecotoopgroepen in Nederland.* (with english summary: Distribution and nature value of ecotope groups in the Netherlands). Special series 'Onderzoek Effecten Grondwaterwinning' report no. 6, Bilthoven.

Up-to-date information on nature quality for environmental management in Flanders   11

Geert de Blust, Desiré Paelinckx and Eckhart Kuijken

ABSTRACT - Environmental management and nature conservation policy need up-to-date information on the state of the environment.
To avoid wasting time, the Flemish administration prefers to work with existing databases to describe the biotic environment. In this respect, the Biological Valuation Map and existing flora distribution maps are examined for their applicability for environmental management and nature conservation policy. Their characteristics and shortcomings are discussed.
The Biological Valuation Map is a standardized survey and evaluation of the biotic environment of Belgium. The flora distribution maps summarize the results of a systematic survey of vascular plants.
The Biological Valuation Map is used primarily for the overall description of nature. The map's legend units alone are inadequate to estimate the vulnerability of sites and vegetations to environmental changes. A combination with flora distribution maps and soil characteristics seems to be the most promising approach for such an application. However, some of the legend units are incompletely defined or cover too broad a range of ecological conditions, making them unsuitable for use in environmental impact assessment and the like. This is the case for such an important category as grassland. Consequently, a detailed analysis of the Biological Valuation Map, soil maps and additional field work remain necessary to increase the applicability of the map.
We conclude that for a revision of the Biological Valuation Map, more attention has to be paid to the precise definition of legend units, with respect to their ecological amplitude, and that explicit criteria should be formulated permitting unambiguous mapping.

Introduction

Surveys of environmental quality and bio-diversity have always been important tools for environmental management. However, their content, their range and the methods used are subject to discussion, both by those who provide the information and by those who use the information for policymaking.

In the Environmental Policy Plan (Ministerie van de Vlaamse Gemeenschap, 1990) and the Nature Development Plan (Ministerie van de Vlaamse Gemeenschap, 1991) for Flanders, the objectives, strategies, and time schedule for a newly developed policy on environmental management and nature conservation are outlined. In both documents the permanent need for up-to-date information on the state of the environment is clearly expressed. Whereas the existing databases of water and air quality made up the groundwork of policy development, it was recognized that data on the biotic environment played a minor role in environmental management so far.

As a result of the new policy, the executive administration demands strategies to monitor the state of the environment, with special reference to nature values, and to judge the impact of their management measures. To support the region-oriented policy, methods have to be developed to describe and locate the different regions relevant for environmental management. In this respect, the administration is interested, for example, in the mapping of ecosystems, not as a goal on its own, but as a basis for vulnerability assessment, the definition of deposition standards or the evaluation of the biotic environment. Due to the urgent need, methods have to be based on existing data covering the whole region. Data can be taken from the environmental quality control database for air and surface waters, and from maps of ecosystem components such as vegetation, soil, flora and fauna. Important examples of such data are the Belgian Soil Map (Maréchal and Tavernier, 1974), The Biological Valuation Map of Belgium (De Blust et al., 1985a, b), and the Atlas of the Flora of Belgium and Luxemburg (Van Rompaey and Delvosalle, 1979).

In this contribution we will introduce the Biological Valuation Map and the flora distribution maps as the core components of a biotic database for environmental management. We will describe the data collection, refer to some applications, and discuss their usefulness for environmental management. Learning from their characteristics and shortcomings, we will end with some recommendations for further elaboration and revision of the databases.

The Biological Valuation Map

The Biological Valuation Map is based on a standardized, uniform survey and evaluation of the biotic environment of Belgium (De Blust et al., 1985a, b; Kuijken and Heirman, 1984). The mapping scale is 1: 25,000. Part of such a map is shown in Plate 11.1. The maps are published in sets of eight, covering 640 km^2. With each set of maps a monograph is published, in which the biotic, abiotic and cultural-historical landscape characteristics of the region are described.

In 1978 the minister of Public Health made an appeal to scientific institutes and universities to set up a national mapping project for the biotic environment. Within a year, 10 universities and institutes cooperated in forming a steering-committee. Although the minister's cabinet thought the project could be finished in a couple of years, it is still unfinished. The project was hampered by short-term contracts, interruptions, and ever-changing deadlines.

The project ended as a national one in 1986. Since then, the two regions were supposed to finish the remaining fieldwork and to update and publish the remaining maps. In the Walloon region, there was no continuation until now. In Flanders, the Institute of Nature Conservation carried on with the project; the fieldwork was finished in 1988. The institute is now digitizing the maps and incorporating them in a GIS (Paelinckx et al., 1991).

Figure 11.1 and 11.2 show the present state, respectively for publication and digitization. Almost half the maps of Flanders and Belgium are published.

The field survey was carried out using a uniform legend for the whole country. This legend was composed on the basis of existing knowledge on the vegetation types of Belgium and on general landscape ecological characteristics. New data, such as phytosociological relevés or species lists collected exclusively for this purpose, were not used for the development of this legend. The expert knowledge of project leaders and surveyors was decisive for identifying homogeneous legend units. It is important to mention that project leaders and surveyors were biologists, agronomists, or forest ecologists. Specialists in abiotic disciplines, such as physical geographers or soil scientists, did not participate in the project.

The field survey started in 1978 with a provisional legend and a final legend was prepared in 1979 on the basis of the earlier experiences (Noirfalise et al., 1985). The legend units are specified in the Appendix to this chapter.

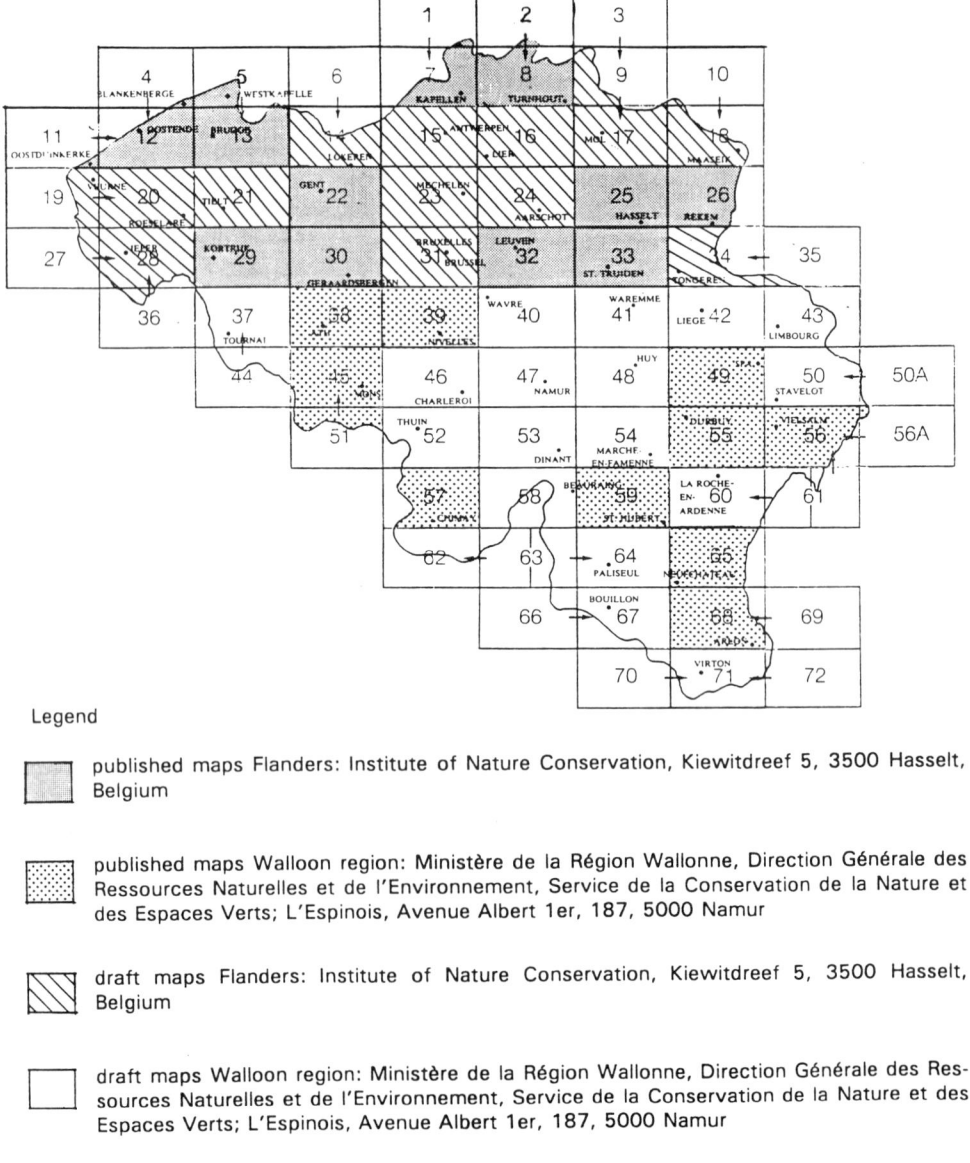

Figure 11.1 The state of publication of the Biological Valuation Map (1.1.93)

INFORMATION ON NATURE QUALITY IN FLANDERS 227

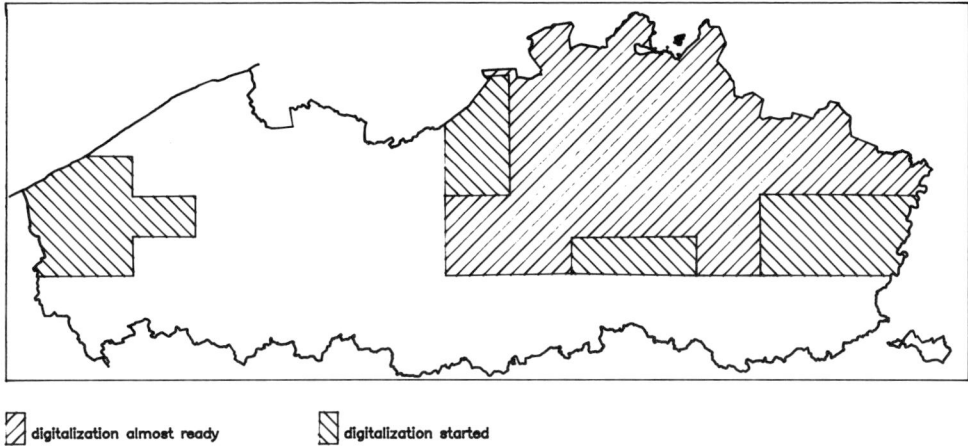

Figure 11.2 The state of digitization of the Biological Valuation Map of Flanders (1.1.93)

The legend units are grouped in 17 classes that largely correspond with formations, as distinguished in phytosociology (Table 11.1). The classes of grassland, shrub, mixed deciduous woodland and azonal woods were divided in subclasses according to soil conditions or agricultural practice. All classes and subclasses contain several basic units. There are 128 units and 34 variants, giving a total number of 162 legend units.

A syntaxonomical equivalent and a list of diagnostic plant species characterize most of the legend units. The legend, however, does not fit into an existing syntaxonomical hierarchy. If we consider all classes, except 'other surveyed elements' and 'urban areas', about half the number of the units match with an order, alliance, or association *sensu* Braun-Blanquet (Table 11.1). Two-thirds of all associations concern woodland units. Grasslands, on the other hand, correspond mainly with alliances. Dunes and salt marshes cannot be referred to syntaxonomically.

These differences can be explained mainly by the choice of developing an easy and pragmatic mapping system and partly by a lack of recent phytosociological surveys.

Table 11.1 The classes of the legend units of the Biological Valuation Map of Belgium; the number of legend units per class and the number of legend units with reference to existing syntaxa. Ass. = Association; All. = Alliance

		units	variant	corresponding to Ass.	All.
A.	Stagnant waters	7	4		1
M.	Marshes	8			5
H.	Grassland	18	4	1	9
C.	Heathland	7	4	4	
T.	Moorland	2			1
D.	Dunes, mudflats, and salt marshes	6	2		
S.	Scrub	10	1	4	4
F/Q.	Mixed deciduous woods	15		13	
E.	Woods of steep slopes and escarpments	2		2	
V.	Woods of alluvial soils, fens, and bogs	8		7	
R.	Ruderal woodland	1		1	
P.	Conifer plantation	2	12		
L.	Poplar plantation	2	5		
N.	Other plantation	1			
B.	Arable land	6			
K.	Other surveyed elements: small landscape elements, 'cultural landscape'	25	2		
U.	Urban areas	8			
		128	34		
		162			

The surveyors worked out a landscape classification before starting the fieldwork. Insight in the region was gained by using topographic, soil, hydrographic, and historical maps, especially the 18th century map of De Ferraris. This information was used to divide the region in so-called homogeneous areas. On the scale of 1: 50,000, these areas are homogeneous for geological, pedological, and hydrological conditions, and for the cultural and morphogenetic landscape evolution. The classification is hierarchically structured. This approach proved to be very useful for understanding the presence and distribution of legend units. Ideas of coincidence and relationships between legend units and abiotic conditions were generated. Nevertheless, much work is still to be done in this field. Most of the findings of this approach were used for the description of the different areas in the monographs. A systematic analysis of the topological and chorological relations, however, was not completed.

Mapping in the field was carried out on a scale of 1: 10,000 or 1: 25,000. Areas smaller than a quarter of a hectare were not mapped. These were usually incorporated into a complex with surrounding units. Complexes were also defined if it was impossible to distinguish individual mapping units or if there was a more or less regular mixture, for example, of arable land and grassland. Small landscape elements were rarely mapped separately but, instead, added to the complexes.

If the original vegetation was still present, although changed or degraded by plantation or other forms of cultivation, this was indicated by adding a particular symbol for the planted tree species or by using a slash bar, indicating the remnants of the original vegetation still present in the actual vegetation. So 'Lhi/Va' for example, is a poplar plantation with an undergrowth, originating from *Ulmo-Fraxinetum* woodland. These notations made it possible to describe a different degree of degradation.

Nature conservation policy insisted very strongly that an evaluation should accompany the inventory. To this end, a classification in three classes regarding the nature conservation value was added: very high biotic value, high biotic value and little biotic value. The criteria used were rarity, vulnerability, naturalness, and replaceability (De Blust et al., 1985a, b). The evaluation was done without using the data collected during the survey. So even for criteria that, in principle, can be quantified, the evaluation is a purely qualitative one, based on best professional judgement.

Because the project was intended to give an overview of the biotic environment of the country, evaluating the legend units alone was unsatisfactory. As we made clear, legend units were largely defined on the basis of vegetation, land use and sometimes geomorphology. Information on fauna was lacking. To fill this gap, information on rare birds and large mammals occurring in the studied region was collected from naturalist groups. For the final nature conservation value map, these additional data were also taken into account to define important areas. This resulted in an upgrading of the value of some of the mapped areas.

Applications

Until now, the main application of the maps and database lies in the determination of the nature conservation value of a particular site. As such, they are used for environmental impact assessment, physical planning, land re-allotment, and nature conservation policy. The Biological Valuation Map is also referred to in jurisdiction. In certain designated areas, e.g. Special Bird Protection Areas of the EC-Directive 79/409, the owners or occupiers of parcels with a vegetation classified as having high nature conservation value, have to get a permit for potentially damaging operations.

In nature conservation, rarity is a very important evaluation criterion. The rarity of the legend units of the Biological Valuation Map was defined on the basis of best professional judgement. With the present GIS database it is now possible to determine rarity more precisely by using the actual area of the various legend units. For this application it is necessary to tackle the problem of mapping units that are named as complexes. We solved this problem by weighing each unit in the complex so that a certain part of the total area of that complex was assigned to each composing unit. Different weighing, e.g. 50% - 30% - 20% or 70% - 20% - 10%, did not significantly affect the results for most legend units when working in a large study area. The higher the frequency of the unit, the less the influence of the weighing.

Beyen (1993) has also done an exercise on this. For 116 km^2 grid cells, he analyzed the relation between area of stagnant water and reed on the one hand, and the number of marsh birds on the other hand. In those cells where stagnant water and reed were mapped as a complex, their precise surface was determined in the field. Figure 11.3 shows the rather large difference between the estimated area of both legend units, using the weighing factor and their real area. However, when applying either the estimated figures or the measured figures in regression analyses of surface area of legend unit against occurrence of marsh birds, no significant difference between both regressions was obtained (Figure 11.4).

INFORMATION ON NATURE QUALITY IN FLANDERS 231

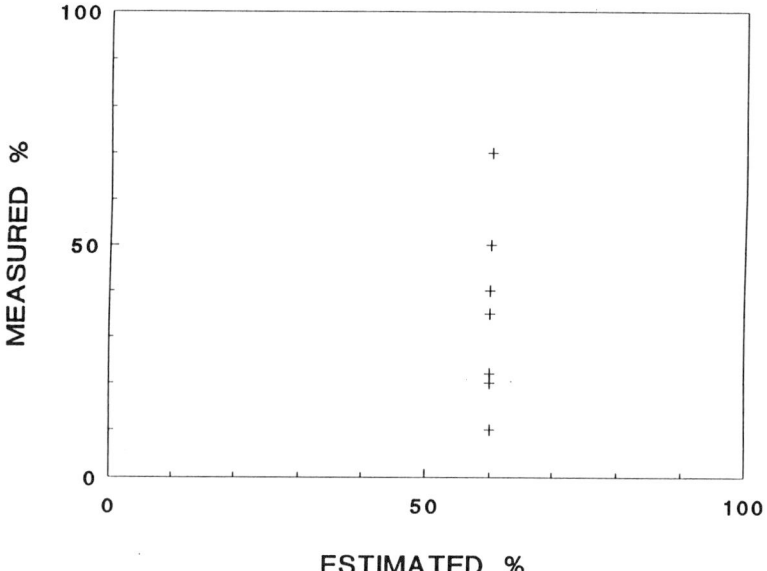

Figure 11.3 The surface area of the first, i.e. dominant, constituent of complex legend units according to an estimate based on the Biological Valuation Map in comparison to the surface areas measured in the field (Beyen, 1993)

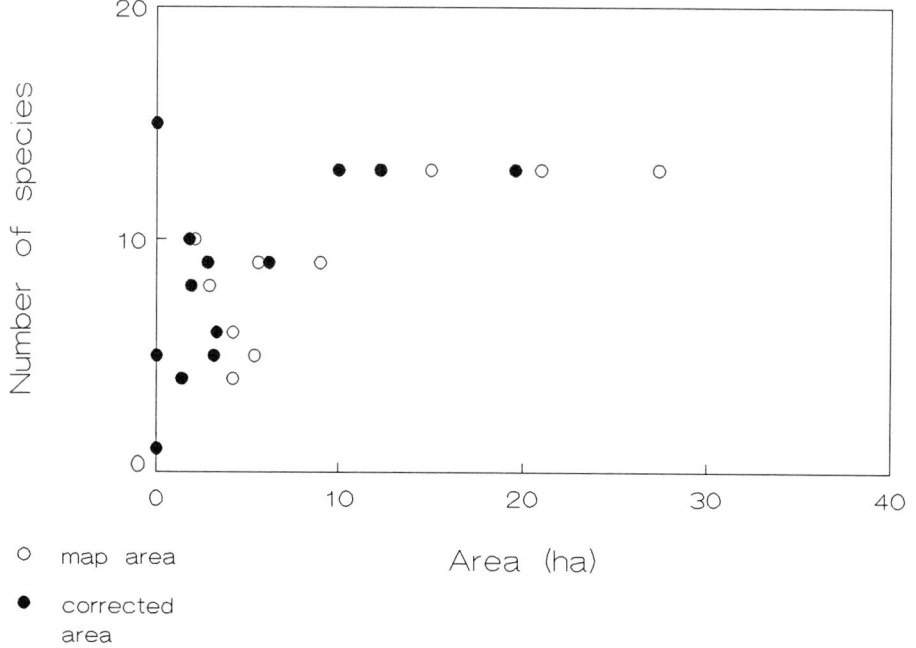

Figure 11.4 The relation between the number of marsh bird species and the surface area of mesotrophic water (Am) as estimated from the Biological Valuation Map (open circles), or alternatively measured in the field (black circles) (Beyen, 1993)

Discussion on the applicability

It is important to consider the limitations of maps and databases. After many years of experience with applications of the Biological Valuation Map we can mention a number of limitations.

Primarily, although it was the intention to have a uniform survey and map, a certain heterogeneity can be detected in the final maps. This has different reasons: the experience between surveyors differed, and with survey time being limited to only a single visit, this may have led to misinterpretations.

Secondly, the scale of 1: 25,000 often made it necessary to map complexes. Whether or not a given situation was mapped as a complex depended very much on the personal appreciation of the surveyor, the accessibility of the site and the time pressure. The scale also caused an underestimate of the importance of small landscape elements.

We have already pointed at the heterogeneity of the legend units. Some units are very broad or have a rather trivial set of characteristic species. Although they are mostly named by one syntaxon, it is clear that during the survey a specific legend unit was used for many more syntaxa than those mentioned in the list. This means a third limitation — that the ecological amplitude of a legend unit as it is used, is much more variable than one would expect from the description of the legend units. Table 11.2 illustrates this for eutrophic waters.

Table 11.2 Comparison of the syntaxonomical description of a legend unit for eutrophic waters and the syntaxonomical units attributed to it by various surveyors in the field survey

legend unit	attributed syntaxonomical units:	
Ae (*Nympheion*)	*Lemnetea*:	*Lemnion minoris*
	Potametea:	*Magnopotamion*
		Nymphaeion
		Parvopotamion
		Hydrocharition
		Callitricho-Batrachion
	Phragmitetea:	*Glycerio-Sparganion*
		Apion nodiflori
		Cicution

This puts important limits on the applicability of the map as a tool for comprehensive ecological interpretations and descriptions of the environment on the basis of the legend unit alone, especially for those who are unfamiliar with the way the surveyors interpreted the legend units.

Also, during the project, surveyors experienced that it was often very difficult or even impossible to attribute a vegetation type in the field to a legend unit. Important shortcomings were found for pioneer vegetations, for species-rich weed vegetations, and for grasslands. For meadows and agricultural grasslands the legend provided only a very limited number of suitable units. This contrasts with the considerable variety of grassland types that can be distinguished in agricultural areas and also with the importance of these grasslands for nature conservation. It is clear that the compilers of the list paid more attention to the semi-natural vegetations than to the vegetations of intensively used land. In Flanders, however, agricultural land occupies much more space than the more natural vegetations.

Finally, the individual maps differ in their time base. The survey extended over more than five years. As a result, an analysis of the complete set of maps that takes into account a certain point in time, cannot be done accurately.

Flora distribution maps

A systematic survey of vascular plants started in Belgium in the late '30s. In 1939 the Institute of the Floristics of Belgium was founded as a society of volunteers, both professionals and amateurs. In 1972, the first edition of the Atlas of the Flora of Belgium and Luxemburg was published (Van Rompaey and Delvosalle, 1972). This atlas contains distribution maps of 1626 Spermatophytes and Pteridophytes from the period 1930 to 1970. For some rare species, the occurrence before 1930, if known, is also given.

The survey was done in grid cells of 1 km^2. On the distribution maps, grid cells have a size of 16 km^2. In 1972, the total number of 1 km^2 species lists was about 12,000. This means one species list per 4 km^2.
The second edition of the atlas was published in 1979 (Van Rompaey and Delvosalle, 1979). To prepare this edition, surveys were concentrated in the regions with inadequate data. The result was 30,288 new finds of species and

an increase of the mean species number from 220 to 230 species per km² (Boon, 1981).
The first edition of the atlas exists in digital form (Boon, 1979).

Applications

Research on the applicability of flora distribution maps has a long tradition (Van der Maarel, 1971, Haeupler, 1974). Widely used applications are the evaluation of sites on the basis of the rarity of species and the relative contribution of different socio-ecological groups, as well as an indication of site conditions by means of the indicative values of plant species (Gommes and Froment, 1978, Stieperaere, 1980, Stieperaere and Fransen, 1982, Van der Meijden et al., 1983). Rankings according to the rarity of species or socio-ecological groups of species exist for Belgium and for Flanders (Cosyns, 1992, Hermy et al., 1990, Stieperaere and Fransen, 1982).

As a basis for regional environmental policy and physical planning for Flanders, De Baere et al. (1986) gave an overall appreciation of the environmental quality of Flanders, analysing the occurrence of a selection of plant species from the atlas.
To this end, 12 socio-ecological groups, representative for more or less natural environments, were defined, each with a set of indicative species. The species were exclusive for each group. Most of them were very rare. The number of species per grid cell (16 km²) was calculated for each socio-ecological group giving a relative degree of the development of that vegetation type. It must be remarked that one is always talking about potential development. After all, occurrence in the same grid cell does not necessarily imply that the species are actually growing together. A classification of the grid cells, using the scores of the groups as variables, was carried out. This resulted in 23 clusters. By combining related and spatially linked clusters, 5 groups were obtained. They show clear zones on the map (Figure 11.5). These zones could be interpreted in terms of natural physiographic background and in terms of degree of human influence and the resulting degradation of characteristic vegetations.

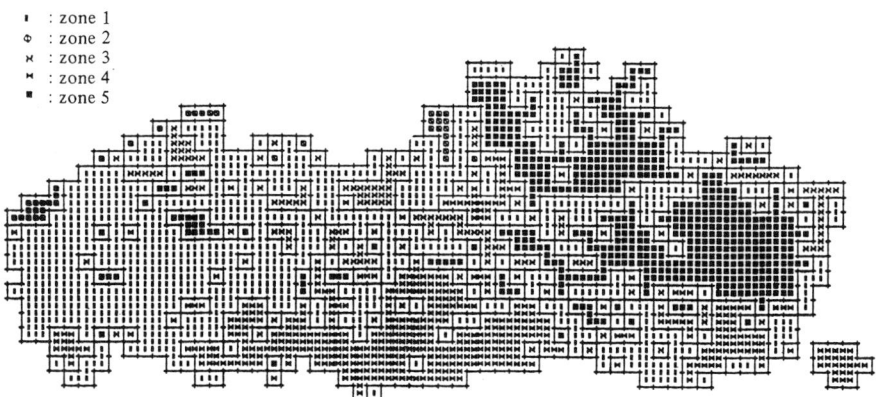

LEGEND

zone 1: strong human influence; no typical or dominating vegetation types
The average number of vegetation types per grid cell is very low. When a vegetation type is present, it is only weakly characterized by a few character species. This situation is typical in regions with intensive agriculture. The spatial distribution is independent from abiotic regional characteristics.

zone 2: well-developed dune and salt marsh vegetations
The spatial distribution is limited to the coastal zone and, only scarcely, to marine polders. In the latter case it is mostly replaced by group 1 and/or 3.

zone 3: more or less well-developed vegetation types of alluvial regions (marsh and grassland vegetations of wet sites)
Only the vegetation of alluvial sites has a presence of more than 10%. The eutrophic site types dominate. Development of the vegetation is low. This type is scattered over Flanders. It has to be interpreted as an impoverished form of the next zones. It is typical for alluvial regions (more or less strong human influence) or agricultural areas with small alluvial sites.

zone 4: presence of either well-developed deciduous woodlands on mesotrophic or acid sites, or Calthion vegetations
This group extends for a big part in the Loam region.

zone 5: many well-developed vegetation types; vegetations typical for well-developed alluvial systems and heaths are often present.
This type is typical for the Campine district and for some smaller sandy regions.

Figure 11.5 Differentiation of Flanders based on the presence of plant species in grid cells, and described in terms of vegetation types and human influence. An application of flora distribution maps (De Baere et al., 1986)

Discussion on the applicability

Even for the second atlas, we cannot state that the survey was done completely systematically (Cosyns, 1992). Interesting sites were visited more frequently than the boring agricultural areas. This remains a weak point of the whole project and limits its applicability (Boon, 1979, Leten, 1991).

Although this atlas is very useful for various analyses, it remains a spatial generalisation of the existing data. The much greater detailed information of the original 1 km^2 lists is not available in digital format. Digitization of these original lists has many advantages, but is laborious. In Flanders only some minor initiatives were taken in this field. At present no systematic digitization is planned.

The integration of the Biological Valuation Map and plant distribution maps

From both the Biological Valuation Map and the plant distribution maps descriptions of the state of the biotic environment of a region can be derived. The Biological Valuation Map gives the precise location, area, and shape of legend units, but neglects their actual species composition. The plant distribution maps give the flora of a grid cell in the magnitude of square kilometres, but tell nothing about the precise location of species.

It is clear that a combined analysis of both types of data may be promising for landscape description and classification. Moreover, insight is obtained in the ecological potential of a region for nature conservation measures.
So far, only a comparative study of the Biological Valuation Map and the flora distribution maps was carried out. Van den Abeele (1986) examined the possibility of elaborating spatial classifications using the Biological Valuation Map and/or plant species lists.
In a study area that covered a significant part of Flanders, 235 cells of 1 km^2 were selected (Plate 11.2 and 11.3). For all the cells a recent flora inventory was available. Two TWINSPAN classifications of the cells were carried out, using the legend units of the Biological Valuation Map on the one hand and the floristic composition on the other. Unfortunately, the map database was simplified by attributing complexes to the first unit of the complex. Thus,

much information was lost, especially on the occurrence of units occupying small areas. Plates 11.2 - 11.5 give some results of the two classifications.
Classification on the basis of floristic data already gives ecologically meaningful groups after only two divisions (Plate 11.2). The differentiation is due to abiotic master factors, such as soil moisture and trophic state, which are associated with large geographical regions. A further division gives an even more precise regional differentiation (Plate 11.3).
The classification of the units of the Biological Valuation Map gives a totally different result. Even after three divisions, most cells are put in one cluster, characterized by a dominance of grasslands and poplar plantations (Plate 11.4). A geographical differentiation, as caused by natural environmental characteristics and reflected in plant species distribution, is faded-out on the level of vegetation units. Here, land use determines the composition of the clusters. The results of the classification with the Biological Valuation Map illustrate the extreme uniformity of the landscape, due to intensive agriculture, urbanisation, and abandonment of traditional land use.

We may conclude that it is most valuable to use the two sets of data in combination. With the valuation map one gets a clear picture of the present landscape: the dominating land use and the presence of more natural vegetations. The flora dataset adds to this:
- the composition of the vegetation as an important quality criterion; and
- the potential of the landscape: the fact that plant species of natural vegetation types still survive in small landscape elements and contribute to the regional diversity.

Further elaboration of the Biological Valuation Map

The flora distribution maps and the Biological Valuation Map are being used with increasing frequency to describe, in an indirect way, the site conditions in a region. The use of socio-ecological groups to interpret plant species in terms of site conditions has had wide interest. The indicative value of plant species on which it is based has become well established (Ellenberg et al., 1992).

When using the Biological Valuation Map, one starts by assuming that the legend units give indications of the environmental qualities. In fact, this has yet to be proven.

Research priority is now being given to this question, because the majority of the applications of the Biological Valuation Map lies in this field. Environmental impact assessment for land re-allotment and groundwater extraction projects, for instance, makes use of the Biological Valuation Map to derive the vulnerability of vegetations and sites due to environmental changes. It has to be examined whether the Biological Valuation Map and its legend units are suitable for this purpose.

Until now we followed different approaches. First of all, legend units were ranked according to their vulnerability by best professional judgement. Scientists of the Institute of Nature Conservation and the Institute of Land Development, who were familiar with field conditions, independently prepared lists of decreasing vulnerability for desiccation and eutrofication. The resulting vulnerability maps of a selected study area showed remarkably low similarity (Van Ghelue et al., 1993). Only 50% of the mapped area corresponded. The vague definition of many of the legend units, leaving much to be interpreted by the user, is largely responsible for this disagreement.

To improve its usefulness for site descriptions, the Biological Valuation Map has to be upgraded by combining it with other data. As stated above, a combination with floristic data would be very promising.

In the exercise of Van den Abeele (1986), the Biological Valuation Map and plant species lists were linked on the basis of grid cells. When phytosociological relevés would be assigned directly to legend units, a more precise species list per unit would be obtained.

Such relevés can be interpreted in terms of site conditions by means of the indicator values of the individual species (Ellenberg et al., 1992, Londo, 1975, 1988). With a set of relevés covering the whole definition of a legend unit, the amplitude and mean value of that legend unit can then be calculated for a specific ecological parameter. In this way Heyrman and Verheyen (1991) obtained a relatively good and reliable ecological interpretation for about 50% of the legend units of the Biological Valuation Map.

The other half of the legend units could not be used to derive site conditions. This was partly due to the fact that they could not be defined in terms of vegetations at all. But, in addition, many units appear to have an ecological amplitude that is too broad. And because many legend units with a narrow ecological amplitude and a clear phytosociological description often concern more natural vegetations, the greater part of a survey area remains without sufficient ecological interpretation. This greatly reduces the applicability of this approach.

We hope to partially solve this problem by combining the biotic data with soil maps and hydrological maps. Van Ghelue et al. (1993) analysed the relation between legend units and soil types. To this end, a digitized Biological Valuation Map and a digitized soil map were compared, using a grid of 20 x 20 metres. From the soil map, among other factors, soil texture, drainage classes, and profile development were taken into account. For some 440,000 points, the observed combinations of the 119 Biological Valuation Map legend units and the 176 soil types were compared with the theoretical distribution, using a chi square test. It can be concluded that there was an overall dependence between the legend units of the Biological Valuation Map and the soil types.

The sum of the difference between theoretical distribution and observed distribution of all the legend units in relation to a particular soil parameter was used as a measure. Thus could we find the most discriminant parameter classes.

For example, for parent material it turned out that peat and, to a lesser degree, loam were characterized by a rather clear set of legend units. As for soil drainage, the very dry and very wet conditions seemed to have their own assemblages of legend units. The intermediate drainage classes had less influence on the distribution of the legend units. Finally, by analysing the distribution of the legend units in relation to the soil profile development, it became clear that Entisols influenced distribution the most, far more than Inceptisols or Spodosols.

At a later stage, correlations between legend units of the Biological Valuation Map, vegetation relevés with known geographical position, and characteristics of the underlying soil derived from soil maps will be analysed. This will enable a subdivision of the legend units with broad ecological amplitudes, such as poplar and pine plantations or intensively used grassland, into more specifically defined local subunits.

A first attempt of such a combined analysis (Nagels, 1990), indicates a possibility of describing another 30% of the legend units in ecological terms.

Final conclusions and recommendations

The Biological Valuation Map of Belgium and the plant distribution maps are two very detailed databases of the biotic environment. Environmental management can gain a lot of efficiency from them. Both databases, however, show some shortcomings. This poses important restrictions on the applicability of especially the Biological Valuation Map.

The original purpose of the Biological Valuation Map was to provide a database of the biotic environment of Belgium. The adopted legend units, however, were based on a phytosociological approach with a bias toward vegetations important for nature conservation. Additional data on fauna had to complete the map.

As it turned out, the overall picture of the map is not very consistent. The information present is far too heterogeneous. This considerably reduces the applicability for environmental management.

In the first place, the shortcomings of the Biological Valuation Map concern the legend units:

- many legend units are insufficiently defined, leaving much to be interpreted by the surveyor. When the surveyor does not make the criteria for his choices explicit, this may lead to mistakes in the assignment that cannot be rectified afterwards. Consequently, the comparability of the individual map sheets is low.
- many legend units contain a wide range of vegetations and thus have a broad ecological amplitude. As a result, the usefulness for indicating environmental parameters through topological relations is limited.

With the published maps and data, it is impossible to improve the first shortcoming. One has to accept an unknown number of mistakes and uncertainties on the map.

The second shortcoming can be dealt with through detailed analysis of additional biotic and abiotic data. On the basis of differences in soil type, groundwater, and species composition, legend units can be subdivided into more homogeneous subunits that can be related to known plant communities.

The existence of complex legend units causes additional, serious problems to applications, especially when these concern surface areas. However, the use of complexes is inevitable. Complexes do not have to result in an unacceptable loss of information, but this requires that criteria on when to use them (the maximum surface area of the complex, the maximum surface area of the composing mapping units) are more obvious, and that the arrangement of the units in the complex is unambiguous.

These conclusions lead to the first important recommendation that, in the case of a revision of the Biological Valuation Map or if a similar project would be planned, much attention must be given to the strict and unambiguous definition of the legend units. They must have a set of diagnostic parameters that allow the surveyor to make his assignment explicit.

A second recommendation concerns the ecological amplitude of the legend units. For the purpose of environmental management, it is of the utmost importance that they are as narrow as possible, i.e., as far as permitted by the scale of the map. To make a uniform application possible, it is required that the legend units belong to the same level of a typification hierarchy.

Finally, we want to emphasize that environmental management gains a great deal if comprehensive maps, such as the Biological Valuation Map, are combined with detailed, monothematic databases. The comprehensive map serves the overview, i.e., the integration needed for general environmental policy, whereas additional monothematic data, which are often site-specific, are suitable for handling well-defined or site-related issues.

This requirement for mutual consolidation must direct the preparation of a new Biological Valuation Map. In the future, the Biological Valuation Map will be part of a pluriform and interrelated tool-kit for environmental management.

Acknowledgements

All people who worked on the Biological Valuation Map, especially J. Heirman and the cited authors, contributed to the development of the ideas in this article. M. Blokken, G. Vannijlen, and M. Rombouts digitized the maps and helped us with the figures. D. Van Straaten and T. Van Thilborg introduced us to GIS and the advanced computer world.

References

Beyen, S., 1993. *Relatie tussen broedvogelsamenstelling en vegetatiekenmerken in Limburgse moerasgebieden.* Eindverhandeling Hoger Instituut der Kempen, Geel.

Boon W., 1979. Enkele kritische bemerkingen over de Atlas van de Belgische en Luxemburgse Flora. *Dumortiera* 11: 14-34.

Boon W., 1981. Enkele kwantitatieve gegevens over de tweede editie van de Atlas van de Belgische en Luxemburgse flora. *Dumortiera* 19/20: 13-21.

Cosyns E., 1992. *Voorlopige standaardlijst van de Vlaamse flora.* Vrije Universiteit Brussel.

De Baere, D., G. de Blust and R.F. Verheyen, 1986. Een globale beschrijving van natuur in Vlaanderen voor de ruimtelijke planning. *Landschap* 3/2: 140-148.

De Blust, G., A. Froment, E. Kuijken, L. Nef and R. Verheyen, 1985a. *Biologische Waarderingskaart van België, Algemene Verklarende Tekst*. Die Keure, Brugge.

De Blust, G., A. Froment, E. Kuijken, L. Nef and R. Verheyen, 1985b. *Carte d'Evaluation Biologique de la Belgique, Texte explicatif general*. Die Keure, Brugge.

Ellenberg, H., H. Weber, R. Düll, V. Wirth, W. Werner and D. Paulissen, 1992. *Zeigerwerte von Pflanzen in Mitteleuropa*. Scripta Geobotanica 18, Göttingen.

Gommes, R. and A. Froment, 1978. L' evaluation des sites dans les remembrements rureaux, l'exemple du remembrement du Germy (Rochefort). *Bull. Soc. Roy. Bot. Belg.* 111: 193-206.

Haeupler, H., 1974. *Statistische Auswertung von Punktrasterkarten der Gefässpflanzenflora Süd-Niedersachsens*. Scripta Geobotanica 8.

Heyrman H. and R.F. Verheyen, 1991. *Operationalisatie van de Biologische Waarderingskaart van België voor het Provinciaal beleid inzake leefmilieu*. Universitaire Instelling Antwerpen.

Hermy, M., L. Vanhecke en M. Leten, 1990. *Een checklist van de Vlaamse flora*. Instituut voor Natuurbehoud - Nationale Plantentuin van België, intern document.

Kuijken, E. and J. Heirman, 1984. The Biological Evaluation Map of Belgium: an applied ecological inventory program. In: D. Bickmore (ed.), *Further Examples of Environmental Maps*. International Geographical Union, Madrid, pp. 47-50.

Leten M., 1991. Hoeveel twijfel is toegestaan? De problematiek van betwistbare floristische gegevens. *Dumortiera* 49: 22-35.

Londo G., 1975. *Lijst van de Nederlandse hydro-, freato- en afreatofyten*. R.I.N., Leersum.

Londo G., 1988. *Nederlandse freatofyten*. Pudoc, Wageningen.

Maréchal, R. and R. Tavernier, 1974. *Pedologie. Atlas van België - Bladen 11A en 11B; Kaarten + Verklarende tekst* (64pp.). Brussel.

Ministerie van de Vlaamse Gemeenschap, 1990. *Milieu-beleidsplan en Natuurontwikkelingsplan voor Vlaanderen*. Voorstellen voor 1990-1995. Brussel.

Ministerie van de Vlaamse Gemeenschap, 1991. *De Groene Hoofdstructuur van Vlaanderen. Richtnota*. AMINAL, Dienst Natuurbehoud en Instituut voor Natuurbehoud, Brussel.

Nagels A., 1990. *Bruikbaarheid van de Biologische Waarderingskaart voor effectvoorspelling van grondwaterstandsdaling in de karteringseenheidsklasse 'populierenaanplanten'*. Licentiaatsthesis, Universitaire Instelling Antwerpen.

Noirfalise, A., H. Stieperaere and L. Vanhecke, 1985. Liste des unites cartographiques. In: De Blust, G., A. Froment, E. Kuijken, L. Nef and R. Verheyen, *Carte d'Evaluation Biologique de la Belgique, Texte explicatif general*. Die Keure, Brugge, pp. 61-90.

Paelinckx, D., H. Heyrman, R. Verheyen and E. Kuijken, 1991. The GIS database Biological Evaluation Map for Flanders: construction and applications. In: D.A. Ondaatje (ed.), *Proceedings EGIS'91*. Second European conference on geographical information systems, Brussels, pp. 826-832.

Stieperaere, H., 1980. The species-area relation of the belgian flora of vascular plants, and its use for evaluation. *Bull. Soc. Roy. Bot. Belg.* 112: 193-200.

Stieperaere, H. en K. Fransen, 1982. Standaardlijst van de Belgische vaatplanten, met aanduiding van hun zeldzaamheid en socio-oecologische groep. *Dumortiera* 22: 1-41.

Van den Abeele, V., 1986. *Studie van de relatie tussen flora, fytogeografie, bodem en ecologische kartering in Midden-Vlaanderen*. Licentiaatsverhandeling Universiteit Gent.

Van der Maarel, E., 1971. Florastatistieken als bijdrage tot de evaluatie van natuurgebieden. *Gorteria* 5: 176-188.
Van der Meijden, R., E.J.M. Arnolds, F. Adema, E.J. Weeda and C.L. Plate, 1983. *Standaardlijst van de Nederlandse Flora*. Rijksherbarium Leiden.
Van Ghelue, P. and K. Decleer, G. De Blust, D. Paelinckx and E. Kuijken, 1993. *Aanzet tot een regionaal landschapsecologisch model (RELEM) voor het gebruik in de landinrichting*. Rapport Instituut voor Natuurbehoud, A93.91
Van Rompaey, E. and L. Delvosalle, 1972. *Atlas van de Belgische en de Luxemburgse Flora: Pteridofyten en Spermatofyten*. 1st edition. Nationale Plantentuin van België, Meise.
Van Rompaey, E. and L. Delvosalle, 1979. *Atlas van de Belgische en de Luxemburgse Flora*. 2nd edition. Nationale Plantentuin van België, Meise.

Appendix

Legend units of the Biological Valuation Map of Belgium

* = legend unit characterized by a list of species

A. STAGNANT WATERS

Ah	brackish waters *
Ae	eutrophic waters (*Nympheion*) *
Aer	newly created (mineral soil)
Aev	well established (mud)
Am	mesotrophic waters without permanent vegetation
Ao	oligotrophic waters (e.g. *Littorellion*)*
Ap	deep or very deep waters (sand pits, ...)
Apo	with gentle slopes
App	with steep slopes
Ad	sedimentation basin
Ab	barrage lake

M. MARSHES

Mr	reedland (*Phragmition*)*
Mz	vegetations of *Scirpus maritimus* *
Mm	vegetations of *Cladium mariscus* *
Mc	tall sedge vegetations (*Magnocaricion*) *
Md	quaking fen
Ms	acid fens (*Caricion curto-nigrae*) *
Mk	alkaline fens (*Caricion davallianae*) *
Mp	dune slack calcareous fens

H. GRASSLANDS
1. semi-natural humid grassland
variants: ...b with shrubs and trees

Hc	moist moderately fertilized meadow (*Calthion*) *
Hj	moist moderately fertilized meadow, dominated by *Juncus* *
Hf	moist tall herbaceous vegetation with *Filipendula ulmaria* *
Hfc	presence of *Cirsium oleraceum*
Hft	presence of *Thalictrum flavum*
Hm	unfertilized wet meadow (*Molinion caerulea*) *
Hmo	oligotrophic subtype *
Hmm	mesotrophic subtype *
Hme	eutrophic subtype *

2. dry grassland
variants: ...b with shrubs and trees

Ha	unfertilized dry grassland (*Thero-Airion*) *
Had	of decalcified dunes
Hn	species rich *Nardus* grassland (*Violion caninae*) *
Hk	dry calcareous grassland (*Brometalia erecti*) *
Hd	xeric grassland on calcareous sand (*Galio-Koelerion*) *
Hv	calaminarion grassland (*Violion calaminariae*) *
Hz	grassland on polluted soils

3. mesophilic agricultural grassland

Hu	mesophilic hay meadows (*Arrhenatherion elatioris*) *
Hp	species rich permanent pasture (e.g. *Cynosurion*) *
Hp*	transition to wet meadow
Hpr	permanent pasture with ditches or microrelief
Hx	species poor pasture
Hr	abandoned pasture and meadow

C. HEATHS
variants: ...b with shrubs and trees

Cg	dry heath (*Calluno-Genistetum*) *
Ce	atlantic wet heath (*Ericetum tetralicis*) *
Ces	with species of raised bogs *
Cm	degraded heath, dominance of *Molinia caerulea*
Cd	degraded heath, dominance of *Deschampsia flexuosa*
Cp	degraded heath, dominance of *Pteridium aquilinum*
Cv	dry heath with *Vaccinium* (*Calluno-Vaccinietum*) *
Ct	heath of raised bogs (*Vaccinio-Ericetum*) *
Ctm	with dominance of *Molinia caerulea*

T. RAISED BOGS

T	active raised bog (*Sphagnion atlanticum*) *
Tm	degraded raised bog, dominance of *Molinia caerulea*

D. DUNES, MUDS and SALT MARSHES

Dz	sand bank
Dl	beach
Ds	mudflat
Da	salt marsh *

Dd	coastal dunes with *Ammophila* (white dunes)
Dm	inland drift sands

S. SCRUBS
1. scrubs on dry soils

Sg	broom scrub (*Sarothamnion*)
Sgu	with *Ulex europaeus*
Sp	thorn thicket (*Rubion subatlanticum*) *
Sk	scrub on calcareous soils (*Berberidion*) *
Sx	stable *Buxus* scrubs (*Berberidion*)
Se	scrubs of clearings (Epilobietalia) *
Sd	dune scrubs (*Hippophaetum*) *
Sz	scrubs on abandoned land

2. scrubs on wet soils

Sm	scrubs with *Myrica gale* (*Myricetum gale*) *
So	willow scrub on acid soils, bogs (*Saliceto-Franguletum*) *
Sf	willow scrub on mesotrophic to eutrophic soils (*Salicetum triandrae-viminalis*) *

F. and Q. MESOPHILIC FORESTS
1. forests on acid soils

Qb	acidophilous oak wood (*Querco-Betuletum*) *
Fb	acidophilous beech wood (*Querco-Betuletum*)
Qs	mesotrophic acidplilous oak wood (*Fago-Quercetum*) *
Fs	mesotrophic acidophilous beech wood (*Fago-Quercetum*) *
Ql	oak wood with *Luzula luzuloides* (*Luzulo-Quercetum*) *
Fl	beech wood with *Luzula luzuloides* (*Luzulo-Fagetum*) *
Ff	beech wood with *Festuca altissima* *
Qd	wood of coastal dunes
Qx	xerophilic oak wood on slate *

2. forests on neutral soils

Qa	oak-hornbeam wood (*Stellario-Carpinetum*) *
Fa	beech wood with *Anemone* (*Milio-Fagetum*) *
Qe	oak-hornbeam wood with *Endymion* (*Endymio-Carpinetum*) *
Fe	beech wood with *Endymion* (*Endymio-Fagetum*) *
Fm	beech wood with *Melica* (*Melico-Fagetum*) *

3. forests on calcareous soils

Qk   calcareous oak-hornbeam wood (*Ligustro-Carpinetum*) *
Fk   calcareous beech wood (*Cephalanthero-Fagetum*) *

E. ESCARPMENT FORESTS

Ek   escarpment wood on calcareous soil (*Tilio-Aceretum*) *
Es   escarpment wood on acid soil (*Ulmo-Aceretum*) *

V. WOODLAND of ALLUVIAL SOILS, FENS and BOGS
1. woods on alluvial soils (*Alno-Padion*)

Va   alluvial ash-elm wood (*Ulmo-Fraxinetum*) *
Vf   alder-oak wood
Vb   mesotrophic ash-alder wood of fast-flowing rivers (*Stellario-Alnetum*) *
Vn   tall herb alder wood (*Macrophorbio-Alnetum*) *
Vc   alder-ash wood of springs and spring rivulets (*Carici-Fraxinetum* and *Cardamini-Alnetum*) *

2. bog woodland

Vm   mesotrophic alder wood with sedges (*Carici elongatae-Alnetum*) *
Vo   oligotrophic alder wood with *Sphagnum* (*Sphagno-Alnetum*) *
Vt   birch bog woodland (*Vaccinio-Betuletum pubescentis*) *

R. RUDERAL FORESTS

Ru   elm wood (*Violo odoratae-Ulmetum*) *

P, L, N. PLANTATIONS

Pp   plantations of *Pinus*
Ppi   young plantation
Ppa   dense plantation, without undergrowth
Ppm   older plantation with undergrowth
Ppmh   with grasses
Ppms   with dwarfshrubs
Ppmb   with shrubs and small trees
P   other conifer plantations (except *Pinus*)
Pi   young plantation
Pa   dense plantation, without undergrowth

Pm	older plantation with undergrowth
Pmh	with grasses
Pms	with dwarfshrubs
Pmb	with shrubs and small trees
L	poplar plantations
Lh	poplar plantation on wet soils
Lhi	with grasses or tall herbs
Lhb	with shrubs or small trees
Ls	poplar plantation on dry soils
Lsh	with grasses and herbs
Lsi	with tall herbs
Lsb	with shrubs or small trees
N	other plantations of broad-leaved trees

B. ARABLE LAND

Bs	arable land on sand
Bl	arable land on loam
Bu	arable land on clay
Bg	arable land on stony loam
Bk	arable land on calcareous stony loam
Bc	arable land on chalk

K. INDIVIDUAL ELEMENTS

Kn	watering place
Kb	row of trees
Kh	hedge
Khw	wooded bank
Ks	abandoned railway or important railway verge
Kw	sunken road
Km	old wall or ruines with important vegetations
Kt	talus
Kd	dike
Kr	cliff
Kra	acid
Krc	calcareous
Kv	pingo
Kk	karst
Ka	duck decoy
Ku	undefined pioneer vegetation
Ku*	on calcareous raised grounds
Kc	quarry
Ko	dumping ground
Kg	rubble heap

Kf	abandoned fort
Ki	airfield
Kj	tall trees orchard
Kl	low trees orchard
Kp	park and/or graveyard
Kpk	castle park
Kpa	arboretum
Kq	nursery or greenhouse
Kz	raised ground or industrial ground

U. URBAN AREAS

Ud	densely built up areas
Ua	residential areas with gardens
Un	residential areas in 'green environment'
Ur	buildings in agricultural area
Ui	industrial plants
Uv	recreation site
Uc	camping site

Monitoring 'small biotopes' 12

Jesper Brandt, Esbern Holmes and Dorthe Larsen

ABSTRACT - Small uncultivated areas within the agricultural landscape, in Denmark called 'small biotopes', have attracted growing attention because of their importance for wildlife and their recreational and aesthetic value. The ecological role of these small, uncultivated areas can be demonstrated by regarding them as the lowest level in a hierarchy of ecological networks. But they also have to be understood as integrated functional parts of the agricultural land use system. During the development of a monitoring system for small biotopes in Denmark we have encountered many problems and found a number of solutions. We discuss some of the issues concerning data collection, classification, database construction and the informational context that is necessary for the practical use of such a monitoring system.

Introduction

In Europe there is a rapidly increasing awareness of the importance of the landscape's spatial composition. One aspect of this is the development of the concept of ecological networks. This initially purely academic concept now plays an increasingly important role in landscape valuation, planning and management. The ecological network of a landscape is probably more correctly viewed as a hierarchy of ecological networks, where the highest level is composed by the larger core areas and corridors at the scale of regional planning (see Figure 12.1a) and the lowest level is composed by the small, uncultivated areas within and between the fields of the agricultural landscape (see Figure 12.1b).

Until now, emphasis has been given to the higher levels of a hierarchy of ecological networks and monitoring systems (e.g. CORINE) for this type of area are being developed (Jongman, 1992). This chapter deals with some of the problems related to monitoring the lowest level of the hierarchy, composed of what we call *small biotopes* because they have been studied primarily as small sites (topos) for wild plants and animals (bio).

Although unassuming, small biotopes are of ecological importance, especially in intensively utilised agricultural landscapes, such as the Danish, because they provide about one-third of the total habitat for wildlife. In that sense we are classifying and monitoring types of nature for wildlife. But classifying small biotopes for our purpose is only secondarily a matter of nature types (like ponds, heaths and meadows), since they are primarily a function of agricultural and other anthropogenic landscape-forming processes (Agger and Brandt, 1988). A monitoring system should take into account this dual status of the small biotopes.

The aim of this monitoring system is to:
- map the structure and dynamics of the small biotopes;
- describe and analyse the relation between agriculture and small biotope pattern; and
- provide a knowledge base for education, information campaigns, development of new methods for regulation and economic incentives among farmers and other land users to promote ecologically sound management of the agricultural landscape.

This chapter presents some of the issues concerning data collection, classification, database construction and the informational context that is necessary for the practical use of such a monitoring system. The chapter is divided into two parts. The first part is a general introduction to the small biotope monitoring system and the second part is a series of lessons learned from ten years of experience with small biotope monitoring.

MONITORING 'SMALL BIOTOPES' 253

A

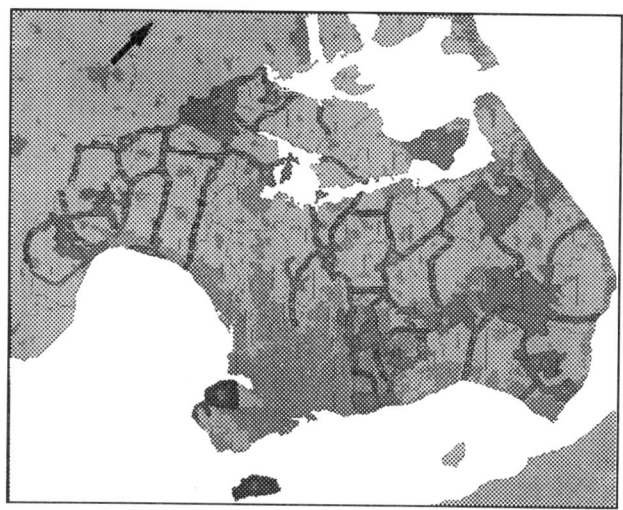

B

Figure 12.1A High level ecological networks have, since the end of the 70's, been increasingly used by Danish regional planning authorities. The map shows a proposal for larger core areas and corridors as found in the regional plan for the Greater Copenhagen area (Hovedstadsrådet, 1982)

Figure 12.1B An actual, low level ecological network of small biotopes as found around the village Tågerup, indicated with an arrow in Figure 12.1A

History of the small biotope monitoring system in Denmark

A small biotope monitoring system was set up in Denmark in the late '70s, initially with 13 test sites. This was expanded to a coverage of 32 sites of 4 km² each, during campaigns in 1981, 1986 and 1991 (see Figure 12.2).
Basically the small biotope monitoring consists of:
1. Detailed field registrations of all linear and areal biotopes less than 2 ha.
2. Interviews with farmers concerning agricultural practice as well as the functions of and plans for the small biotopes.
3. Historical registration in 5 test sites based on topographical maps and aerial photographs.

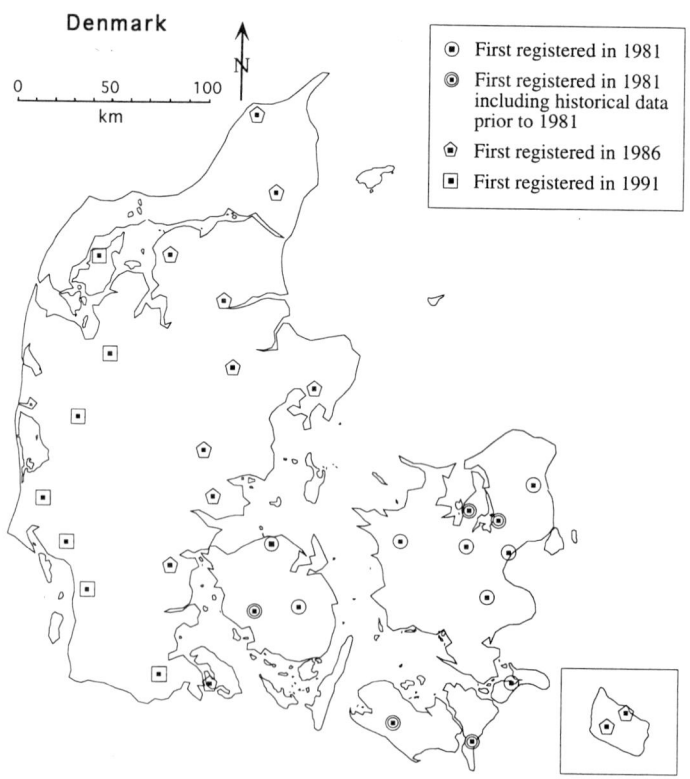

Figure 12.2 The 32 test sites of 4 km² surveyed in the monitoring programme

The motivation for the 1981 campaign was the impression of a rapid decrease in number and quality of small biotopes following the concentration, specialisation and industrialisation of Danish agriculture (Biotopgruppen: Agger et al., 1986).

The 1986 campaign was a main source of information on the status and development of marginal land within the intensively-used Weichsel moraine landscapes in Denmark. Here, the dynamics of small biotopes was considered an indicator for the intensification/extensification process within agriculture (Agger and Brandt, 1987). In 1986, spontaneous marginalisation of agricultural land was observed: nine abandoned fields were registered compared to none five years earlier.

The 1991 campaign was carried out in co-operation with the Ministry of Environment as part of the national monitoring programme for wildlife: a monitoring programme not only for small biotopes (Agger et al., 1992, Brandt, 1991) but also for other (larger) types of habitats and selected animal and plant species (Agger and Owesen, 1990).

Impact on policy

An important goal for the project has been to influence policy and decision makers, by changing the focus of conservation interests to incorporate threatened everyday nature values. This goal has been achieved in the sense that the term small biotopes is now an everyday concept in Danish environmental debate. It has fundamentally influenced a new nature protection act since June 1992. This replaced the former nature conservation act, including §43, stating a list of nature types under 'general protection', which means areas that cannot be altered without permission, although no compensation is given. The new act expands the list of nature types regulated by the general protection. In addition, the minimum size of landscape elements regulated by the law has also been lowered considerably to a mere 100 m^2 for small lakes and ponds and 2500 m^2 for most other biotopes (see Table 12.1).
These landscape elements are included in the Nature Protection Act mainly because of their importance as habitats for threatened animal and plant species. Even the incorporation of the smallest lakes and ponds, down to 10 x 10 metres, is primarily justified by their importance for threatened species, especially the 14 existing species of amphibians (most threatened is the Fire-

Table 12.1 The history of general protection — without compensation — of biotopes in the Danish agricultural landscape according to the Nature Conservation Act (1937, 1972, 1978, 1984 §43) and the Nature Protection Act (1992 §3,4 and 12) (Min. size in m^2)

	1937	1972	1978	1984	1992
Barrows	all	all	all	all	all + 2 m buffer zones
Other archaeological sites					most types + 2 m buffer zone
Water courses		> 1.5 m	> 1.5 m + specially selected	> 1.5 m + specially selected	high priority + 2 m buffer zones
Lakes and ponds		all natural lakes	> 1 000	> 500	> 10
Bogs			> 5 000	> 5 000	> 2 500
Heaths				> 50 000	> 2 500
Salt meadows				> 30 000	> 2 500
Fresh meadows					> 2 500
Commons					> 2 500
Stone and earth dikes					all on topographical maps (provisionally) + 2 m buffer zones

bellied toad, *Bombina bombina*, but also *Hyla arborea*, *Triturus alpestris*, *Bufo viridis* and *Rana ridibunda* are threatened). The protection of threatened plants and birds is the reason for including more of the extensively used agricultural types of land, such as small meadows, heaths, and commons. But also the protection of historical heritage has been strengthened in the legislation: the first landscape element under general protection was barrows from the Iron and Bronze Ages. Now the legislation has been widened to include almost all recognizable archaeological features and their near surroundings. Also, rather recent historical cultural elements, such as stone and earth dikes from the 19th century, have been incorporated based on a true mixture of cultural and nature protection arguments (Skov- og Naturstyrelsen, 1992).

Consequences for future monitoring

The tightening of the legislation toward still smaller landscape elements not only protects important wildlife habitats, but also less important landscape elements. These can be regarded as elements of an ecological network in the local agricultural landscape. The major part of the ecological network in the agricultural landscape — especially the flat, 'non-exotic', intensively-used landscape — is, however, still unregulated. This is particularly the case for linear features like hedgerows, field divides, brooks and road verges, but also thickets, woodlots, solitary trees, dry marl-pits, and other small areas, mostly of rather recent (agri-)cultural origin. A protection of many of these smaller areas is doubtful, primarily because it would impede a flexible management of the agricultural landscape. The existing law has probably already gone too far in this direction. Instead, we advocate a more flexible legislation based on the role the element plays in the ecological network and in relation to other uses of the agricultural landscape. Thus, these landscape elements should be managed by regarding them not as individual biotopes, but as part of such functional networks.

It is important to keep in mind that the existence of these local networks is not only physically, but also economically linked to agriculture. These ecological networks, therefore, cannot be managed without simultaneously influencing agriculture and vice versa.

A small biotope monitoring system must take this dual linkage into account. The system must enable a continuous evaluation of the effects of the legislation. It must also support the development of new methods for regulation by offering flexible tools for the analysis of the mechanisms behind changes of the small biotope network.

Lessons of small biotope monitoring

In the following sections we will discuss some of the lessons that can be learned from our ten years of work with small biotope monitoring. Not all of the proposed solutions have yet been implemented in the Danish small biotope monitoring system and they are, therefore, still of a somewhat theoretical character.

How to define the 'small biotope' concept

The term 'small biotope' was created to enable us to study the 'rapid decrease in number and quality' of the small, uncultivated areas within the agricultural landscape of Denmark. 'Small biotopes' are defined as '*uncultivated* areas that are *permanently covered with vegetation* (or water) and situated *within* the agricultural areas'. Furthermore, a small biotope must be smaller than 2 ha and either larger then 10 m^2 or longer than 10 m with a width of more than 0.1 m (Agger and Brandt, 1984).

In this definition the small biotopes are regarded as part of the land use, but contrasting to the cultivated areas. Thus, the small biotopes are *not* defined in terms of natural landscape structure, e.g. physiotopes and their chorological extensions.

Based on our small biotope definition, only approximately 1/4 of the small biotopes of Danish agricultural landscapes can be considered to be of natural origin and, even then, are often highly transformed. The rest can be traced back as being manmade features, primarily related to present or former agricultural land use such as dikes, marl pits, etc.

To reflect the anthropogenic nature of the small biotopes, we have chosen to integrate the small biotope classification into the general land use classification. Furthermore, we have chosen to use everyday terms for the small biotope classification, however, giving these everyday terms a precise definition. For instance, a hedge is defined as 20 metres of a linear biotope, of which a minimum of 50% is covered with trees or shrubs, and where the surface is between 0.25 m under and 0.75 m above the surrounding fields. If the surface had been higher we would have had the small biotope type 'hedge on dike'.

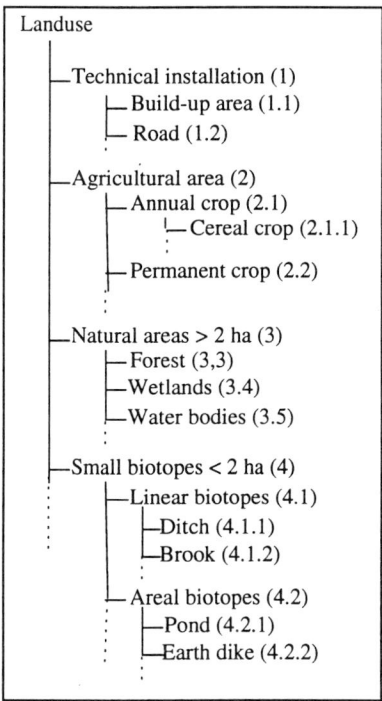

Figure 12.3 Part of the small biotope classification in a CORINE inspired hierarchically-structured land use classification

A major problem of the small biotope definition is that we have stressed that the small biotopes must be within or between agricultural fields. Hence would-be small biotopes within and directly adjacent to farmsteads and urbanised areas are not registered. Restricting the study to a certain matrix is related to the landscape ecological tradition where biotopes are regarded as patches and corridors embedded in a certain matrix, in our case, an agricultural matrix. This does, however, have a major drawback, particularly in connection with the historical analysis, where small biotopes can come into being and disappear as a result of changes solely in the surrounding matrix. For instance, the dismantling of an agricultural holding might involve the upcoming of a ruderate, thickets, hedgerows and ponds, although they always existed as biotopes related to the former garden. In a monitoring system it would certainly also be relevant to follow how existing, small biotopes can be properly embedded in an urban or recreational environment related to an urbanisation process.

Another problem arises from the maximum area clause of the small biotope definition. This clause has been introduced to enable us to concentrate on the smaller biotopes (< 2 ha) that are more liable to changes in the agricultural matrix. The stability of the larger biotopes is often induced by the sharing of ownership among several agricultural holdings. The drawback of this maximum area clause is that changes in the small biotope population may be wrongly interpreted. Small biotopes may, for instance, arise from larger biotopes; when a bog is drained it often results in several small bogs remaining in the lowest parts of the area, consequently, the drainage activity results in an increased number of small biotopes. On the other hand, the amalgamation of two small biotopes resulting in the total area surpassing the 2 ha-limit causes a decrease in the number of small biotopes. These problems can only be handled by regarding the small biotopes as part of the general land use, thus supporting the integration of the small biotope classification with the general land use classification.

How to handle nature types within the context of small biotopes

Although we do not include the concept of nature types in the definition of small biotopes we are well aware that relating small biotopes to nature types may reveal important information. In earlier studies we grouped the small biotope classes based on their dominant nature type. This grouping, which is illustrated in Figure 12.4, is relevant for a functional description of different types of patches and corridors comprising a local ecological network. Based on this grouping of small biotope classes, we have observed changes in the population of the different groups of small biotope classes as shown in Table 12.2.

In Table 12.2 we have grouped the small biotopes by common nature type characteristics, namely, wet or dry small biotopes. In comparison to the more homogeneous agricultural production areas the small biotopes are characterised by a high degree of internal heterogeneity. In the monitoring system this internal heterogeneity has been represented as four main land cover types: open water, reed vegetation, herbaceous vegetation and woody vegetation.

Experience has, however, shown that the addition of some non-vegetational land cover categories is appropriate. An extended, but still very simplified land cover classification contains eight classes, as shown in Figure 12.5.

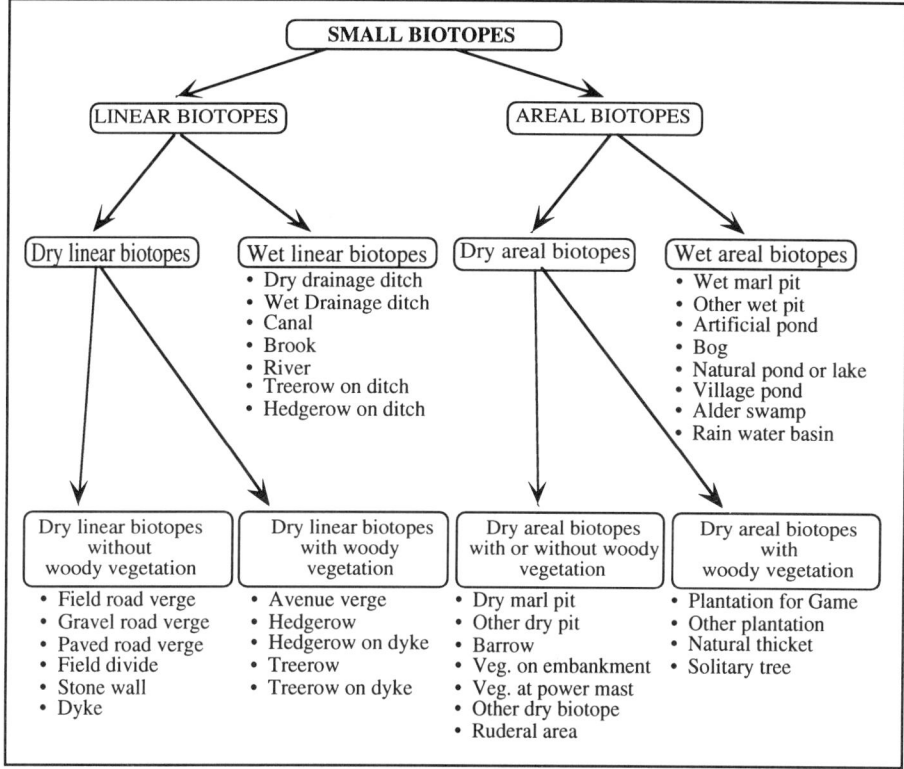

Figure 12.4 A nature type-oriented classification of small biotopes used in the Danish small biotope monitoring system

This allows the representation of internal heterogeneity of land cover within the individual small biotopes. Besides this, the land cover differentiation allows us to link the primarily anthropogenic oriented classification of small biotopes to the more traditional classification system of nature types for conservation purposes (see Figure 12.6).

In this role the land cover classification has shown its worth in connection with a special paragraph in the new Nature Protection Act stating that bogs less than 2500 m^2 are included if they contain an area of open water (including a related reed vegetation) of more than 100 m^2. Based on our land cover data we are able to estimate that this rule will double the number of small bogs covered by the general protection.

How to incorporate agricultural information

Socially induced processes are probably the dominant factor in the production and removal of small biotopes. We believe that these processes should be studied within a context of what Neef (1984) calls *action fields*, e.g. agricultural fields, holdings or owner associations. Until now, we have concentrated on the level of farm holdings, thus relating small biotope data to farm size, spatial configuration, ownership, specialisation, introduction of machinery and game-orientation.

Table 12.2 Table showing the development of small biotopes in Denmark in 1981-91, based on a nature type-oriented classification

DEVELOPMENT OF SMALL BIOTOPES IN DENMARK 1981 - 1991* NATURE TYPE-ORIENTED GROUPING		1981-86 (% per year)	1986-91 (% per year)
13 TEST SITES IN EASTERN DENMARK (52 km^2)	Wet linear	-0.1	-1.1
	Dry linear	-0.1	+0.2
	All linear	-0.1	0.0
	Wet areal	-1.8	-0.8
	Dry areal	+0.9	+2.0
	All areal	-0.6	+0.6
10 TEST SITES IN EASTERN JUTLAND (40 km^2)	Wet linear		+3.2
	Dry linear		0.0
	All linear		+0.4
	Wet areal		+2.4
	Dry areal		+4.7
	All areal		+3.7
25 TEST SITES IN DENMARK (100 km^2)	Wet linear		+0.3
	Dry linear		0.0
	All linear		+0.1
	Wet areal		+0.3
	Dry areal		+2.6
	All areal		+1.5

* Indicated as % annual change on average for all test sites; the linear in % of length; the areal in % of number.

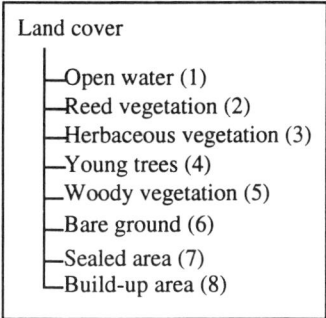

Figure 12.5 Land cover categories for the description of internal small biotope heterogeneity

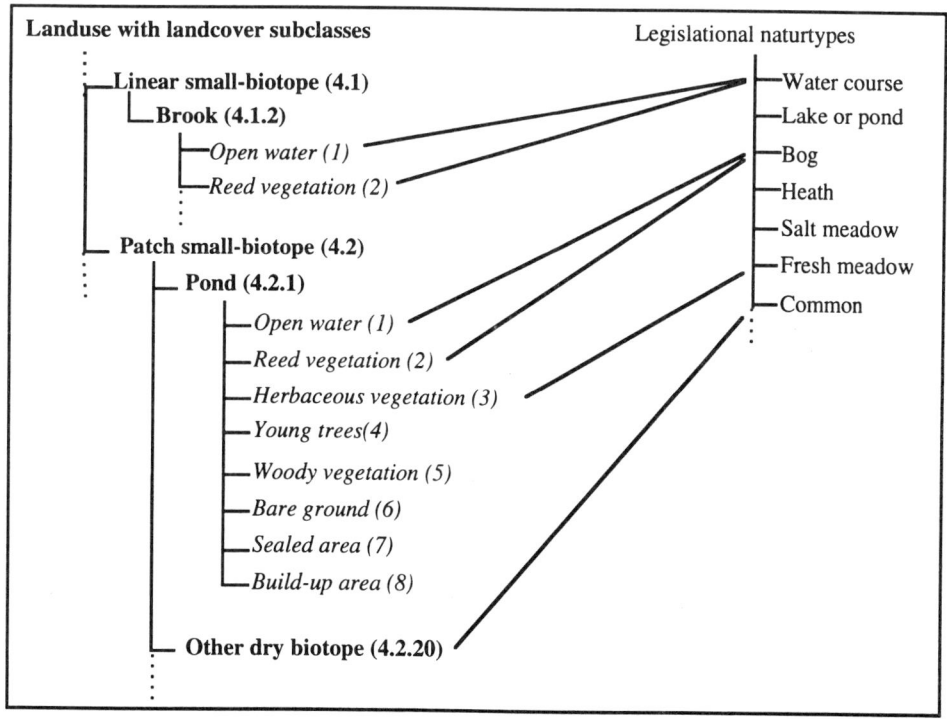

Figure 12.6 The relation between small biotope subclasses and legislational nature types

Signs of an agricultural industrialisation process are of special relevance, as pointed out by Meeus et al. (1990): 'Where intensification, generally speaking, can be seen as the optimal use of the land, closely tied to the local landscape and to existing ecosystems, industrialisation is primarily motivated by technological developments. Agriculture no longer depends on qualities of the existing landscape and its ecological structures, but seems instead to spread out in a disjointed way'.

In addition to relating small biotope data to the agricultural holding, it is also important to examine small biotope data at the field level. For instance, to study how the process of intensification and extensification of agricultural land use influences the small biotope pattern within the single agricultural holding. Ideally this should be based on field-related information on crop rotation, seasonal vegetation cover, use of pesticides, and intensity and distribution of manure.

This combination of field and small biotope related information opens possibilities for an integrated management of agricultural landscapes. Mayrhofer and Schawerda (1991) explicitly relate areal data on small-biotopes to areal data on adjacent fields by the calculation of so-called 'eco-points' (Öko-punkte) as an integrated measure of an overall ecologically sound management of the agricultural landscape.

Another way to study the interrelation of the agricultural land use and the small biotope is to examine the fluctuations in the small biotope population based on the role of the small biotope classes in the landscape. For this purpose we have developed the following grouping of the small biotope classes (see Figure 12.7).

Examining the small biotopes in the view of this grouping we arrive at the results shown in Table 12.3. When comparing these results with Table 12.2 it seems that the explanatory power of the land use functional grouping is greater than that of the nature type grouping. For instance, the increase of the game-oriented small biotopes seems quite significant. This implies that the increase of the dry areal, small biotopes found in Table 12.2 is controlled by these biotopes' role in game care.

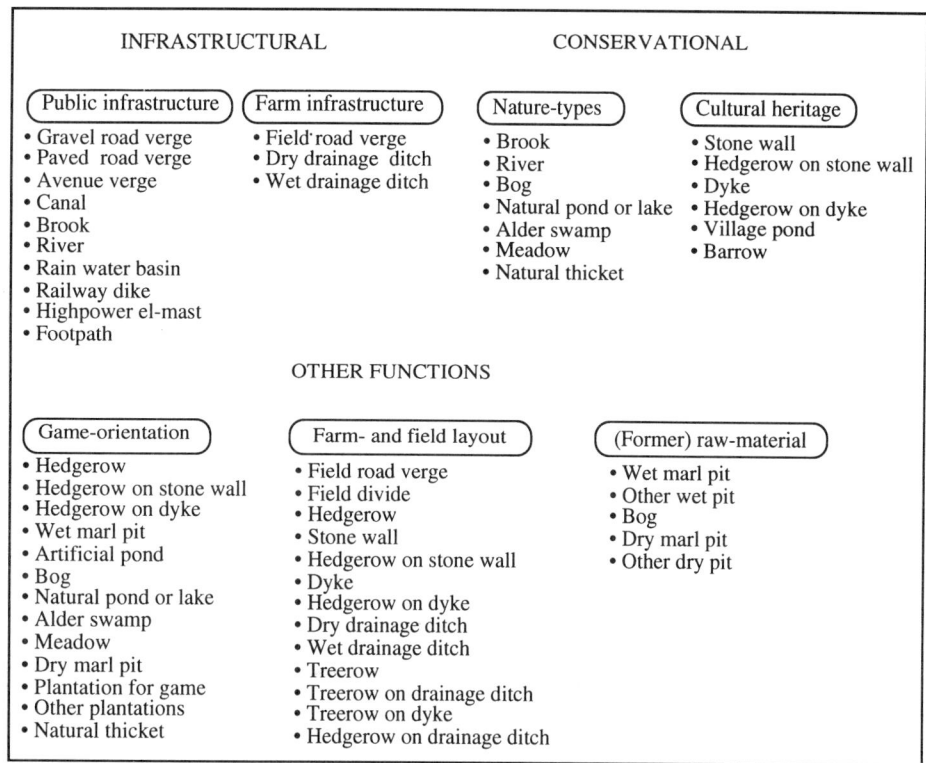

Figure 12.7 A land use functional grouping of small biotopes used in the Danish small biotope monitoring system

How to incorporate geo-ecological parameters

Geo-ecological parameters can be incorporated into the monitoring system based on a holistic landscape classification at the lowest chorological level (nanochore, ecoseries) (Andersen and Jensen, 1991) or as more or less complex single parameters, e.g. soil type, geomorphology, elevation, etc.

A thorough investigation into the statistical correlation between small biotope types and geo-ecological parameters can hopefully help us monitor technological shifts in agricultural land use. Thus, we hope that if there is a technological shift towards a more landscape-adapted land use in the future, this should be unveiled in the correlation between small biotope classes and geo-ecological parameters.

Table 12.3 The development of small biotopes in Denmark in the period 1981-91, based on a land use functional grouping; compare with Table 12.2

DEVELOPMENT OF SMALL BIOTOPES IN DENMARK 1981 - 1991* - A FUNCTIONAL CLASSIFICATION		1981-86 (% per year)	1986-91 (% per year)
13 TEST SITES IN EASTERN DENMARK (52 km²)	Public infrastructure	0.0	+ 0.9
	Farm infrastructure	- 1.2	- 1.7
	Nature types	+ 0.1	+ 0.5
	Cultural heritage	0.0	- 0.1
	Game orientation	+ 0.2	+ 3.2
	Farm and field layout	- 0.6	- 0.6
	(Former) raw material	- 0.2	+ 0.6
10 TEST SITES IN EASTERN JUTLAND (40 km²)	Public infrastructure		+ 0.9
	Farm infrastructure		+ 0.5
	Nature types		- 0.3
	Cultural heritage		- 1.1
	Game orientation		+ 5.3
	Farm and field layout		- 0.1
	(Former) raw material		- 2.3
25 TEST SITES IN DENMARK ** (100 km²)	Public infrastructure		+ 0.8
	Farm infrastructure		- 0.9
	Nature types		+ 0.4
	Cultural heritage		0.0
	Game orientation		+ 3.8
	Farm and field layout		- 0.5
	(Former) raw material		- 0.1

* Indicated as % annual change in average for all test sites; the linear in % of length; the areal in % of number.
** Including 2 test sites on Bornholm in the Baltic Sea

How to perform the data collection

There are a number of different data sources available for the monitoring of small biotopes, such as topographical maps, remotely sensed data — aerial photographs or high resolution satellite images — and last but not least, field survey.

Several surveys of small biotopes in agricultural landscapes are based on topographical maps. These maps are, however, of restricted usefulness because some of the most common and changeable types of small biotopes such as ditches along roads and field divides are rarely present. Furthermore, the surface area of linear and small areal features cannot be deduced from the map because these are only represented as line or point signatures on the map. Topographical maps traditionally stress features of military interest such as restrictions to military transportation or opportunities for military shelter. This gives the topographical map a bias towards certain types of more stable, wet and visually dominating biotopes. Therefore it is almost impossible to retrieve a reliable quantified description of small biotopes based on topographical maps. However, in cases where other sources of data are lacking — particularly for historical investigations — the topographical maps are an indispensable data source.

The use of aerial photographs is more reliable, but still presents some problems. It is, for instance, difficult to judge the width of linear features because of the shadows from woody vegetation. Moreover, a classification of small biotopes based on aerial photographs alone can be extremely difficult due to their internal differentiation. Woody vegetation with a height of less than 3 metres can rarely be distinguished from scrubs and herbaceous vegetation.

This leaves us with field surveys as the most reliable, although time consuming, way of monitoring small biotopes. Field registration is the only way to obtain information on the biological content of small biotopes such as tree species composition and recent anthropogenic influence, e.g. disposal of waste and stones, cutting and scorching, sign of game care and the like. For field survey aerial photographs are a very important tool for geographical positioning of the features in the field as well as for the digitizing process. General land use data were also obtained solely by field survey, because up-to-date aerial photographs were unavailable. An interesting future perspective is to investigate the extent to which such a general land use survey may be supported by employing high resolution satellite data such as SPOT data.

Interviews are necessary for collecting information on the functions, care of and attitudes towards future utilisation of the small biotopes. This can be done through telephone interviews, but personal interviews with the farmers are definitely preferable. The personal interview is a source of information not only for objective facts like location, and type of the holding; it will also supply an impression of the farmers' general attitude towards the small biotopes. Building such a sound, up-to-date understanding of modern agriculturists' strategies and priorities, and how they can be actively involved in the management of the agricultural landscape is indispensable for any strategy for maintaining and developing local ecological networks. It is important to understand that much of the time-consuming activities put into the establishment, maintenance and removal of small biotopes are to be seen as long-term investments put into action in periods of downtime. These activities are directed by the present conception of what is seen as the right strategy for the future and such conceptions are, obviously, changing through time.

How to handle the data of the monitoring system

The data of the monitoring system can be characterised as geographical data which basically consist of two types of data: thematic and spatial. The spatial data concern the geographical location and extent of the feature on the earth's surface, whereas the thematic data compose the (non-spatial) description of the feature, for instance the vegetation of a small biotope or the crop of a field. The term *thematic data* is used as an alternative to the more commonly used *attribute data*, as this term is used in a more general sense in database theory.

The storage and manipulation of the very large amount of data collected for the monitoring system must be handled by some sort of database system. It seems obvious that the system should be computer-based. The core of such a system is a database where the data are stored and a management system that supplies the user with an interface for data input and output — usually in the form of a query language. A geographical information system (GIS) can be regarded as a special-purpose database system able to handle geographical data.

The monitoring system has some crucial demands for the GIS. First of all, the data structure of the thematic data is non-simple, i.e. the structure consists of several related types of features like agricultural holdings, small biotopes, interviews and so on. Further, each of these features is described by a range of attributes — the small biotope is described by type, tree composition, signs of game care, etc. The GIS must, of course, be able to support this complex data structure.

Secondly, an important task for the monitoring system is to perform spatial analyses of different kinds. The query language of the GIS must include a full range of spatial queries supporting area and distance calculations, network analysis, buffer zoning and the like. We will return to the subject of how to represent the spatial data in the GIS to support these queries in the following section.

Thirdly, it is important to be able to perform analyses based on both thematic and spatial data. It must for instance be possible to answer a query like 'select all small biotopes with more than 100 m^2 of open water and located next to a field of permanent grass'. The first part concerns thematic data, namely an attribute of the small biotopes, whereas the second part includes a 'neighbourhood' query on the spatial data.

These demands do not seem unreasonable. It has been, however, very difficult to find a commercial GIS that is actually able to fulfil them. These GIS's can be identified as belonging to one of two main types.

One type of GIS is primarily concerned with handling spatial data. Representation of thematic data is typically restricted to attaching a single attribute to each feature. This type of GIS is, obviously, uncapable of handling the complex thematic data structures of the monitoring system data.

The other type of GIS is based on a dual architecture. The thematic data are handled by a general-purpose database system able to store and manipulate complex data structures, whereas the spatial data are still represented in a separate, dedicated database system. This solution fulfils the first two requirements as it handles thematic and spatial data very efficiently. The important drawback is that spatial and thematic data can only be integrated with great difficulty. Integrated queries are often only possible to a very limited degree, and desirable features of the database management system (like query optimisation, integrity, and transaction control) are not applied to these integrated queries. For example, the spatial description of an object may still exist in the spatial database, while its thematic description has been deleted from the thematic database (Vijlbrief et al., 1992). A typical example of a GIS with dual architecture is ARC/INFO.

The solution that we found to this problem is representing both thematic and spatial data in a general-purpose relational database system. This allows access to both thematic and spatial data through a common query-language enabling integrated queries. The relational database system certainly supports complex data structures, enabling representation of both thematic and spatial data. This system does have a severe drawback in that the handling of spatial data is not optimised as in the dedicated database system above and thus shows rather low performance. For the purpose of the small biotope monitoring system this is, however, compensated by the flexibility of both data representation and access.

How to support spatial analysis

The spatial characteristics of features and, in particular, the spatial interrelationships of features are key elements in most of the analyses the monitoring system must cater to. Therefore, the database system must supply flexible tools for spatial analysis. The design of the spatial data structure is the basic system hook — it determines which analyses are possible and which are not.

The features of the monitoring system — small biotopes, fields, property units and the like — are characterised by being distinct features with sharply defined boundaries as opposed to a continuously varying natural landscape. Consequently, it is important to be able to handle these features and the boundaries between them as coherent objects. This is the main reason for choosing a vector representation of polygons, lines and points over the raster-cell structure, which is often regarded more suitable for spatial analysis.

The database must include sufficient spatial information to enable analysis on characteristics such as calculation of area and length. This implies that the extent of each feature should be represented. However, this is not feasible for all types of features. In the field, the spatial data of the small biotopes are registered on maps in a scale of app. 1: 10,000. Since the smallest features registered are areal biotopes of 10 m^2 and linear features 10 metres long by 0.1 metres wide, the scale used is insufficient to enable a satisfying areal representation of these small features. In general, at this scale it is impossible to represent the width of most linear features and very small areal features (less than approximately 250 m^2). As the linear features on average constitute about one-half of the total area of small biotopes and the main part of non-protected small biotopes, the representation of their area is important — not least from a management point of view.

The benefit of using a mapping scale small enough to cater to such detail does not balance the additional work on field registration nor the digitizing that this would cause. Instead, the width of the linear features and the area of small areal features are represented by 'sketch'-geometry, meaning that these spatial characteristics are treated as thematic data. The spatial characteristics of a linear feature are represented as a row of points defining a line and an attribute describing the width of the feature. Consequently, only a mean width of the feature can be represented. This is, however, not a serious limitation as most linear features, such as field divides and hedgerows, are characterised by a fairly uniform width. The very small areal features, also named 'point' features, are represented by an area attribute attached to a single point.
'Sketch'-geometry is also used to represent linear features which are close together and parallel. A road and ditch with parallel courses are, again due to the scale, linked to the same line on the map.

A subset of the spatial data which is of vital importance for many landscape ecological analyses is the topology (math: non-metric geometry). Within the mathematical tradition, topology describes the spatial relations between features, such as neighbourhood and intersection. The design of the spatial data structure in the database determines which topological relationships are represented and thus accessible.

When existing GIS's claim to support *'full topology'* they refer to the ability of handling neighbourhood analysis, i.e. determining which features are next to each other on a single map overlay (this includes handling of islands). This, for example, enables the study of how the maintenance of small biotopes is related to the agricultural management of the surrounding fields.

Our definition of 'full topology', however, also includes the handling of spatial relationships across map overlays, i.e. handling not only the neighbourhood of features but also spatial overlapping. This is accomplished in our database system by integrating the spatial data of all map overlays into a single overlay, so that all thematic features are referencing the same map. This enables, for instance, an analysis of the spatial correlation between small biotopes and property boundaries. This analysis is based on determining the spatial overlap of small biotopes and property boundaries originally belonging to a land use map and a land property map, respectively.

How to cope with changes through time

The database now includes data from several historical registrations and three registrations based on field surveys. A major part of our work has been to analyse the status of the small biotopes for each individual registration year and monitor the overall changes. We are, however, convinced that much is to be learned by monitoring the history of individual small biotopes.

This poses the problem of defining rules for the historical relationship between the small biotopes registered through the years in a reasonable manner. On the one hand, the definition must be loose enough to allow the small biotope to change through time and still be registered as the same, although changed, small biotope. On the other hand, it must be tight enough to enable us to identify substantial changes or new small biotopes when they emerge. The question is: When must a small biotope be registered as a new small biotope, as opposed to merely being a more or less changed version of a small biotope registered in a previous year? This question is not a trivial one (Langran, 1992; Brandt et al., 1990) and we have been unable to formulate fool-proof rules for the answer.

As a rule of thumb, we say a small biotope is to be regarded as an incarnation of an earlier registered small biotope if there is a large degree of spatial correlation between the two. This rule usually overrides a causal relationship between small biotope incarnations. Thus, we do not consider a small biotope to be an incarnation of one previously registered just because its appearance explains the disappearance of the other small biotope.

On the other hand, even though there is no causal connection between the disappearance of one small biotope and the appearance of another, we define them as being incarnations of the same small biotope if there is a large degree of spatial correlation between them. For instance, this is the case if a hedgerow is removed and a ditch then dug at the same location. The ditch is then considered to be a reincarnation of the hedgerow. This is also the case if the ditch is not dug until many years after the hedgerow was removed. Consequently, we may find a small biotope that was registered in 1981, disappeared in 1986 and reappeared in 1991 as a different type.

Conclusion

The small biotopes form the lowest level of a hierarchy of ecological networks in the agricultural landscape. They should be considered an integrated part of the agricultural land use and, therefore, they must be managed in accordance with their close relation to the agricultural structure.

A monitoring system for small biotopes should reflect these realities by emphasizing the relationship between small biotopes and agriculture and enabling detailed statistical analysis of the spatial and thematic aspects of this relationship. Important lessons learned are:

- The small biotopes must be regarded as an integrated part of the general land use.
- It should be possible to handle an internal land cover differentiation of the small biotopes.
- The monitoring system should include general information on the agricultural practice, with special emphasis on land use intensity and industrialization.
- The system should allow for a spatial analysis of the relation between small biotope pattern and geo-ecological parameters and classifications.
- The data collection should be based on field surveys including personal contact with the farmers as a source of information, not only on location and type of the holding but, also, concerning the farmers' actual and future utilisation of the small biotopes.
- The system must support a fully developed topology enabling spatial analysis.
- The system should enable time series analysis.

References

Agger, P., E. Andersen, J. Brandt, E. Holmes, J.E. Jensen, D. Larsen and J. Rasmussen, 1992. *Udviklingen i agerlandets småbiotoper 1981-91*. Roskilde Universitet og Skov- og Naturstyrelsen.

Agger, P. and J. Brandt, 1984. Registration methods for studying the developement of small-scale biotope structures in rural Denmark. In: J. Brandt and P. Agger (eds.), *First international seminar of IALE on methodology in landscape ecological research and planning in Roskilde, Denmark*. Geo-Ruc, pp. 61-72.

Agger, P. and J. Brandt, 1987. *Småbiotoper og marginaljorder*. Skov- og Naturstyrelsen. (Miljøministeriets projektundersøgelser 1986). Teknikerrapport 35.

Agger, P. and J. Brandt, 1988. Dynamics of small biotopes in Danish agricultural landscapes. *Landscape Ecology*, 227-240.

Agger, P. and C.H. Owesen, 1990. Monitoring wildlife - an example of a programme set up in Denmark. *Ekologia/CSSR* 9/3: 303-314.

Andersen, E. and J.E. Jensen, 1991. A proposal for a nanochoretypification of Danish landscapes. In: J. Brandt (ed.), *European seminar on practical landscape ecology in Roskilde, Denmark*. International Association for Landscape Ecology (IALE), pp. 143-146.

Biotopgruppen: P. Agger, J. Brandt, E. Byrnak, S.M. Jensen and M. Ursin, 1986. Udviklingen i agerlandets småbiotoper i Øst-Danmark. Institut for Geografi, Samfundsanalyse og Datalogi, Roskilde Universitetscenter. Fors.rapp. No. 48.

Brandt, J., 1991. Land use, landscape structure and the dynamics of habitat networks in Danish agricultural landscapes. In: J. Baudry, F. Burel and V. Hawrylenko (eds.), *Comparisons of landscape pattern dynamics in European rural areas. 1991 Seminars in Ukraine, Normandy*. French and Ukrainian MAB Committee, UNESCO, pp. 213-229.

Brandt, J., E. Holmes and D. Larsen, 1990. Design of a regional database for landscape-ecological studies. In: *Third scandinavian research conference on Geographical Information Systems in Helsingør*. Danish Geological Survey, pp. 74-87.

Hovedstadsrådet, 1982. *Forslag til udpegning af fredningsinteresseområder*. Planlægningsdokument PD354, Copenhagen.

Jongman, R., 1992. *Landscape ecology and spatial organisation Europe*. Draft version coll nr 100-311. Dept. of physical planning and rural development.

Langran, G., 1992. Time in Geographic Information Systems. In: D.J. Peuquet and D.F. Marble (eds.), *Technical Issues in geographic Information Systems*. Tayler & Francis, London etc.

Mayrhofer, P. and P. Schawerda, 1991. *Die Bauern, die Natur & das Geld. Modell Ökopunkte Landwirtschaft*. Verein zur Förderung der Landentwicklung und intakter Lebensräume (LiL).

Meeus, J. H. A., M. P. Wijermans and M. J. Vroom, 1990. Agricultural landscapes in Europe and their transformation. *Landscape Urban Plan*. 18/3-4: 289 - 352.

Neef, E., 1984. *Applied landscape research*. Paper dedicated to the participans of The First International Seminar of The International Association for Landscape Ecology (IALE).

Skov- og Naturstyrelsen (ed.), 1992. *Naturbeskyttelsesloven. Lov nr. 9 af 3 januar 1992. Forberedelser, folketingsbehandling samt andet materiale vedrørende loven*. Skov- og Naturstyrelsen, Hørsholm.

Vijlbrief, T. and P. Van Oosterum, 1992. The GEO++ System: an Extensible GIS. In: *Proceedings of the International Symposium on Spatial Data Handling, Charleston, South Carolina*, pp. 40-50.

The use of floristic data to establish the occurrence and quality of ecosystems 13

Kees (C.) L.G. Groen, Ruud van der Meijden
and Han (J.) Runhaar

ABSTRACT - For environmental management, reliable information is needed on the occurrence and quality of ecosystems. A complete mapping of ecosystems in combination with an exhaustive survey is often expensive, especially on a nationwide scale. Therefore, it is worthwhile investigating whether floristic surveys can be a useful alternative.
In this chapter the applicability of a nationwide floristic database for the Netherlands on 1-km^2 grid cells called FLORBASE is discussed. It is compared with a database on land-cover upgraded with information from vegetation relevés, which is set up in the context of the Landscape Ecological Mapping of the Netherlands: LKN.
It appears that floristic information can be used as a rough estimate of the occurrence of ecosystems in terms of presence in grid cells, and that this information is especially relevant for establishing the quality of these ecosystems. Floristic databases have obvious advantages, the most important of which is the fact that they can be filled and updated relatively easily with information gathered by both amateur and professional naturalists.

Introduction

Due to the growing awareness of the importance of protecting and enhancing nature values among national and regional authorities, there is a growing demand for information on the *quantity* and *quality* of ecosystems. What is the surface area covered by certain ecosystem types and what is the quality of these ecosystems in terms of species richness?

The most direct way to obtain information on the surface area and quality of ecosystems is by mapping them. Maps of ecosystems or vegetation as the most manifest part of ecosystems, with added information on the presence and abundance of species, may contain all the needed information. However, a complete nationwide mapping of ecosystems, in combination with an exhaustive survey on a sufficiently detailed mapping scale, is hardly a realistic option at present because of the costs. Moreover, frequent and recurrent updating of such maps and surveys will certainly not be feasible. Therefore, we have to find other means to obtain the necessary information for environmental planning and for evaluating the national nature conservation policy.

One possibility is to use stratified sampling as an alternative to exhaustive survey, as discussed by Bunce (see chapter 8 in this book). In this chapter, however, we shall discuss another alternative, namely the use of floristic data on grid cells as a means for assessing the occurrence and quality of ecosystems in these grid cells.

We shall discuss the advantages and disadvantages of the use of floristic data as stored in a nationwide database for the Netherlands concerning 1-km^2 grid cells. To this end, we shall make a comparison with the use of data on the surface area of ecosystem types as stored in another nationwide database for 1-km^2 grid cells, which has been compiled in the context of the Landscape Ecological Mapping of the Netherlands (LKN). The main question of comparison is whether floristic data provide a good source of information on the occurence and quality of ecosystems. We shall also discuss the use of floristic data for monitoring ecosystem quality. But first we shall give some background information on the databases we used.

Databases

Two nationwide databases contain relevant information on the (possible) occurrence of ecosystems and ecosystem quality, namely, the database compiled in the context of the Landscape Ecological Mapping of the Netherlands, LKN, and the floristic database FLORBASE.

The *Landscape Ecological Mapping of the Netherlands* (LKN) was started in 1984. Its objective was to bring together in one geographical database all information collected in different projects that may be relevant for landscape ecological studies on national and regional scales (Canters et al., 1991). The project is being financed by the ministries of Housing, Physical Planning and Environmental Management (VROM) and Agriculture, Nature Conservation, and Fisheries (LNV), which consider LKN to be a cornerstone for environmental and nature management policy. The project will be finished in 1994, but component information from the database has already been used for regional and national planning projects (Bolsius et al., 1992; see also Claessen et al. in chapter 10 in this book).

LKN contains information on different abiotic and biotic attributes, such as geomorphology, soil, groundwater, land cover and landscape elements, vegetation, and vertebrate fauna. The information is stored per 1-km^2 grid cell. Each grid cell may contain various records on each attribute. The number of records depends on the detail of the underlying maps or inventories and the internal heterogeneity of the grid cell.

Information on the vegetation is mainly retrieved from inventories carried out by different provincial agencies. The type of data collected varies per province. Some made vegetation relevés, others compiled species lists per landscape element or composed a vegetation map. For transforming and unifying the variety of floristic and phytosociological data into information on the occurrence of ecosystem types, we used the ecological species groups as specified in the ecotopes' classification described by Runhaar and Udo de Haes in this book (see chapter 7). This transformation is based on the indicative value of the entire species composition for ecotope types. In the LKN database the information is stored in terms of the surface area of each ecotope type per 1-km^2 grid cell.

Apart from problems due to the non-comparability of data from different sources, the LKN database suffers from the fact that parts of the country are not covered by provincial inventories. As the majority of provincial agencies are decreasing their efforts towards monitoring vegetation, it is very unlikely that we shall be able to update the LKN database on vegetation in the near future. Consequently, LKN is not suited for monitoring changes.

Another source of information is the nationwide floristic database FLORBASE, which has recently been set up in co-operation by the National Herbarium and the Centre of Environmental Science, both of Leiden University. This database contains records on the occurrence of plant species in 1-km^2 grid cells, from 1975 onwards. This database is the sequel to the ATLAS database (see Mennema et al., 1980; 1985; Van der Meijden et al., 1989), which used grid cells of 25 km^2 and contains records from 1900-1980. The

FLORBASE data come from different sources, from both amateur botanists and the provincial inventories already mentioned above (Groen et al., 1992).
Presently FLORBASE contains more than 3,500,000 records on the occurrence of plant species. Plate 13.1 shows the number of plant species per grid cell recorded so far. It illustrates that the inventories of the western part of the Netherlands, as well as of the northeastern province of Drenthe, are relatively good, both quantitatively and qualitatively, with more than 150 species/km^2 on average. In other parts of the country, the average number of species per grid cell is much smaller, because the inventories were confined to only relatively rare species with high indicative value, or only specific landscape elements were sampled.
The prospects for updating FLORBASE are good, since there is an active organisation of amateur botanists who provide more than 200,000 new records yearly. In addition, the costs of floristic inventories are relatively low compared to the more detailed inventories carried out by the provincial agencies. The main question, however, is whether floristic data provide sufficient information on the occurrence of ecosystem types and their quality for national and regional environmental management.

Estimating ecosystem occurrence from floristic data

For the eco-hydrological model DEMNAT (see Claessen et al., chapter 10), FLORBASE was used to establish the distribution of wet and moist ecosystems in the Netherlands. The ecosystems were classified according to the *ecotope classification* as described by Runhaar and Udo de Haes in chapter 7. We did, however, make some generalizations into *ecotope groups* and, also, distinguished aquatic ecotope types according to an earlier approximation of this ecotope classification. We shall, as did Claessen et al. in chapter 10, continuously speak of ecotope types for reasons of simplicity.
For the ecotope classification, site factors are used as classification characteristics, and species composition is used as the main mapping characteristic. The fact that in the classification itself the species composition is directly linked to the site factor classes is an advantage for environmental impact assessment (see Runhaar and Udo de Haes in chapter 7).

Witte and Van der Meijden (1990, 1992) developed a method to determine the occurrence of ecotopes of different types in grid cells, as well as their species richness from the floristic data. They first selected indicator species for each

THE USE OF FLORISTIC DATA TO ESTABLISH ECOSYSTEM QUALITY 279

ecotope type. The ecological species groups, as specified for each ecotope type (Runhaar et al., 1987; Groen et al., 1993), form the basis for this selection.

The indicator species were then screened as to their indicative value. Species that would often be identified incorrectly and species that may occur in many different ecotope types, and are thus not very indicative, have been left out.

For each selected species an indicator value has been determined inversely related to the number of ecotope types in which it may be found. Species that occur in one ecotope type only were attributed the highest indicative value. Table 13.1 specifies the indicative values of species belonging to ecotope type K21 ('herbaceous vegetations on wet, nutrient-poor, acid sites').

Table 13.1 Indicative value W of plant species for ecotope type K21, 'herbaceous vegetations on wet, nutrient-poor, acid sites'. When the underlined species are found in a grid cell, the completeness of K21 is 5.17; With threshold values of 2.0 and 7.0 the completeness factor is 0.63.

W	Plant species
1.0	Andromeda polifolia
0.5	Carex curta
1.0	Drosera intermedia
0.67	Drosera rotundifolia
1.0	Erica scoparia
0.5	Erica tetralix
0.33	Eriophorum angustifolium
1.0	Eriophorum vaginatum
0.5	Gentiana pneumonanthe
0.5	Juncus tenageia
1.0	Lycopodium inundatum
1.0	Narthecium ossifragum
0.5	Oxycoccus macrocarpos
1.0	Oxycoccus palustris
0.5	Polygala serpyllifolia
1.0	Rhynchospora alba
1.0	Rhynchospora fusca
1.0	Vaccinium uliginosum
1.0	Wahlenbergia hederacea

The sum of the indicative values of the species recorded in a grid cell is used to calculate a measure of 'completeness' of an ecotope of a certain type. If, for instance, the underlined species from Table 13.1 are found in a grid cell, the sum amounts to 5.17.

This sum cannot yet be regarded as a measure for completeness, because the number of species to be expected differs for the various ecotope types. For example, the number of vascular plant species in oligotrophic bogs is relatively small, whereas calcareous grasslands are often very rich in species. These differences are reflected in the number of species assigned to the ecological species groups per ecotope type. In order to account for these differences, the sum of indicative values has been 'normalised' to a *completeness factor* ranging from zero to one. Below a certain *threshold value* the ecotope type is assumed to be absent and the completeness factor is set at zero. This threshold value is specific for the ecotope type; it is lower for K21, which comprises oligotrophic bogs and wet heathlands, than it is for K43, which comprises calcareous grasslands. Above another threshold value the relative species richness is assumed to be maximal, and the completeness factor is set at 1. This threshold value is also specific for the ecotope type, as some ecotope types may be more rich in species than others. Completeness factors between zero and one are calculated by linear interpolation between the two threshold values.

Plate 13.2 and Plate 10.2 exemplify the use of this method to establish the distribution of two ecotope types: type A17 and K27 respectively. Type A17 ('aquatic vegetations in fresh, moderately nutrient-rich water') is very common in the western parts of the Netherlands on fluvial clay and lowland peat soils, whereas K27 ('herbaceous vegetations on wet, moderately nutrient-rich sites') is much less common, and predominantly occurs in the eastern part of the Netherlands on sandy and peat bog soils.

Reliability of the completeness factor as an ecosystem indicator

The distributions as shown on the maps of Plates 13.2 and 10.2 are consistent with what is generally known about the distribution of these ecotope types. However, they give little insight into the reliability of the use of completeness factors to establish the occurrence of ecotopes in individual cells. Does a completeness factor between zero and one mean that an ecotope of this very type is present and does a completeness factor of zero imply its absence? And, if so, what is the relation between the completeness factor and the surface area of the respective ecotope in a grid cell?

To answer these questions we compared the completeness factors we calculated for all grid cells with the information on the presence and surface area of ecotopes in the LKN database. This comparison is slightly hampered by the fact that the LKN database contains information on only the surface area of ecotopes of different type, but does not give any information on the species richness of these ecotopes.

First, we shall examine the first question, the one on the correspondence between a completeness factor non-zero or zero, respectively, and the presence/absence of an ecotope of identical type according to the LKN database. For this comparison only grid cells were used with more than 100 recorded species in FLORBASE *and* with at least 4 ecotopes recorded in LKN.

In Plate 13.3 the presence of ecotopes of type A17 ('aquatic vegetations in fresh, moderately nutrient-rich water'), as established from LKN and/or FLORBASE, is presented, demonstrating the degree of overlap. The plate presents the data on the western part of the Netherlands, which part is best covered in both databases. In only 40% of the grid cells did this ecotope type's presence follow from both LKN and FLORBASE. In 47% of the cases A17 was indicated only by the LKN database, and in 13% of the cases only by the completeness according to FLORBASE.

The grid cells in which A17 is only indicated by the LKN database are concentrated in ecodistricts (see Klijn in chapter 5 of this book for the ecodistrict classification) with sand or marine clays as dominant parent material. In these ecodistricts, ecotopes of type A17 are mostly poor in species. In contrast, in the lowland peats and fluvial clay areas, where the correspondence is much better (51-59%), ecotopes belonging to type A17 are often much richer in species. This indicates that on the basis of floristic information we probably miss poorly developed ecotopes.

Other ecotope types characterized by a (moderately) high nutrient availability show a similar pattern. This is illustrated in Table 13.2 for ecotope types A18 and K27.

Table 13.2 Overlap of ecotopes of various types in the Netherlands in percentages, as derived from LKN and as derived from FLORBASE (FB)

Ecotope type	FB + LKN	FB	LKN
A17	40	13	47
A18	56	3	41
K27	41	13	46
K28	62	21	17
K21	50	46	4
K22	14	70	16
K41	44	48	8
K42	8	78	14

Ecotopes with low nutrient availability (types K21, K22, K41 and K42), however, show a reverse pattern (see Table 13.2). These are indicated more often on the basis of the floristic data. This may be caused partly by the fact that the provincial agencies focused their inventories on the rural areas, which results in nature reserves being ill-represented in the LKN database. Another explanation of this difference may be the retarded response of the presence of plant species to intensified land use. Although vegetations of nutrient poor sites have largely disappeared from the rural areas due to excessive eutrophication, individual plants or small populations of certain long-living species may have survived in places. In species lists concerning homogeneous sites they will be recognized immediately as relicts of the past. After lumping species lists per km^2, however, the presence of a number of these species may give the false impression that an ecotope of the type concerned is still present.

Secondly, we investigated whether there is a relation between the completeness factors as derived from FLORBASE and the surface areas of ecotopes of the same type according to LKN. This analysis was carried out for only those grid cells that contained ecotopes of a certain type according to the above described method. No relation was found.

For a justified interpretation of the results of the above comparisons, we should take into account that the LKN database cannot be considered a 100% reliable source of information concerning the occurrence of ecotope types. A variety of many different sources of information was used to determine the presence and surface area of ecotopes in the grid cells. Occasionally, the same information was used for the LKN database as for FLORBASE, although more detail has been retained for LKN, namely, species lists per land cover type or landscape element within a grid cell, instead of the mere species lists per grid cell for FLORBASE. Therefore, we have to be careful in drawing conclusions from this comparison.

The results suggest that floristic data give important information on the *quality* of ecotopes in terms of species richness, but not on the *quantity* of ecosystems in terms of surface area.

The use of floristic data for monitoring changes in environmental quality

Floristic data are not only relevant for establishing the present situation for environmental impact assessments, but can also be used for monitoring changes in ecosystem quality in terms of changing species composition.

One of the main problems for applications of this kind is due to the fact that past inventories have not always been sufficiently systematic. This obliges us to realise that an increase in the number of grid cells in which a species is recorded is not necessarily caused by an actually increased occurrence, but may be caused by increased search intensity, enhanced knowledge on the identification of 'difficult' species, or changes in the interest botanists show for certain ecotope types.

For the Netherlands, the latter phenomenon can be exemplified in the case of aquatic ecotope types. The scarcity of records of plant species of aquatic ecosystems from the beginning of the century is very likely due to the fact that botanists were not interested in these ecosystems at that time (Witte and Van der Meijden, 1992). As a result, the records in the floristic database indicate a considerable increase of aquatic species, whereas in fact a decrease is known to have occurred in many cases, at least since 1970 (Clausman and Groen, 1988).

Still, floristic data can be very valuable for monitoring environmental changes, as long as we are well aware of possible pitfalls connected with this type of data. We shall demonstrate this by an example aimed at quantifying biotic parameters for environmental quality assessment by using data from FLORBASE on species presence only.

As one of the quality parameters especially relevant for lowland peat areas, we selected a group of 16 plant species that are indicative for the highly valued but threatened ecotope type G27: 'grasslands on wet sites with moderately high nutrient availability'. These species should be easy to identify, so that we can be relatively certain that they were not overlooked in the inventories. The selected species group (Table 13.3) is called the 'Marsh Marigold group' after its most outstanding and appealing representative, the Marsh Marigold (*Caltha palustris*).

We compared the occurrence distribution of the Marsh Marigold group before and after 1983, not only for the ecodistrict called the Lowland Peat area (see Klijn in chapter 5 for the ecodistrict classification and map) but for the entire country. Only those grid cells were taken into account which were well investigated in both periods, a demand which was met by 2824 grid cells.

For these cells we compared the number of species of the Marsh Marigold group found in the period 1975-1983 with the number found in the period 1984-1990. Plate 13.4 illustrates the results of this comparison for the western part of the Netherlands. Figure 13.1 shows the numbers of cells with either increasing or decreasing species numbers. On closer examination we see that the distribution in this figure is skewed: there are more grid cells on the left

Table 13.3 The Marsh Marigold group

Achillea ptarmica
Caltha palustris
Carex disticha
Carex vesicaria
Cirsium palustre
Epilobium palustre
Hydrocotyle vulgaris
Hypericum quadrangulum
Juncus conglomeratus
Lotus uliginosus
Lychnis flos-cuculi
Mentha aquatica
Ranunculus flammula
Senecio aquaticus
Stellaria palustris
Valeriana dioica

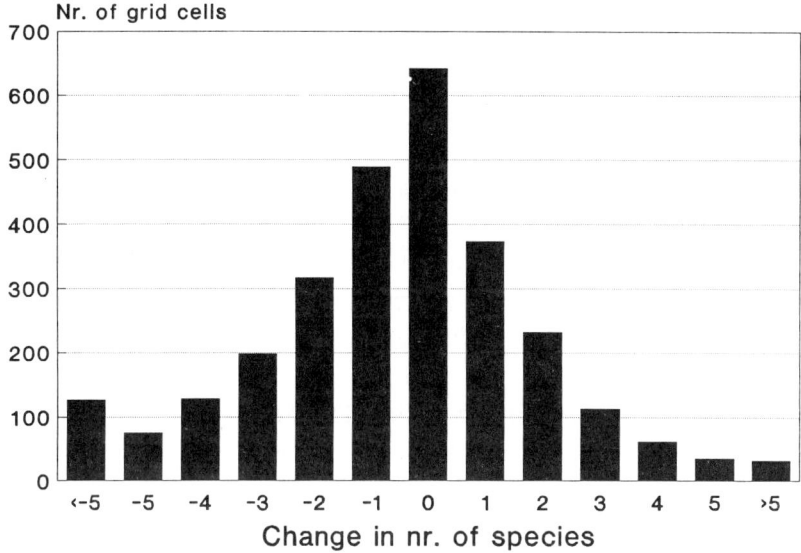

Figure 13.1 Numbers of grid cells with change in species numbers of the Marsh Marigold group between the survey periods 1975-1983 and 1984-1990. Source: FLORBASE

then on the right. In 47% of the grid cells we see a decrease, in 30% an increase. The decrease in species numbers is largest in cells with more than eight species of this group before 1983 (see Table 13.4).

The majority of the grid cells with rapidly decreasing species numbers is found within the Lowland Peat area. This can mainly be explained by the intensified agriculture in this area, which has led to eutrophication and a sub-

Tabel 13.4 Changes in the distribution of the Marsh Marigold group between the survey periods 1975-1983 and 1984-1990 relative to the number of species in the first period

	Initial nr. of species			
Change	0	1-8	9-16	Total
Increase	102	632	114	848
No change	137	417	88	642
Decrease	0	892	442	1334
	239	1941	644	2824

sequent decrease in species characteristic for mesotrophic conditions (Clausman and Groen, 1988; Melman, 1991; Van Strien, 1991).The above example resembles the procedure to quantify biotic quality parameters for the Lowland Peat area followed by Latour et al. (1991). The main difference is that they used data on both the presence *and abundance* of a *smaller* group of species, viz. 12, which they derived from provincial inventories in the western part of the country, carried out between 1974 and 1983.

The results we achieved in the above example based on *presence only* very closely resemble the results of Latour et al. This suggests that information on the abundance of plant species within grid cells does not add much.

A second example concerns the changes in the '*Littorella* group' in the eastern part of the province of Brabant, as based on two different inventories. The *Littorella* group consists of species characteristic for ecotope type W12, i.e., 'nutrient-poor, weakly acid water' (Table 13.5).

Ecotopes of this type were never very common, but in the last 15 years a rapid decrease has occurred resulting from acidification caused by atmospheric pollution. In the most recent inventory, in only 1 cell out of 48 an increase in species number was recorded, whereas in 33 cells there was a decrease.

Table 13.5 The Littorella group

Apium inundatum
Callitriche palustris
Echinodorus ranunculoides
Echinodorus repens
Eleocharis acicularis
Hypericum elodes
Littorella uniflora
Lobelia dortmanna
Ludwigia palustris
Myriophyllum alterniflorum
Lythrum portula
Pilularia globulifera
Ranunculus ololeucos
Scirpus fluitans

Discussion

The question we posed in the beginning of this chapter was whether floristic data in a 1-km^2 grid would provide a good source of information for establishing the occurrence and quality of ecosystems for national policy purposes.

As to the occurrence of ecotopes of various types, we have the impression that floristic data do not always yield sufficiently complete pictures of their spatial distribution. When using floristic data for environmental impact assessments, we probably miss information on less-developed ecotopes. Also, we may conclude an ecotope to be present on the basis of merely a number of species that are in fact relicts occurring at great mutual distance within a grid cell.

In addition, we want to emphasize that it may be practically impossible to trace ecotopes of all distinguished types on the basis of indication by the present ecological species groups, because some ecotope types contain too few vascular plant species. Establishing the occurrence of ecotopes of such types by means of biotic indicators can only be achieved by also taking into account other taxonomic groups, such as mosses, lichens, or aquatic or soil macro-fauna.

For monitoring we are mainly interested in changes in the *quality* of ecosystems, as expressed by changes in the presence of species characteristics for the ecotope types under concern. The fact that floristic data do not always give a complete picture of the occurrence distribution of ecotopes of a certain type is then of little relevance. We maintain, therefore, that floristic data provide an outstanding means for monitoring changes in ecosystem quality due to human impacts.

The examples given in the previous section showed how floristic data can be used for monitoring changes in environmental quality. In both examples we worked with groups of plant species instead of with individual species. We prefer this, for two reasons.

In the first place, the reliability of the method increases by using more species. Individual species always pose the inherent risk that a better recognition of the species or a changed interest in the species among botanists leads us to drawing the wrong conclusions. By working with groups of species, this risk is diminished, because such non-systematic errors are levelled out.

Another advantage of working with species groups is that the interpretation of the observed changes is easier, i.e., if the groups have an ecologically indicative value. In the examples given, the groups consist of species that share a preference for certain site conditions, so that observed effects can be linked directly to changes in site factors. If we want to analyse the spatial distribution of a single species, we must realize that its occurrence is determined by many factors, including some of which we may have very little knowledge, for example, on factors determining seed dispersal or germination. For the same reasons, we doubt whether it is possible to predict effects on the species level, as advocated by Latour et al. in chapter 9 of this book.

FLORBASE presently contains data from a wide variety of inventories carried out according to different methods and aimed at different purposes. This is no problem as long as we are only interested in establishing the present situation concerning species occurrence, because we may correct for 'inventory effects' due to the different methods (Witte and Van der Meijden, 1992).
For monitoring purposes, however, it is a different case. By not standardizing the survey method, an additional source of uncertainty is introduced that negatively affects the sensitivity of the measure used for monitoring. Of course, the reliability of comparisons over time can be enhanced substantially by standardizing the method. However, we should realize that most of the records originate from amateur botanists, implying that it is not very likely that such a standardization can be achieved in practice. That is, if we want to continue maximally exploiting the enthusiastic efforts of the large number of amateur botanists in the future as well.
Thus, we should remain very careful when interpreting recorded changes based on secondary data. But an aid for a more sound interpretation of observed changes may also be found in setting up a monitoring *network* of grid cells throughout the country. This should be investigated at regular intervals according to a standardized method, with the objective of constituting a reference data set. Data from such a reference set can then be used to check whether trends found in other grid cells are likely to be real trends or merely a coincidence.
Monitoring changes in the species composition within ecotopes may best be assessed by means of a stratified sample concerning the occurrence of only a number of selected species from the ecological species groups within ecotopes of the specified type (see Bunce in chapter 8 of this book). We are currently planning such an enterprise as a sequel to composing the FLORBASE database.
Including information on the *abundance* of species *within* grid cells may be a means within close reach to enhance the sensitivity of a nationwide floristic monitoring project. However, as was demonstrated in the previous section, the

profit of such an effort is doubtful. We ask for closer investigations to assess whether it is worthwhile.

Finally, we want to stress that whoever aims at establishing the 'present' situation should be well aware of the fact that data, and especially floristic data, age, because of the huge impact of human activities. To account for various undesired effects we should, for nationwide analyses, explore the possibilities of combining accurate spatial data on more stable ecosystem components or characteristics, such as land use, soil, hydrology, etc., with spatially less differentiated data on ecosystem quality in terms of species composition.

References

Bolsius E.C.A., J.P. Chardon, C.L.G. Groen, W.B. Harms, F. Klijn, J.L. Mulder, Th. Niessen, C.H.J. Van Oijen, Th.M.F. Peterbroers, J.G.M. Schouffoer, E.P. Sterling, B.J. Vreeken, R.W. de Waal and M. Van 't Zelfde, 1992. *Op weg naar een landsdekkend databestand.* LKN fase 3 Nederland: Interimrapport. LKN rapport 3, Rijksplanologische Dienst, The Hague.

Canters, K.J., C.P. Den Herder, A.A. de Veer, P.W.M. Veelenturf and R.W. de Waal, 1991. Landscape-ecological mapping of the Netherlands. *Landscape Ecology* 5/3: 145-162.

Clausman, P.H.M.A. and C.L.G. Groen, 1988. *Veranderingen in de vegetatie in de Alblasserwaard en de Vijfheerenlanden in de periode 1977-1984.* Rapport Dienst Ruimte en Groen provincie Zuid-Holland, The Hague.

Groen, C.L.G., M. Gorree, R. Van der Meijden, R. Huele and M. Van 't Zelfde, 1992. *FLORBASE; een bestand van de Nederlandse flora, periode 1975-1990.* CML-rapport 91, Leiden/ Rapport Onderzoek Effekten Grondwaterwinning nr. 4. RIVM, Bilthoven.

Groen, C.L.G., R.A.M. Stevers, C.R. Van Gool and M.E.A. Broekmeijer, 1993. *Uitwerking ecotopensysteem III; herziene landelijke typologie en vertaalsleutels voor Overijssel, Gelderland, Noord-Brabant en Limburg.* CML-mededelingen 49, Leiden.

Latour, J.B., C.L.G. Groen and M. Van 't Zelfde, 1991. *De milieukwaliteit van de ecodistricten Het Laagveengebied en De Kalkrijke Duinen; bijlagen.* RIVM-rapp. 711901002, Bilthoven/ CML-mededelingen 73b, Leiden.

Melman, Th.C.P.M., 1991. *Slootkanten in het veenweidegebied; mogelijkheden voor behoud en ontwikkeling van natuur in agrarisch grasland.* Thesis RU Leiden.

Mennema, J., A.J. Quené-Boterenbrood and C.L. Plate (eds.), 1980. *Atlas van de Nederlandse Flora, deel 1.* Kosmos, Amsterdam.

Mennema, J., A.J. Quené-Boterenbrood and C.L. Plate (eds.), 1985. *Atlas van de Nederlandse Flora, deel 2.* Bohn, Scheltema en Holkema, Utrecht.

Runhaar, J., C.L.G. Groen, R. Van der Meijden and R.A.M. Stevers, 1987. Een nieuwe indeling van plantesoorten in ecologische soortengroepen binnen de Nederlandse flora. *Gorteria* 13: 276-359.

Van der Meijden, R., C.L. Plate and E.J. Weeda, 1989. *Atlas van de Nederlandse flora 3.* Leiden.

Van Strien, A.J., 1991. *Maintenance of plant species diversity on dairy farms.* PhD thesis, Leiden University.

Witte, J.P.M. and R. Van der Meijden, 1990. *Natte en vochtige ecosystemen.* Wet. med. KNNV nr 200, Utrecht.

Witte, J.P.M. and R. Van der Meijden, 1992. *Verspreiding en natuurwaarden van ecotoopgroepen in Nederland.* Rapport Onderzoek Effekten Grondwaterwinning nr. 6. RIVM, Bilthoven.

Index

assessment, risk 187
balance, material 122
biodiversity v, 180, 190, 224
biogeocoenosis 142
biotope 56, 252, 258
boundary 33-4, 92
budget, proton 125-7
carrying capacity 135
cause-effect relations 122-3
characteristic, classification 94-6, 126, 143-4
characteristic, diagnostic 26-7, 39, 41, 95-6
characteristic, mapping 95-6
chorological relations 13, 161
classification characteristic 94-6, 126, 143-4
classification, floristic 61
cluster analysis 33, 158
community 139, 158
components 88
conservation, nature 224, 230, 255-6
correlative complex 91
critical load 125-6, 189
data management 268
database 207-8, 224, 268, 276,
database, floristic 208, 233, 276-7
diagnostic characteristic 26-7, 39, 41, 95-6
district 76, 98, 100

dose-effect function 209
ecochore 12-3, 45, 59, 265
ecological network 251-2, 260
ecological species group 146-8, 208, 279
ecosystem v, 1, 3, 24, 50, 85, 119, 139
ecosystem response 184, 202, 204, 209
ecotope 11-2, 24, 40-1, 45, 53, 55-9, 77-8, 94, 102, 140, 142, 204-5, 277
environmental hazard 103-4, 119, 191
environmental impact assessment 7-8, 155, 157, 230
environmental management 63, 85, 200, 224, 264
environmental quality 4
environmental sphere 52, 87
environmental stress 183
floristic classification 61
floristic database 208, 233, 276-7
Gaia v, 10
grid 174, 207, 233, 276
guideline 124
guiding principle 26-8, 140
hazard, environmental 103-4, 119, 191
hierarchy 8-10, 38, 44, 50, 87, 89, 119, 251

hierarchy, spatial 69-72, 80, 90
hierarchy, temporal 90
impact assessment, environmental 7, 8, 155, 157, 230
indicator 187
indicator, site factor 151, 234, 279-80
land attribute 24, 28-9, 40-1, 43, 265, 277
land cover 176-7, 179-80
land evaluation 4, 6
land unit 12, 29, 39, 58, 93, 142, 174
landscape 13-4, 51-2, 55, 76
levels, organizational 9, 50, 119-20
levels, scale 10, 38
levels, system 8
load, critical 125-6, 189
management, data 268
management, environmental 63, 85, 200, 224, 264
management measures 133, 135
management, water 200
mapping 30-2, 86, 94, 97, 225, 229, 252
mapping characteristic 95-6
material balance 122
measures, management 133, 135
model, multiple stress 184-5
modelling, predictive 201
monitoring 177-9, 224, 252, 257, 272, 283-4
multiple stress model 184-5
multivariate analysis 39-40, 173-5, 180
nature conservation 224, 230, 255-6
nature valuation 5, 211, 225, 229, 276, 284-5

nested system 71-2
network, ecological 251-2, 260
occurrence probability, species 185
ordination 33, 37, 176
organizational levels 9, 50, 119-20
physiotope 55, 59
phytosociology 61, 158, 175, 227
predictive modelling 201
properties 26
proton budget 125-7
region 75, 94, 98-9
response, ecosystem 184, 202, 204, 209
response, species-response curve 185-6
risk assessment 187
sampling 34-5, 174-5, 254, 266-7, 276
scale levels 10, 38
scales, spatial 50, 86, 92
scaling 64, 122
site 77, 101, 142, 260
site factor 119, 141, 143-4, 186, 202, 205
site factor indicator 151, 234, 279-80
societal system 2, 63
spatial scales 50, 86, 92
spatial hierarchy 69-72, 80, 90
species composition 145-6, 184, 188, 202
species group, ecological 146-8, 208, 279
species number 234, 278, 284-5
species occurrence probability 185
species-response curve 185-6
species, target 135, 184, 188
stress, environmental 183
stress model, multiple 184-5
subsystems 88

susceptibility 4-5, 102, 105-8
sustainability v
system 25
system levels 8
target species 135, 184, 188
temporal hierarchy 90
topological relations 13

typification 33, 44-5
valuation, nature 5, 211, 225, 229, 276, 284-5
vulnerability 5-6
water management 200
zone 71-2, 94, 97-8

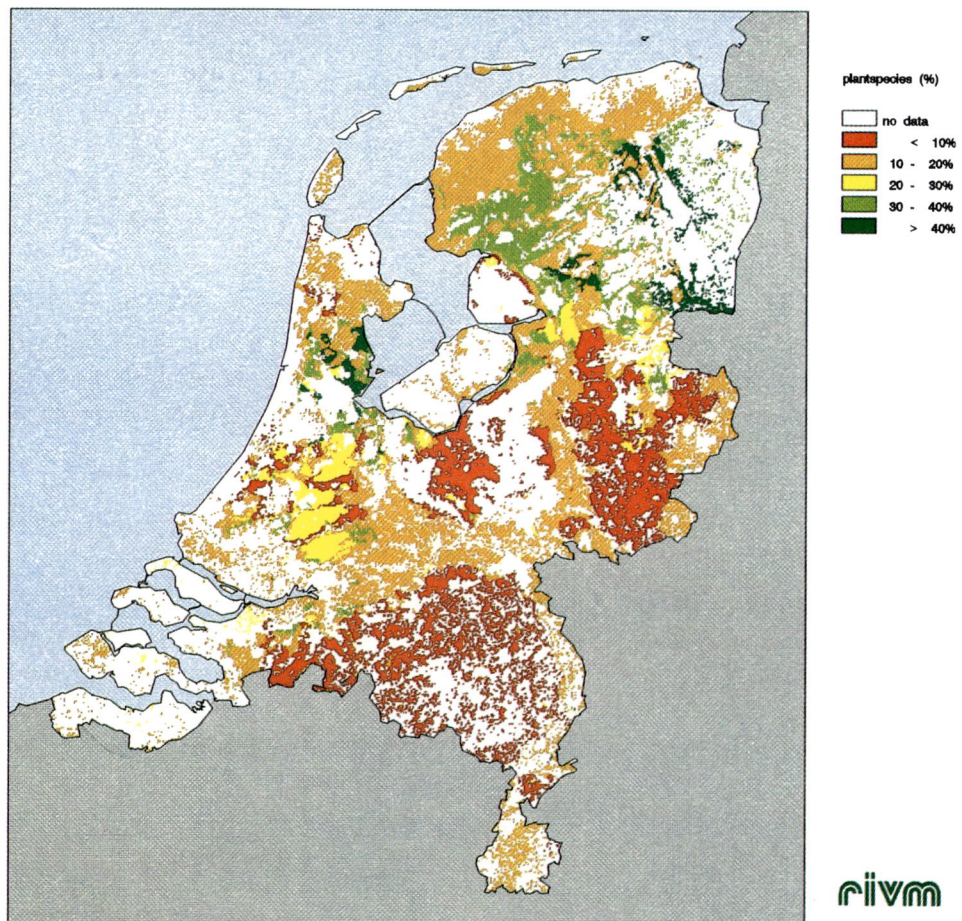

Plate 9.1 The potential species diversity of grasslands in the Netherlands with N loads from fertilizer and manure based on a set of 275 ecological response curves for plant species in the province of South Holland. It is assumed that response curves for South Holland do not differ systematically from ecological response curves for other areas in the Netherlands.

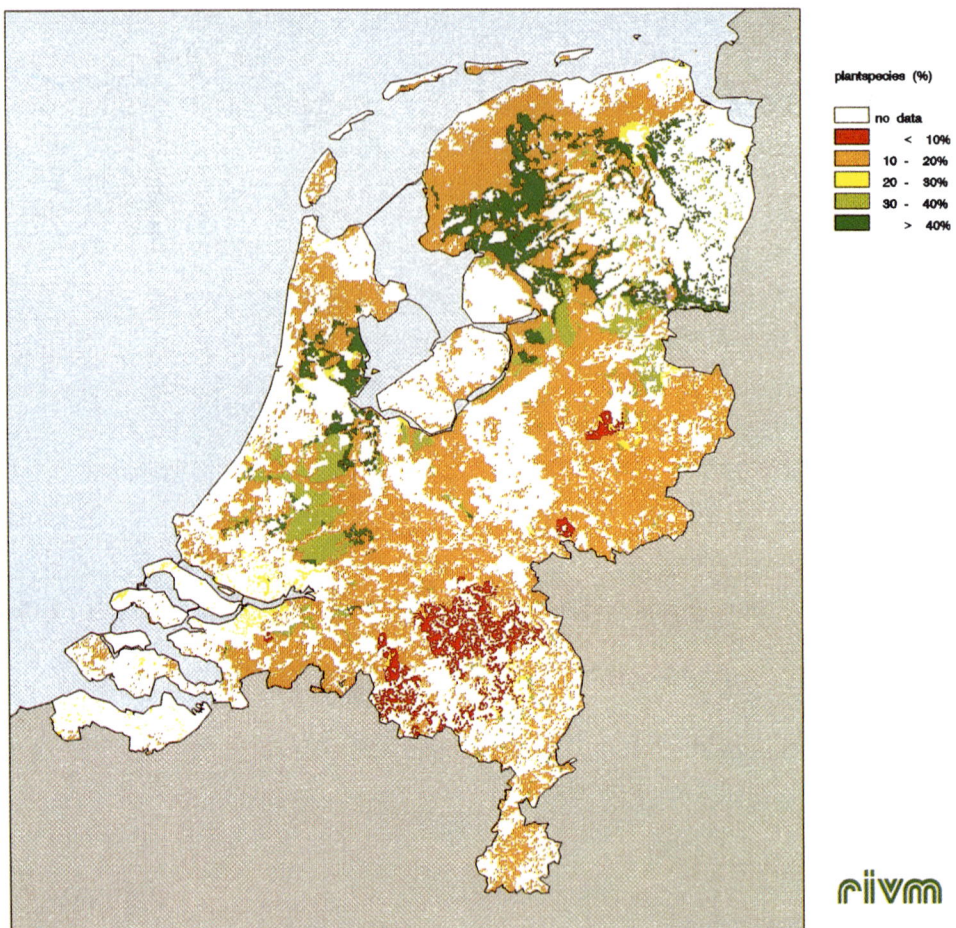

Plate 9.2 As Plate 9.1, but for the year 2010 using a scenario in Maas (1991).

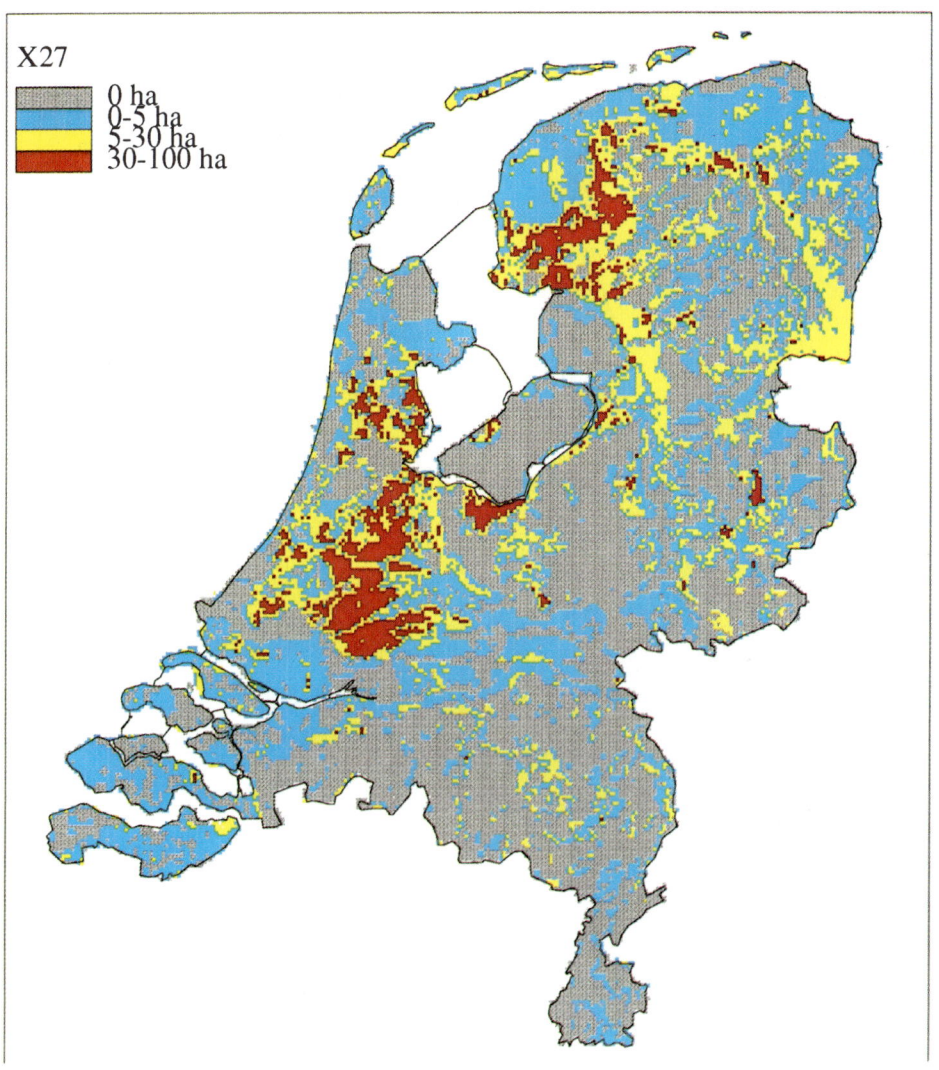

Plate 10.1 Map of potential occurrence of abiotic site type X27 as derived from ecoseries, i.e., wet, moderately rich sites. Surface areas refer to gridcells of 1 km² (Klijn et al., 1992)

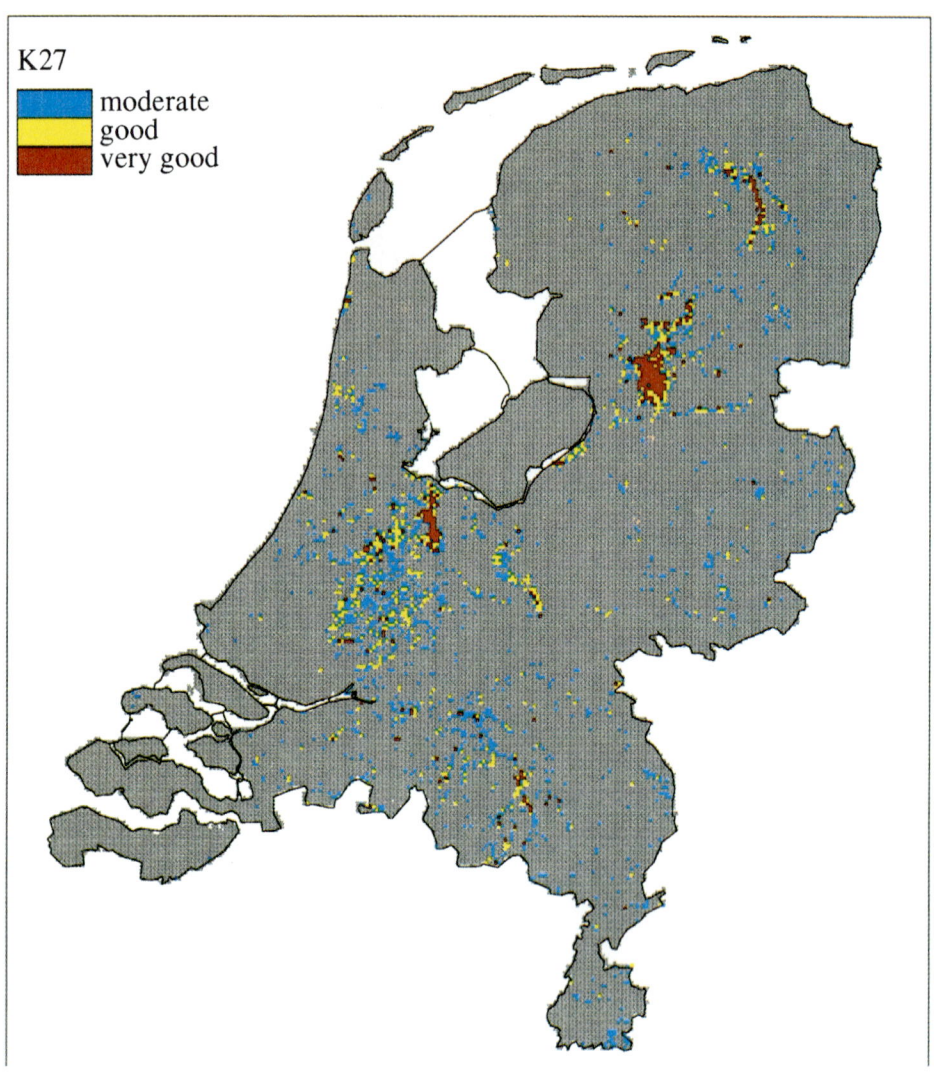

Plate 10.2 Actual distribution and relative species richness, in four classes, of ecotope type K27, i.e. herbaceous vegetations on wet, moderately rich sites (Witte and Van der Meijden, 1992)

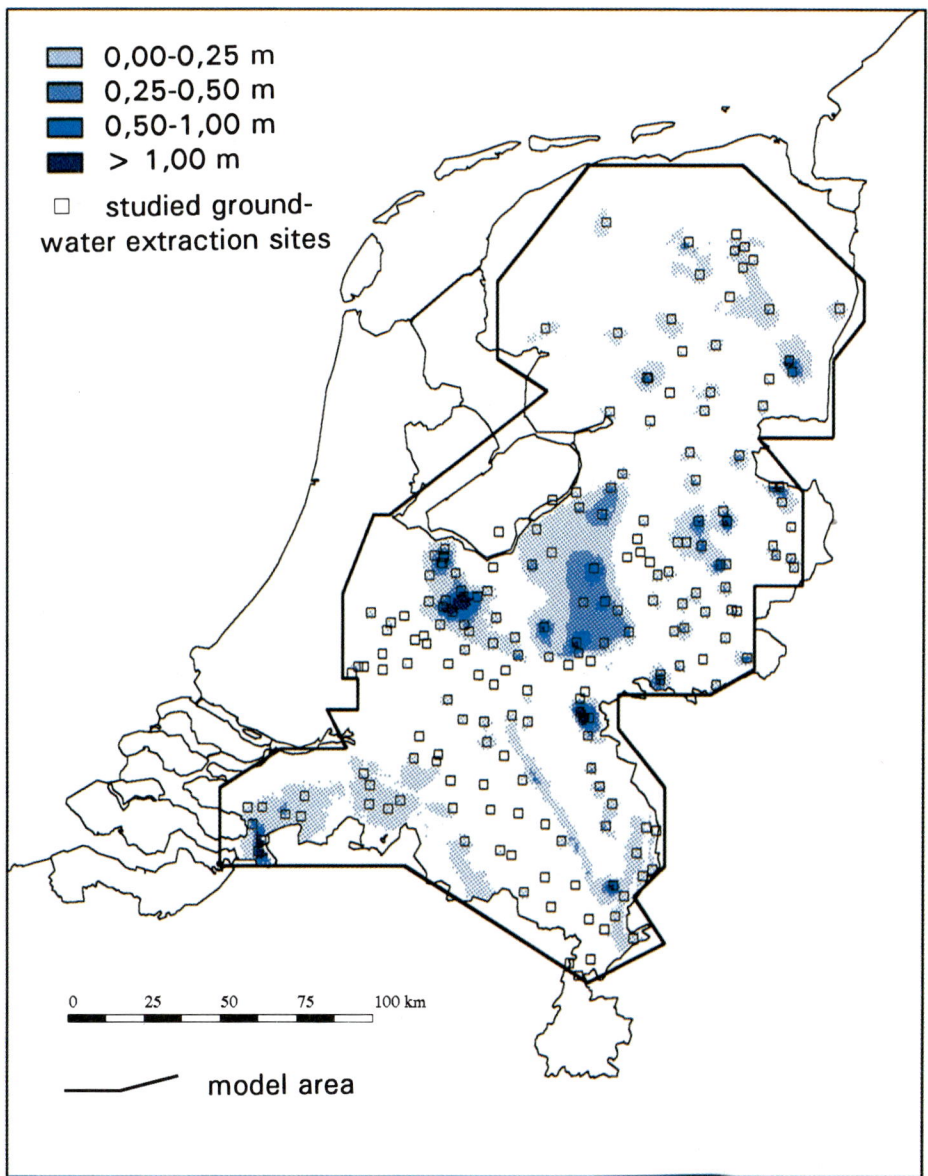

Plate 10.3 The rise of the groundwater level [m] calculated for scenario 3, i.e., a 50% decrease of drinking water extraction. The figure also shows the extension of the modelled area covering about 75% of the Netherlands (Beugelink et al., 1992).

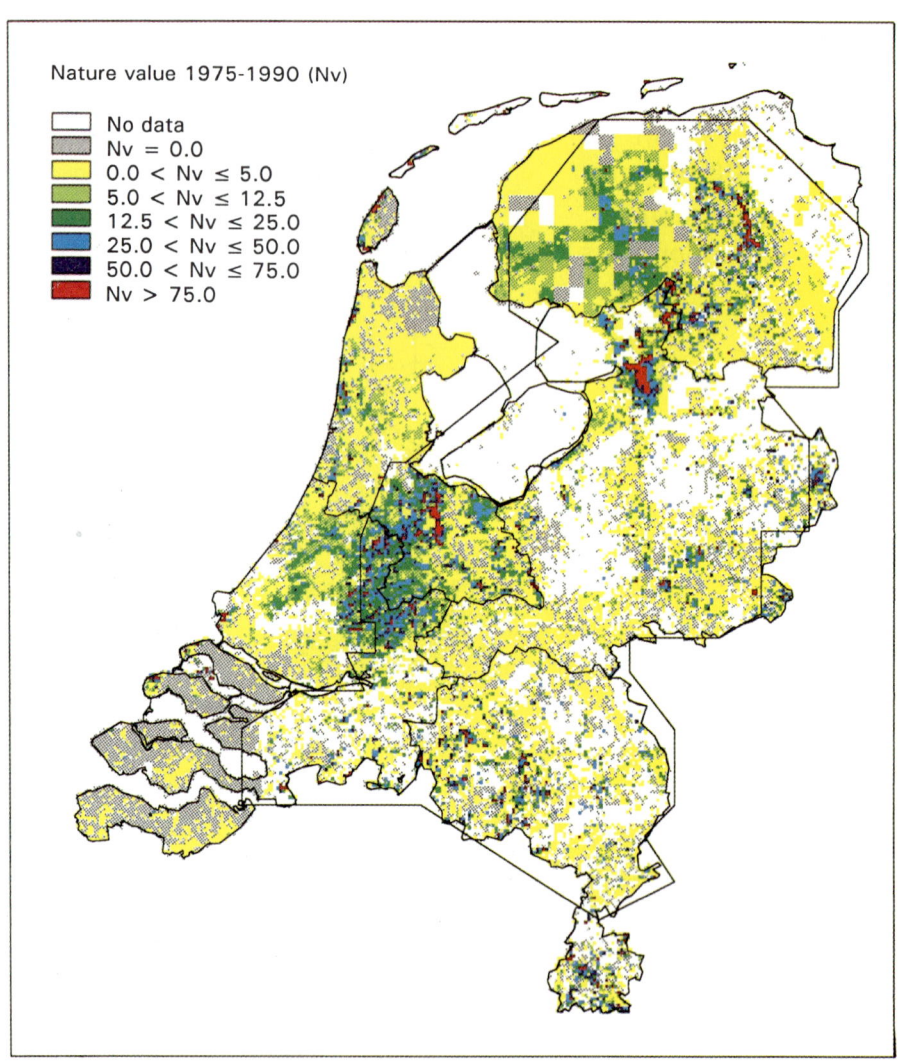

Plate 10.4 Present nature value derived from floristic species finds in the period 1975-1990, totalled for 15 ecotope types (Witte and Van der Meijden, 1992).

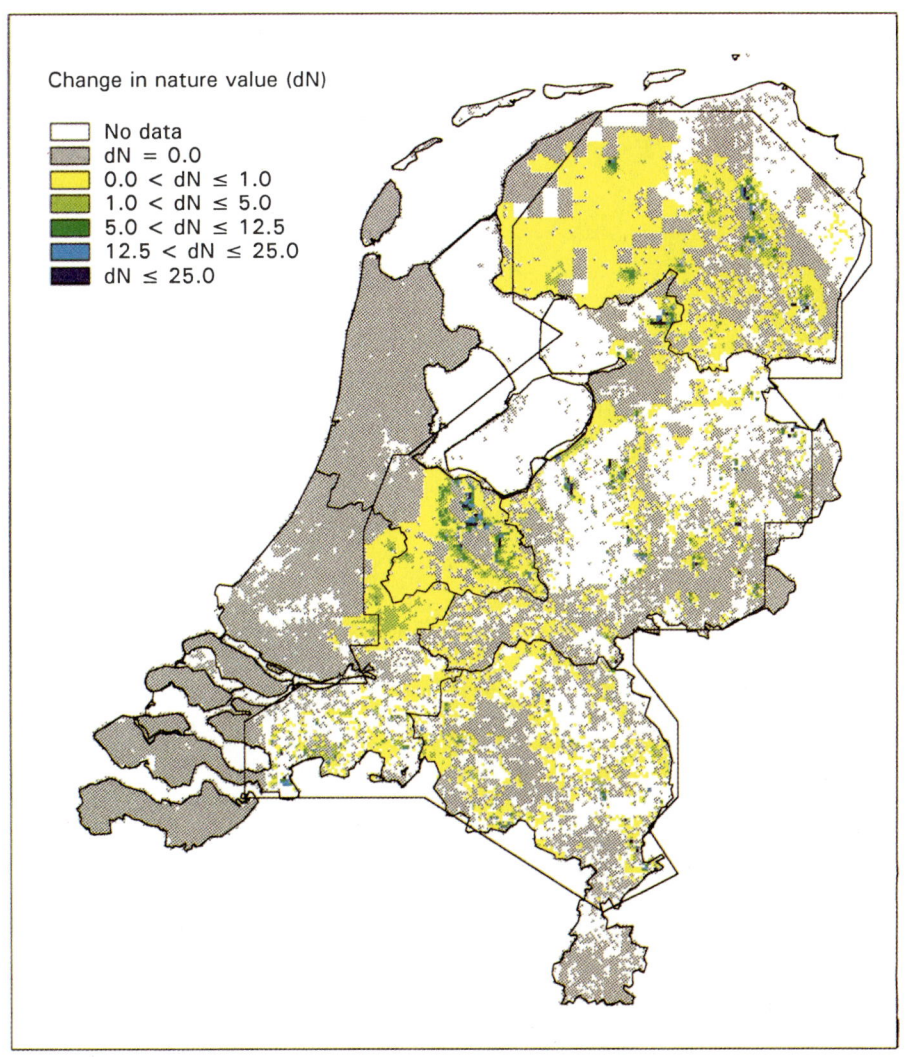

Plate 10.5 Change in the total nature value of the 15 ecotope types, calculated for scenario 3, relative to the present situation (Beugelink et al., 1992).

very high biotic value

high biotic value

little biotic value

Plate 11.1 Part of the Biological Valuation Map of Belgium. For the codes of the legend units: see annex 1.

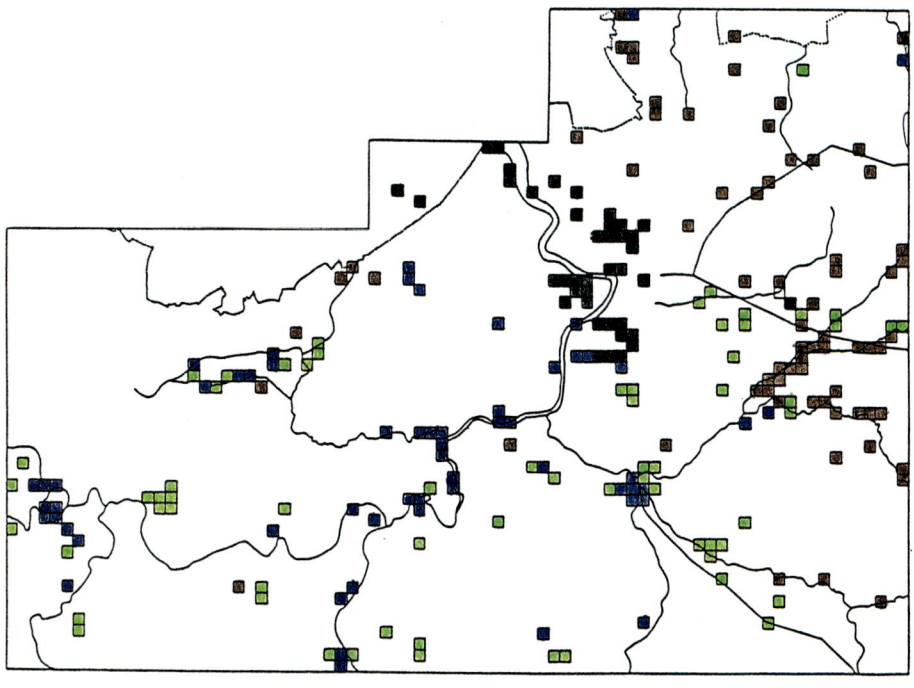

	10	species from wet eutrophic tall herbaceous vegetations, growing mainly in alluvial plains
	11	species characteristic for pioneer ecosystems and for polder areas; they are associated with the industrial area around Antwerp and the lower Scheldt river.
	00	species characteristic for acid, nutrient poor soils (Campine region)
	01	species characteristic for nutrient rich deciduous woodland

Plate 11.2 An ecological characterization of grid cells classified by means of TWINSPAN analysis of plant species lists, i.e. floristic data (Van den Abeele, 1986). Plate 11.2: 2nd division level; Plate 11.3: 3rd division level

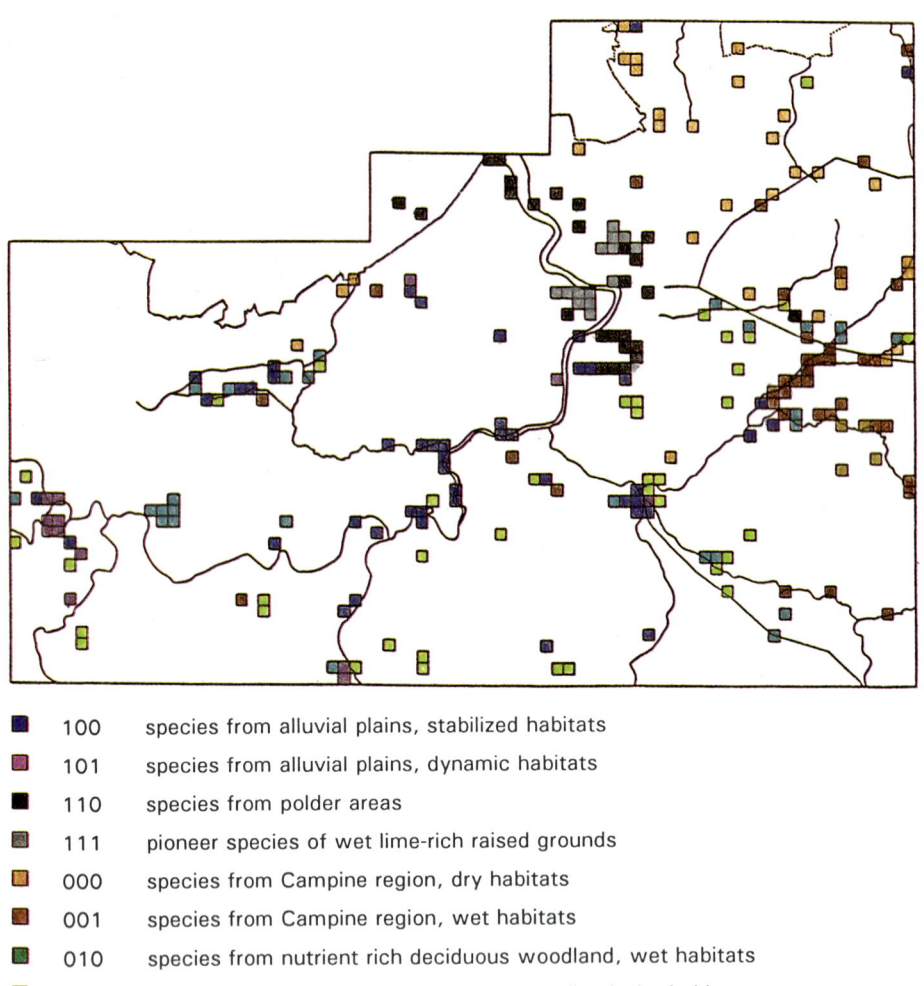

	100	species from alluvial plains, stabilized habitats
	101	species from alluvial plains, dynamic habitats
	110	species from polder areas
	111	pioneer species of wet lime-rich raised grounds
	000	species from Campine region, dry habitats
	001	species from Campine region, wet habitats
	010	species from nutrient rich deciduous woodland, wet habitats
	011	species from nutrient rich deciduous woodland, dry habitats

Plate 11.3 For explanation, see Plate 11.2

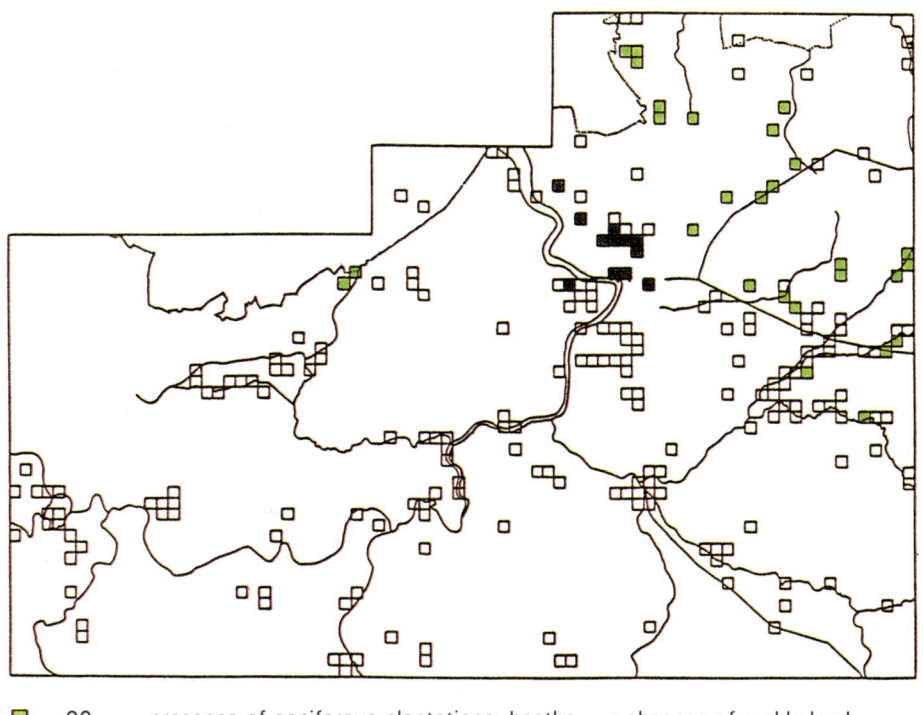

	00	presence of coniferous plantations, heaths, ...; absence of arable land
	01	presence of arable land and absence of coniferous plantations, heaths, ...; dominance of grasslands and poplar plantations
	1	dominance of industrial or raised grounds

Plate 11.4 An ecological characterization of grid cells classified by means of TWINSPAN analysis using the legend units of the Biological Valuation Map (Van den Abeele, 1986). Plate 11.4: 2nd division level; Plate 11.5: 3rd division level

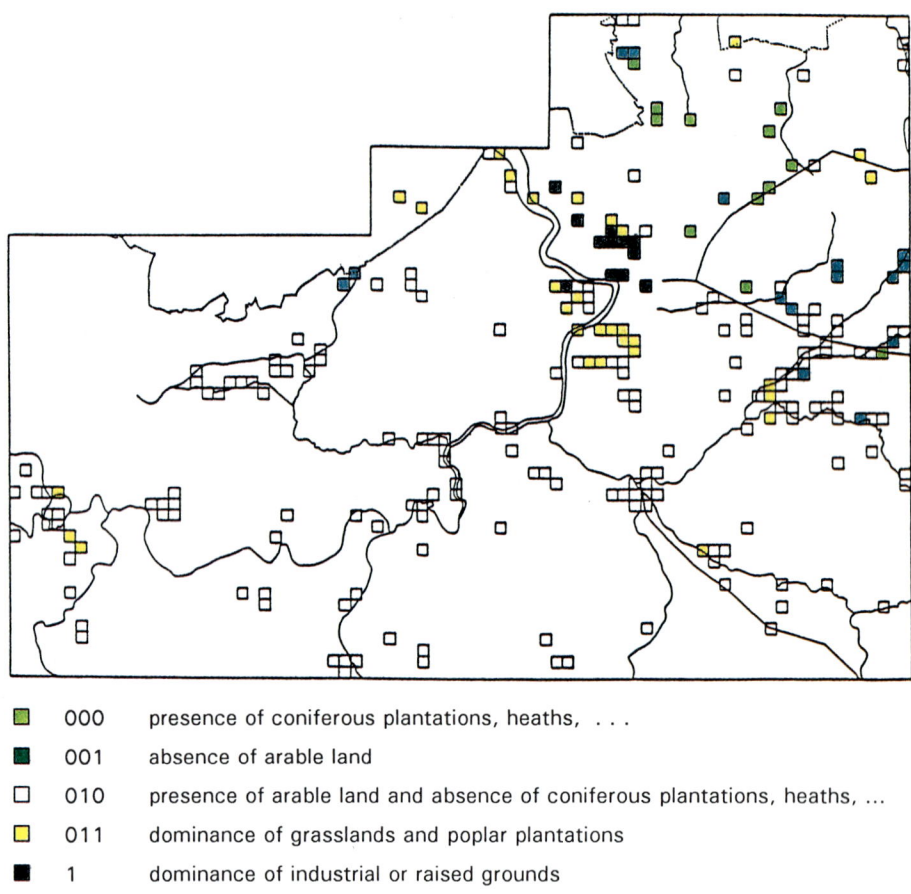

	000	presence of coniferous plantations, heaths, ...
	001	absence of arable land
	010	presence of arable land and absence of coniferous plantations, heaths, ...
	011	dominance of grasslands and poplar plantations
	1	dominance of industrial or raised grounds

Plate 11.5 For explanation, see Plate 11.4

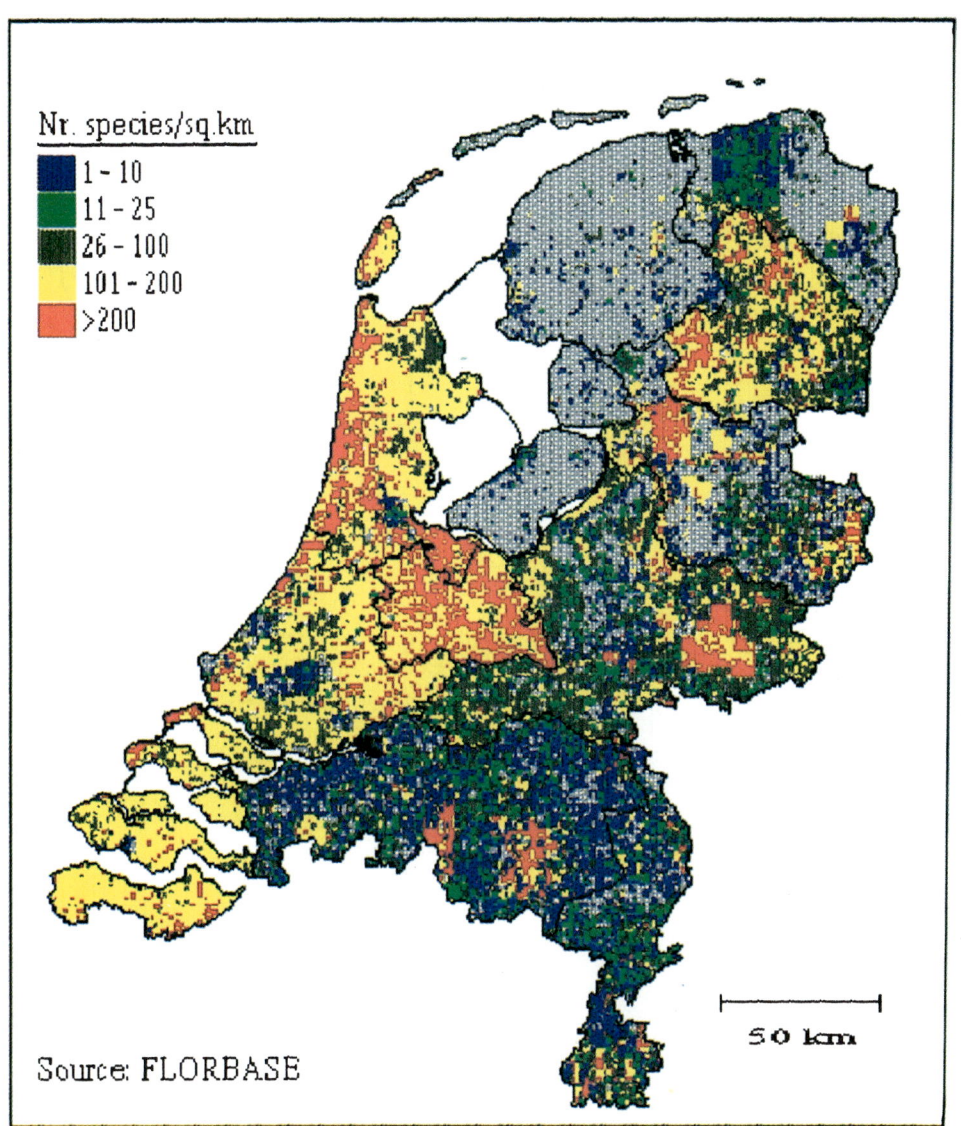

Plate 13.1 The number of plant species recorded in FLORBASE for each square kilometre. Provincial boundaries are given in black.

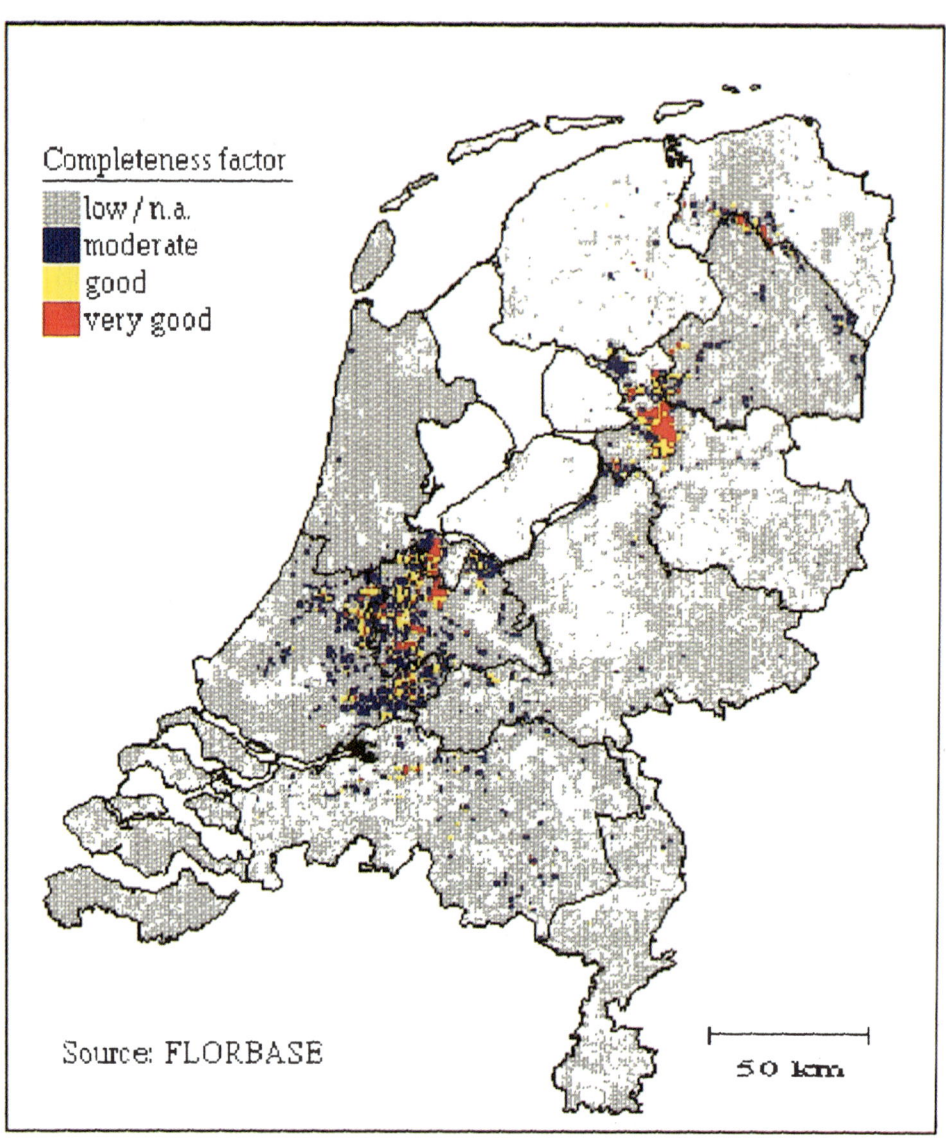

Plate 13.2 The presence of ecotopes of type A17, 'aquatic vegetations in fresh, moderately nutrient-rich water', as derived from FLORBASE (after Witte and Van der Meijden, 1992). Grid cells are indicated in grey, which have a completeness factor of 0. Of the white grid cells no data are available.

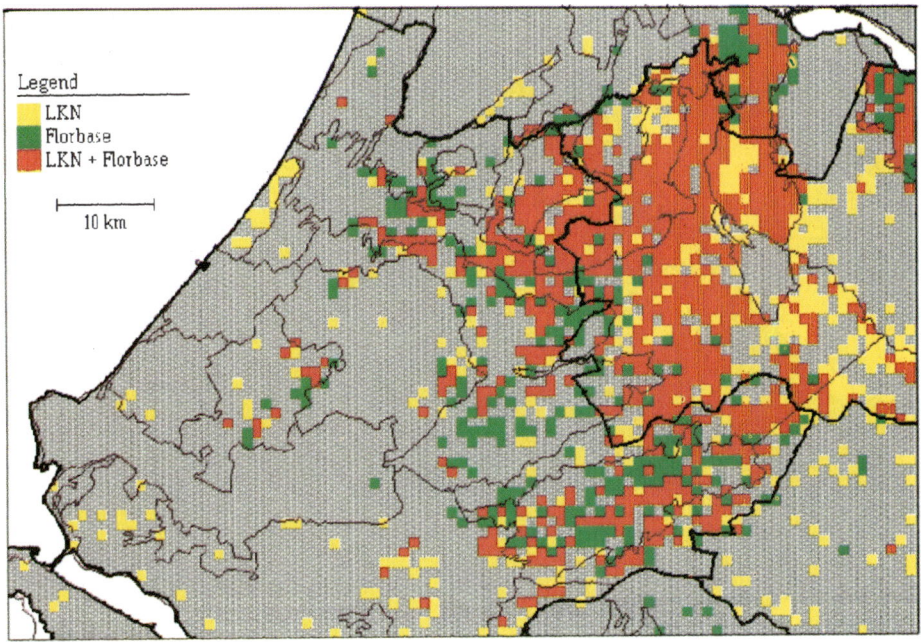

Plate 13.3 The presence of ecotopes of type A17, as derived from LKN and/or FLORBASE. The figure concerns only the western part of the Netherlands. The boundaries of ecodistricts and provinces are indicated (see chapter 5)

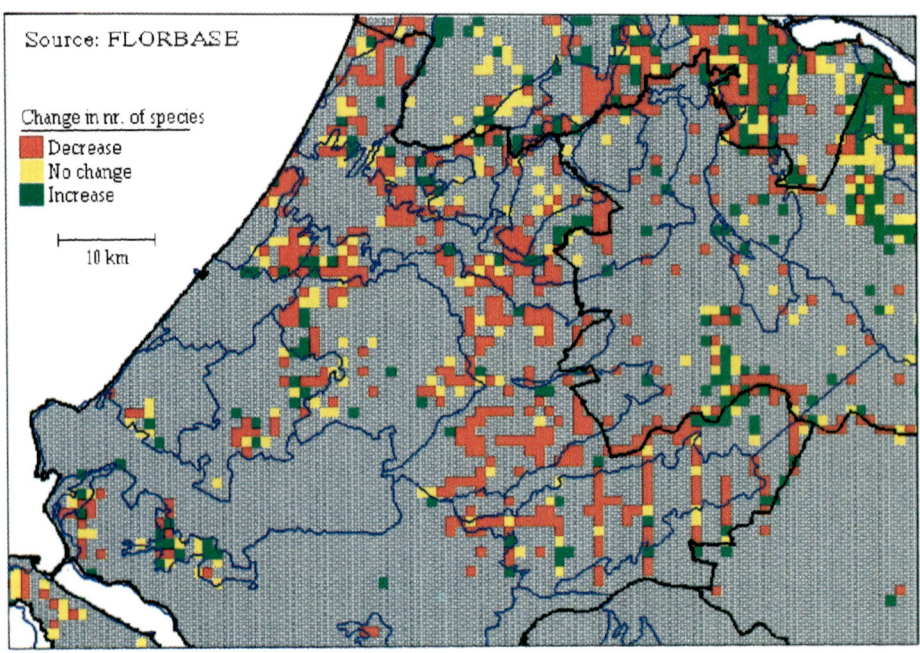

Plate 13.4 Change in species numbers of the Marsh Marigold group between the survey periods 1975-1983 and 1984-1990. The boundaries of ecodistricts and provinces are indicated (see chapter 5).